T0190492

Probability Theory and Stochastic Modelling

Volume 74

Editors-in-chief

Søren Asmussen, Aarhus, Denmark
Peter W. Glynn, Stanford, CA, USA
Thomas G. Kurtz, Madison, WI, USA
Yves Le Jan, Orsay, France

Advisory Board

Joe Gani, Canberra, ACT, Australia
Martin Hairer, Coventry, UK
Peter Jagers, Gothenburg, Sweden
Ioannis Karatzas, New York, NY, USA
Frank P. Kelly, Cambridge, UK
Andreas E. Kyprianou, Bath, UK
Bernt Øksendal, Oslo, Norway
George Papanicolaou, Stanford, CA, USA
Etienne Pardoux, Marseille, France
Edwin Perkins, Vancouver, BC, Canada
Halil Mete Soner, Zürich, Switzerland

The **Probability Theory and Stochastic Modelling** series is a merger and continuation of Springer's two well established series Stochastic Modelling and Applied Probability and Probability and Its Applications series. It publishes research monographs that make a significant contribution to probability theory or an applications domain in which advanced probability methods are fundamental. Books in this series are expected to follow rigorous mathematical standards, while also displaying the expository quality necessary to make them useful and accessible to advanced students as well as researchers. The series covers all aspects of modern probability theory including

- Gaussian processes
- Markov processes
- Random fields, point processes and random sets
- Random matrices
- Statistical mechanics and random media
- Stochastic analysis

as well as applications that include (but are not restricted to):

- Branching processes and other models of population growth
- Communications and processing networks
- Computational methods in probability and stochastic processes, including simulation
- Genetics and other stochastic models in biology and the life sciences
- Information theory, signal processing, and image synthesis
- Mathematical economics and finance
- Statistical methods (e.g. empirical processes, MCMC)
- Statistics for stochastic processes
- Stochastic control
- Stochastic models in operations research and stochastic optimization
- Stochastic models in the physical sciences

More information about this series at http://www.springer.com/series/13205

Erich Häusler · Harald Luschgy

Stable Convergence and Stable Limit Theorems

 Springer

Erich Häusler
Mathematical Institute
University of Giessen
Giessen
Germany

Harald Luschgy
FB IV, Mathematics
University of Trier
Trier
Germany

ISSN 2199-3130   ISSN 2199-3149   (electronic)
Probability Theory and Stochastic Modelling
ISBN 978-3-319-36519-0   ISBN 978-3-319-18329-9   (eBook)
DOI 10.1007/978-3-319-18329-9

Mathematics Subject Classification (2010): 60-02, 60F05, 60F17

Springer Cham Heidelberg New York Dordrecht London
© Springer International Publishing Switzerland 2015
Softcover reprint of the hardcover 1st edition 2015
This work is subject to copyright. All rights are reserved by the Publisher, whether the whole or part of the material is concerned, specifically the rights of translation, reprinting, reuse of illustrations, recitation, broadcasting, reproduction on microfilms or in any other physical way, and transmission or information storage and retrieval, electronic adaptation, computer software, or by similar or dissimilar methodology now known or hereafter developed.
The use of general descriptive names, registered names, trademarks, service marks, etc. in this publication does not imply, even in the absence of a specific statement, that such names are exempt from the relevant protective laws and regulations and therefore free for general use.
The publisher, the authors and the editors are safe to assume that the advice and information in this book are believed to be true and accurate at the date of publication. Neither the publisher nor the authors or the editors give a warranty, express or implied, with respect to the material contained herein or for any errors or omissions that may have been made.

Printed on acid-free paper

Springer International Publishing AG Switzerland is part of Springer Science+Business Media
(www.springer.com)

Preface

Weak convergence of probability measures or, what is the same, convergence in distribution of random variables is arguably one of the most important basic concepts of asymptotic probability theory and mathematical statistics. The classical central limit theorem for sums of independent real random variables, a cornerstone of these fields, cannot possibly be thought of properly without the notion of weak convergence/convergence in distribution. Interestingly, this limit theorem as well as many others which are usually stated in terms of convergence in distribution remain true under unchanged assumptions for a stronger type of convergence. This type of convergence, called *stable convergence* with *mixing convergence* as a special case, originates from the work of Alfred Rényi more than 50 years ago and has been used by researchers in asymptotic probability theory and mathematical statistics ever since (and should not be mistaken for weak convergence to a stable limit distribution). What seems to be missing from the literature is a single comprehensive account of the theory and its consequences in applications, illustrated by a number of typical examples and applied to a variety of limit theorems. The goal of this book is to present such an account of stable convergence which can serve as an introduction to the area but does not compromise on mathematical depth and rigour.

In Chap. 1 we will give a detailed motivation for the study of stable convergence of real random variables and disclose some of its main features. With the exception of one crucial example this introductory chapter contains no proofs, but references to later chapters in which proofs can be found. It will be seen that stable convergence is best thought of as a notion of convergence for conditional distributions of random variables given sub-σ-fields of the σ-field of the underlying probability space on which the random variables are defined. Now conditional distributions are Markov kernels so that the theory of weak convergence of Markov kernels is the proper framework for stable convergence. Since we want to include limit theorems for (continuous-time) stochastic processes later on, it is reasonable to consider from the very start random variables with values in separable metrizable spaces. Therefore, we have to deal with the setting of Markov kernels from sample spaces of arbitrary probability spaces to separable metrizable spaces (which quite often are assumed to be polish). The required facts from the theory of weak convergence of such Markov kernels will be presented in Chap. 2.

In Chap. 3 the material from Chap. 2 is used to describe two approaches to stable convergence of random variables in separable metrizable spaces. In the first approach the limits of stably convergent sequences are always Markov kernels. In the second (essentially equivalent) approach the limit kernels are represented as conditional distributions of random variables. This approach allows for what might sometimes be considered as a somewhat more intuitive description of stable convergence results.

In Chap. 4 we demonstrate the usefulness of stable convergence in different areas. Our focus is on limit points of stably convergent sequences with an application to occupation times of Brownian motion and random index limit theorems as well as the empirical measure theorem and the δ-method.

Chapters 5–10 constitute in some sense the second part of the book in which it is shown that in a variety of known distributional limit theorems the convergence is actually stable or even mixing.

In Chap. 5 we discuss general conditions under which limit theorems in distribution are mixing. In particular, it turns out that the classical distributional limit theorems for centered and normalized partial sums and sample maxima of independent and identically distributed real random variables are automatically mixing.

Chapter 6 is devoted to martingale central limit theorems. Here, stable and mixing convergence is strongly dependent on the filtrations involved and the normalization used. Full stable convergence follows from a nesting condition of the filtrations. Illustrations concern martingales with stationary increments, exchangeable sequences, the Pólya urn and adaptive Monte Carlo estimators.

In Chap. 7 it is shown that the natural extension of Donsker's functional central limit theorem for partial sum processes of independent real random variables to martingale difference sequences holds with stable convergence in the metric space of all continuous real valued functions defined on the nonnegative real axis.

Chapter 8 contains a stable limit theorem for "explosive" processes with exponential rate. Since the increments of these processes are not asymptotically negligible, conditions of Lindeberg-type are not satisfied. Nevertheless, the limits can be normal, but quite different limits are also possible. This result is crucial for deriving stable limit theorems for some estimators in autoregressive processes of order one in Chap. 9 and in Galton-Watson branching processes in Chap. 10. From our point of view, these applications in two classical models of probability theory and mathematical statistics provide once more convincing illustrations of the importance of the concept of stable convergence.

Exercises appear throughout the book. We have supplied solutions of the exercises in Appendix B while Appendix A contains some basic facts about weak convergence of probability distributions, conditional distributions and martingales.

As is apparent from the brief description of its content this book is by no means meant as an encyclopedic account of all major stable limit theorems which have been established in the last 50 years or so. We tried to be reasonably complete in the basic Chap. 3 and in some sense also in Chaps. 4 and 6, but the selection of the material presented in other chapters is quite subjective. As far as our sources are concerned, we tried to give credit where credit is due, but we did not spend much time obtaining definite historical evidence in all cases. In addition to the published sources listed in the References, the first author benefitted considerably from a series of lectures on stable convergence given by David Scott at the University of Munich in the fall semester 1978/79. It is a pleasure to thank Holger Rootzén who made valuable comments on an earlier version of the manuscript. Our thanks also go to a referee for careful reading of the manuscript and for useful suggestions.

Contents

Chapter 1
Why Stable Convergence?

This chapter is of an introductory nature. We make the motivation for the study of stable convergence more precise and present an exposition of some of its features. With the exception of Example 1.2, no proofs are given, only references to later chapters where proofs may be found.

Our starting point is the classical central limit theorem. For this, let $(Z_k)_{k\geq 1}$ be a sequence of independent and identically distributed real random variables, defined on some probability space $(\Omega, \mathcal{F}, P)$. Assume $Z_1 \in \mathcal{L}^2(P)$ and set $\mu = EZ_1$ and $\sigma^2 = \text{Var}\, Z_1$. To exclude the trivial case of almost surely constant variables, assume also $\sigma^2 > 0$. Then the classical central limit theorem says that

$$\lim_{n\to\infty} P\left(\frac{1}{n^{1/2}} \sum_{k=1}^{n} \frac{Z_k - \mu}{\sigma} \leq x\right) = \Phi(x) = \int_{-\infty}^{x} \varphi(u)\, du \quad \text{for all } x \in \mathbb{R},$$

where $\varphi(u) = \frac{1}{\sqrt{2\pi}} \exp\left(-\frac{1}{2}u^2\right)$, $u \in \mathbb{R}$, denotes the density of the standard normal distribution. It is customary to write this convergence of probabilities in a somewhat more abstract way as convergence in distribution of random variables, i.e. as

$$\frac{1}{n^{1/2}} \sum_{k=1}^{n} \frac{Z_k - \mu}{\sigma} \xrightarrow{d} N(0, 1) \quad \text{as } n \to \infty,$$

where $N(0, 1)$ denotes the standard normal distribution, or as

$$\frac{1}{n^{1/2}} \sum_{k=1}^{n} \frac{Z_k - \mu}{\sigma} \xrightarrow{d} N \quad \text{as } n \to \infty,$$

where N is a random variable which "realizes" the standard normal distribution, that is, the distribution P^N of N (under P) equals $N(0, 1)$. To put this notation into a

© Springer International Publishing Switzerland 2015
E. Häusler and H. Luschgy, *Stable Convergence and Stable Limit Theorems*,
Probability Theory and Stochastic Modelling 74,
DOI 10.1007/978-3-319-18329-9_1

broader perspective, recall that for a probability distribution ν on $\mathbb{R}$ and real random variables $(X_n)_{n \geq 1}$ convergence in distribution of $(X_n)_{n \geq 1}$ to ν, written as

$$X_n \overset{d}{\to} \nu \quad \text{as } n \to \infty,$$

is equivalent to

$$\lim_{n \to \infty} Eh(X_n) = \int h \, d\nu \quad \text{for all } h \in C_b(\mathbb{R}),$$

whereas convergence in distribution of $(X_n)_{n \geq 1}$ to a real random variable X, written as

$$X_n \overset{d}{\to} X \quad \text{as } n \to \infty,$$

means $X_n \overset{d}{\to} \nu$ with $\nu = P^X$ and is equivalent to

$$\lim_{n \to \infty} Eh(X_n) = Eh(X) \quad \text{for all } h \in C_b(\mathbb{R}),$$

where $C_b(\mathbb{R})$ is the set of all continuous, bounded functions $h : \mathbb{R} \to \mathbb{R}$. Here it is implicitly assumed that the probability space $(\Omega, \mathcal{F}, P)$ is rich enough to carry a random variable X with distribution ν.

Writing, as usual, $\overline{Z}_n = \frac{1}{n} \sum_{k=1}^{n} Z_k$ for the sample mean of $Z_1, \ldots, Z_n$, an equivalent formulation of the classical central limit theorem is

$$n^{1/2} \left(\overline{Z}_n - \mu \right) \overset{d}{\to} \sigma N \quad \text{as } n \to \infty,$$

which means that $\overline{Z}_n$ considered as an estimator for μ is asymptotically normal, where the asymptotic distribution $N(0, \sigma^2)$ of σN is the centered normal distribution with variance σ^2. If in a statistical setting μ and σ^2 are supposed to be unknown and μ is the parameter of interest and σ^2 is not, i.e. σ^2 is a so-called nuisance parameter, then σ has to be removed from the limit theorem by replacing it by a suitable consistent estimator, if the limit theorem is to be used for statistical inference. The proper tool for doing this is

Theorem 1.1 (Cramér-Slutzky) *Let $(X_n)_{n \geq 1}$ and $(Y_n)_{n \geq 1}$ be sequences of real random variables. If*

$$X_n \overset{d}{\to} X \quad \text{as } n \to \infty$$

for some real random variable X and

$$Y_n \to c \quad \text{in probability as } n \to \infty$$

for some $c \in \mathbb{R}$, then

$$Y_n X_n \xrightarrow{d} cX \quad as \ n \to \infty.$$

A proof of this fundamental result can be found in almost any textbook on asymptotic theory in mathematical statistics. For the sample variance

$$\widehat{\sigma}_n^2 = \frac{1}{n} \sum_{k=1}^{n} \left(Z_k - \overline{Z}_n \right)^2$$

of $Z_1, \ldots, Z_n$ we have $\widehat{\sigma}_n^2 \to \sigma^2$ almost surely as $n \to \infty$ by the strong law of large numbers, and Theorem 1.1 gives

$$n^{1/2} \frac{\overline{Z}_n - \mu}{\widehat{\sigma}_n} \xrightarrow{d} N \quad as \ n \to \infty.$$

This convergence result can now be used in asymptotic statistical inference about μ because it is free from the unknown nuisance parameter σ.

The situation is different in the following setting. Consider the classical super-critical Galton-Watson branching process as a model for exponentially growing populations. For $n \geq 0$ let X_n denote the size of the n-th generation, and α the mean per-capita number of offspring. Here $\alpha > 1$, and if α is unknown, it can be estimated from observed values of $X_0, X_1, \ldots, X_n$ in various ways. For simplicity, we assume here that $\lim_{n \to \infty} X_n = \infty$ almost surely; the general case is considered in Chap. 10. If $\lim_{n \to \infty} X_n = \infty$ almost surely, then the Harris estimator

$$\widehat{\alpha}_n^{(H)} := \frac{\sum_{i=1}^{n} X_i}{\sum_{i=1}^{n} X_{i-1}}$$

is a consistent estimator for α, and

$$\frac{\alpha^{n/2}}{(\alpha - 1)^{1/2}} \left(\widehat{\alpha}_n^{(H)} - \alpha \right) \xrightarrow{d} \sigma M_\infty^{-1/2} N \quad as \ n \to \infty,$$

where σ^2 is the offspring variance (assumed to be positive and finite), where N is a random variable with a standard normal distribution and M_∞ is a positive random variable with positive variance, i.e. M_∞ is a proper random variable and not a constant almost surely, which is independent of N. Since the norming sequence in this limit theorem depends on the unknown parameter α and, more importantly, since the limit distribution is a variance mixture of centered normals with unknown mixing law, the result as it stands is not suitable for asymptotic statistical inference about α. Now

$$\left(\sum_{i=1}^{n} X_{i-1} \right)^{1/2} \frac{(\alpha - 1)^{1/2}}{\alpha^{n/2}} \to M_{\infty}^{1/2} \quad \text{a.s. as } n \to \infty,$$

and we would immediately get

$$\left(\sum_{i=1}^{n} X_{i-1} \right)^{1/2} \left(\widehat{\alpha}_n^{(H)} - \alpha \right) \xrightarrow{d} \sigma N \quad \text{as } n \to \infty,$$

if in Theorem 1.1 the constant limit c could be replaced by a proper random variable. The remaining nuisance parameter σ could then be removed with the help of Theorem 1.1 as it stands and a consistent estimator for σ exactly as in the case of the classical central limit theorem for independent observations discussed before. Unfortunately, as shown by the following example, Theorem 1.1 is no longer true if c is replaced by a proper random variable so that removing the mixing variable M_{∞} from the limit theorem and thereby transforming it into a statistically useful result requires a new tool.

Example 1.2 Consider $(\Omega, \mathcal{F}, P) = \left([0, 1], \mathcal{B}\left([0, 1]\right), \lambda_{[0,1]} \right)$ and set $X_n = 1_{[a_n, a_n + 1/2]}$ for all $n \geq 1$ and some sequence $(a_n)_{n \geq 1}$ of real numbers in $[0, 1/2]$. Clearly, $P^{X_n} = (\delta_0 + \delta_1)/2$ for all $n \geq 1$ so that

$$X_n \xrightarrow{d} X_1 \quad \text{as } n \to \infty.$$

Consider the random variable Y with $Y(\omega) = \omega$ for all $\omega \in \Omega$ and the function $h(u) = (u \wedge 1) \vee 0, u \in \mathbb{R}$. Then $h \in C_b(\mathbb{R})$, and

$$Eh(YX_n) = \int_{a_n}^{a_n + \frac{1}{n}} u \, du = \frac{1}{2} \left(a_n + \frac{1}{4} \right).$$

This shows that for any sequence $(a_n)_{n \geq 1}$ which is not convergent, the sequence $(Eh(YX_n))_{n \geq 1}$ is also not convergent so that the sequence $(YX_n)_{n \geq 1}$ cannot converge in distribution, and in particular not to YX_1. Therefore, Theorem 1.1 does not hold if the limit c in the assumption $Y_n \to c$ in probability as $n \to \infty$ is replaced by a proper random variable. $\qquad \square$

A second example of a more probabilistic nature for a distributional limit theorem in which the limit is a variance mixture of centered normals with non-constant mixing law is as follows (cf. Corollary 6.26). Let $(X_k)_{k \geq 1}$ be a martingale difference sequence w.r.t. an increasing sequence $(\mathcal{F}_k)_{k \geq 0}$ of sub-σ-fields of $\mathcal{F}$. If $(X_k)_{k \geq 1}$ is also stationary and $X_1 \in \mathcal{L}^2(P)$, then the following version of the central limit theorem is true:

$$\frac{1}{n^{1/2}} \sum_{k=1}^{n} X_k \overset{d}{\to} E\left(X_1^2|\mathcal{I}_X\right)^{1/2} N \quad \text{as } n \to \infty,$$

where $\mathcal{I}_X$ is the σ-field of the invariant sets of $X = (X_k)_{k\geq 1}$, N is a random variable with a standard normal distribution and the random variables $E\left(X_1^2|\mathcal{I}_X\right)$ and N are independent. It is important to note that $E\left(X_1^2|\mathcal{I}_X\right)$ is in general indeed a proper random variable so that the limit distribution is a variance mixture of centered normals again. Therefore, though we have

$$\frac{1}{n} \sum_{k=1}^{n} X_k^2 \to E\left(X_1^2|\mathcal{I}_X\right) \quad \text{a.s. as } n \to \infty$$

by the ergodic theorem, we cannot derive

$$\left(\sum_{k=1}^{n} X_k^2\right)^{-1/2} \sum_{k=1}^{n} X_k \overset{d}{\to} N \quad \text{as } n \to \infty$$

by an application of Theorem 1.1 thus removing the mixing variable $E\left(X_1^2|\mathcal{I}_X\right)^{1/2}$ from the limit theorem by a random norming, because for a proper application $\frac{1}{n} \sum_{k=1}^{n} X_k^2$ would have to converge (in probability) to a constant, which is not the case in general (unless the stationary sequence $(X_k)_{k\geq 1}$ is ergodic, of course). Mixed normality in the limit as appearing here and in the Galton-Watson branching process typically occurs in "non-ergodic" or "explosive" models.

As Example 1.2 shows, the concept of convergence in distribution is not strong enough to allow for a version of the Cramér-Slutzky Theorem 1.1 in which the constant factor c in the limit variable is replaced by a proper random variable. There is, however, a stronger notion of convergence for which such a stronger version of the Cramér-Slutzky theorem is true, and this is stable convergence. For a brief exposition of its main features let $(X_n)_{n\geq 1}$ be a sequence of real random variables defined on some probability space $(\Omega, \mathcal{F}, P)$, let $\mathcal{G}$ be a sub-σ-field of $\mathcal{F}$ and let K be a $\mathcal{G}$-measurable Markov kernel from Ω to $\mathbb{R}$. Then the sequence $(X_n)_{n\geq 1}$ is said to converge $\mathcal{G}$-stably to K as $n \to \infty$, denoted by

$$X_n \to K \quad \mathcal{G}\text{-stably as } n \to \infty,$$

if the conditional distributions $P^{X_n|\mathcal{G}}$ of the random variables X_n given $\mathcal{G}$ converge weakly to K in the sense of weak convergence of Markov kernels, i.e. if

$$\lim_{n\to\infty} Efh(X_n) = \int_{\Omega} \int_{\mathbb{R}} f(\omega) h(x) K(\omega, dx) dP(\omega)$$

for every $f \in \mathcal{L}^1(\mathcal{G}, P)$ and $h \in C_b(\mathbb{R})$. In case K does not depend on $\omega \in \Omega$ in the sense that $K = \nu$ P-almost surely for some probability distribution ν on $\mathbb{R}$, then

$(X_n)_{n \geq 1}$ is said to converge $\mathcal{G}$-mixing to ν, and we write

$$X_n \to \nu \quad \mathcal{G}\text{-mixing as } n \to \infty.$$

This means

$$\lim_{n \to \infty} Efh(X_n) = \int f \, dP \int h \, d\nu$$

for every $f \in \mathcal{L}^1(\mathcal{G}, P)$ and $h \in C_b(\mathbb{R})$. Therefore, the weak topology on the set of $\mathcal{G}$-measurable Markov kernels from Ω to $\mathbb{R}$ and the theory of weak convergence of such Markov kernels does provide the proper framework for stable convergence. We will develop this theory (for more general state spaces) as far as necessary in Chap. 2.

To get a feeling for the difference between convergence in distribution and stable convergence, recall that convergence in distribution of random variables X_n towards a distribution ν is in fact weak convergence of the distributions P^{X_n} towards the distribution ν, i.e. the underlying concept is that of weak convergence of probability measures. Now the distributions P^{X_n} may obviously be interpreted as the conditional distributions $P^{X_n|\{\emptyset, \Omega\}}$ of the random variables X_n given the trivial σ-field $\{\emptyset, \Omega\}$. In the concept of stable convergence this trivial σ-field is replaced by some larger sub-σ-field $\mathcal{G}$ of the σ-field $\mathcal{F}$ in $(\Omega, \mathcal{F}, P)$, and the limit distribution ν is replaced by the $\mathcal{G}$-measurable Markov kernel K. Note that $\mathcal{G}$-stable convergence always implies convergence in distribution (take $f = 1$ in the definition of stable convergence).

As for convergence in distribution it can be convenient to "realize" the limit kernel K through a random variable X which satisfies $P^{X|\mathcal{G}} = K$. Such a random variable does always exist on a suitable extension of $(\Omega, \mathcal{F}, P)$. Therefore, if $(X_n)_{n \geq 1}$ and X are real random variables, defined on some probability space $(\Omega, \mathcal{F}, P)$, and $\mathcal{G} \subset \mathcal{F}$ is a sub-σ-field, we say that $(X_n)_{n \geq 1}$ converges $\mathcal{G}$-stably to X as $n \to \infty$, written as

$$X_n \to X \quad \mathcal{G}\text{-stably as } n \to \infty,$$

if X_n converges $\mathcal{G}$-stably to the conditional distribution $P^{X|\mathcal{G}}$. This is equivalent to

$$\lim_{n \to \infty} Efh(X_n) = Efh(X)$$

for every $f \in \mathcal{L}^1(\mathcal{G}, P)$ and $h \in C_b(\mathbb{R})$.

Most useful criteria for $\mathcal{G}$-stable convergence $X_n \to X$ are

$$X_n \xrightarrow{d} X \quad \text{under } P_F \text{ for every } F \in \mathcal{G} \text{ with } P(F) > 0,$$

where $P_F = P(\cdot \cap F)/P(F)$ denotes the conditional probability given the event F, or

$$(X_n, Y_n) \xrightarrow{d} (X, Y) \quad \text{as } n \to \infty$$

for every sequence $(Y_n)_{n \geq 1}$ of real random variables and every $\mathcal{G}$-measurable real random variable Y satisfying $Y_n \to Y$ in probability (cf. Theorems 3.17 and 3.18). In particular, a generalized version of the Cramér-Slutzky theorem about random norming holds under $\mathcal{G}$-stable convergence where full strength is obtained if $\mathcal{G}$ is sufficiently large.

In case X is independent of $\mathcal{G}$ so that $P^{X|\mathcal{G}} = P^X$, $\mathcal{G}$-stable convergence $X_n \to X$ means $\mathcal{G}$-mixing convergence. If X is $\mathcal{G}$-measurable so that $P^{X|\mathcal{G}} = \delta_X$, the Dirac-kernel associated with X, then $\mathcal{G}$-stable convergence $X_n \to X$ turns into convergence in probability just as for $\mathcal{G} = \{\emptyset, \Omega\}$ distributional convergence to a constant means convergence in probability to this constant (cf. Corollary 3.6).

In the two examples discussed above we, in fact, can show that

$$\frac{\alpha^{n/2}}{(\alpha - 1)^{1/2}} \left(\widehat{\alpha}_n^{(H)} - \alpha \right) \to \sigma M_\infty^{-1/2} N \quad \mathcal{G}\text{-stably as } n \to \infty,$$

where $\mathcal{G} = \sigma(X_n, n \geq 0)$ and N is independent of $\mathcal{G}$, and

$$\frac{1}{n^{1/2}} \sum_{k=1}^{n} X_k \to E\left(X_1^2 | \mathcal{I}_X\right)^{1/2} N \quad \mathcal{G}\text{-stably as } n \to \infty,$$

where $\mathcal{G} = \sigma(X_n, n \geq 1)$ and N is independent of $\mathcal{G}$, respectively (cf. Corollaries 10.6 and 6.26). Consequently, the generalized Cramér-Slutzky theorem implies the desired limit theorems

$$\left(\sum_{i=1}^{n} X_{i-1} \right)^{1/2} \left(\widehat{\alpha}_n^{(H)} - \alpha \right) \xrightarrow{d} \sigma N \quad \text{as } n \to \infty$$

and

$$\left(\sum_{k=1}^{n} X_k^2 \right)^{-1/2} \sum_{k=1}^{n} X_k \xrightarrow{d} N \quad \text{as } n \to \infty.$$

As we have seen we can formulate stable limit theorems with Markov kernels as limits or with random variables as limits, if the limit kernels are identified as conditional distributions of these random variables. Both approaches will be developed in Chap. 3 and applied as convenient.

Stable convergence has a number of other interesting consequences and applications beyond the random norming discussed earlier. Let us demonstrate this for the classical central limit theorem. We will see that

$$n^{1/2} \left(\overline{Z}_n - \mu \right) \to N\left(0, \sigma^2\right) \quad \mathcal{F}\text{-mixing as } n \to \infty$$

(cf. Example 3.13). This mixing convergence and the second criterion above applied with $X_n = n^{1/2} \left(\overline{Z}_n - \mu \right)$ and $Y_n = Y$ for all $n \in \mathbb{N}$ imply

$$\lim_{n \to \infty} P\left(n^{1/2} \left(\overline{Z}_n - \mu \right) \leq Y \right) = \int N\left(0, \sigma^2 \right) ((-\infty, y]) \, dP^Y (dy)$$

for any real *random variable* Y on $(\Omega, \mathcal{F}, P)$, whereas convergence in distribution covers only *constants* Y.

Another area in which stable convergence proves its value are limit theorems with random indices, i.e. limit theorems for sequences $\left(X_{\tau_n} \right)_{n \geq 1}$ with random variables X_n and $\mathbb{N}$-valued random variables τ_n with $\tau_n \to \infty$ in probability as $n \to \infty$; see Sect. 4.2. For instance, if $\tau_n / a_n \to \eta$ in probability for some $(0, \infty)$-valued random variable η and $a_n \in (0, \infty)$ satisfying $a_n \to \infty$, then

$$\tau_n^{1/2} \left(\overline{Z}_{\tau_n} - \mu \right) \overset{d}{\to} N\left(0, \sigma^2 \right) ,$$

and this convergence is again $\mathcal{F}$-mixing (cf. Example 4.8). In this context we can also demonstrate the advantage of stable convergence for restrictions to subsets of Ω. Assume that the limiting random variable η is $\mathbb{R}_+$-valued satisfying merely $P(\eta > 0) > 0$. Since by the first criterion $n^{1/2} \left(\overline{Z}_n - \mu \right) \to N\left(0, \sigma^2 \right)$ $\mathcal{F}$-mixing under $P_{\{\eta > 0\}}$, and $P_{\{\eta > 0\}} (\eta > 0) = 1$, we can conclude in this case that

$$\tau_n^{1/2} \left(\overline{Z}_{\tau_n} - \mu \right) \overset{d}{\to} N\left(0, \sigma^2 \right) \quad \text{under } P_{\{\eta > 0\}} .$$

Still another area concerns the fluctuation behavior of stably convergent sequences of random variables; see Sect. 4.1. As for the classical mixing central limit theorem this implies that the set of limit points of the sequence $\left(n^{1/2} \left(\overline{Z}_n - \mu \right) \right)_{n \geq 1}$ coincides with $\mathbb{R}$, the support of $N\left(0, \sigma^2 \right)$, almost surely (cf. Example 4.2).

Historically, the idea of mixing convergence was developed first. Early applications of the concept, not yet in its most general form, can be found in [84, 85, 90, 93, 94]. In the work of Rényi, the idea can be traced back at least to [75] and was developed in its general form in [76, 78]. Therefore, the notion is also known as "Rényi-mixing". More detailed information on the early history of the theory of mixing and its application to random-sum central limit theorems in particular can be found in [21].

Stable convergence originates from [77], where an unspecified limit version of $\mathcal{F}$-stability in the sense of

$$X_n \overset{d}{\to} \nu_F \quad \text{under } P_F$$

for every $F \in \mathcal{F}$ with $P(F) > 0$ and some probability distribution ν_F on $\mathbb{R}$ is used which, however, is equivalent to our definition (cf. Proposition 3.12). The classical limit theory for sums of independent real random variables as well as for maxima of independent and identically distributed random variables provides, in fact, mixing limit theorems (cf. Examples 3.13 (a) and 5.6 (c)). In view of the consequences of stable convergence as outlined above its importance is simply due to the fact that many other distributional limit theorems are stable. The concept of stable convergence played one of its first major roles in the development of the theory of martingale central limit theorems in discrete time; see [41, 82] and the references therein. Later it became important in the theory of limit theorems for stochastic processes; see e.g. the monographs [50, 60]. More recently, stable convergence has appeared as a crucial tool in the investigation of discretized processes [49], the approximation of stochastic integrals and stochastic differential equations [56, 59, 70, 71] and the statistics of high-frequency financial data [1].

As explained in this chapter, stable convergence is in fact weak convergence of conditional distributions, for which weak convergence of Markov kernels is the proper framework. In the next chapter we will therefore present the essential parts of the theory for Markov kernels from measurable spaces to separable metric spaces equipped with their Borel-σ-fields. This somewhat abstract level cannot be avoided if we want to include the convergence of stochastic processes in later chapters, as we will do.

Chapter 2
Weak Convergence of Markov Kernels

As indicated in the previous chapter, stable convergence of random variables can be seen as suitable convergence of Markov kernels given by conditional distributions. Let $(\Omega, \mathcal{F}, P)$ be a probability space and let $\mathcal{X}$ be a separable metrizable topological space equipped with its Borel σ-field $\mathcal{B}(\mathcal{X})$. In this chapter we briefly describe the weak topology on the set of Markov kernels (transition kernels) from $(\Omega, \mathcal{F})$ to $(\mathcal{X}, \mathcal{B}(\mathcal{X}))$.

Let us first recall the weak topology on the set $\mathcal{M}^1(\mathcal{X})$ of all probability measures on $\mathcal{B}(\mathcal{X})$. It is the topology generated by the functions

$$\nu \mapsto \int h \, d\nu, \quad h \in C_b(\mathcal{X}),$$

where $C_b(\mathcal{X})$ denotes the space of all continuous, bounded functions $h : \mathcal{X} \to \mathbb{R}$ equipped with the sup-norm $\|h\|_{\sup} := \sup_{x \in \mathcal{X}} |h(x)|$. The weak topology on $\mathcal{M}^1(\mathcal{X})$ is thus the weakest topology for which each function $\nu \mapsto \int h \, d\nu$ is continuous. Consequently, weak convergence of a net $(\nu_\alpha)_\alpha$ in $\mathcal{M}^1(\mathcal{X})$ to $\nu \in \mathcal{M}^1(\mathcal{X})$ means

$$\lim_\alpha \int h \, d\nu_\alpha = \int h \, d\nu$$

for every $h \in C_b(\mathcal{X})$ (here and elsewhere we omit the directed set on which a net is defined from the notation). Because $\int h \, d\nu_1 = \int h \, d\nu_2$ for $\nu_1, \nu_2 \in \mathcal{M}^1(\mathcal{X})$ and every $h \in C_b(\mathcal{X})$ implies that $\nu_1 = \nu_2$, this topology is Hausdorff and the limit is unique. Moreover, the weak topology is separable metrizable e.g. by the Prohorov metric, see e.g. [69], Theorem II.6.2, and polish if $\mathcal{X}$ is polish; see e.g. [69], Theorem II.6.5, [26], Corollary 11.5.5. The relatively compact subsets of $\mathcal{M}^1(\mathcal{X})$ are exactly the tight ones, provided $\mathcal{X}$ is polish, where $\Gamma \subset \mathcal{M}^1(\mathcal{X})$ is called *tight* if for every $\varepsilon > 0$ there exists a compact set $A \subset \mathcal{X}$ such that $\sup_{\nu \in \Gamma} \nu(\mathcal{X} \setminus A) \leq \varepsilon$; see e.g. [69], Theorem II.6.7, [26], Theorem 11.5.4.

© Springer International Publishing Switzerland 2015
E. Häusler and H. Luschgy, *Stable Convergence and Stable Limit Theorems*,
Probability Theory and Stochastic Modelling 74,
DOI 10.1007/978-3-319-18329-9_2

A map $K : \Omega \times \mathcal{B}(\mathcal{X}) \to [0,1]$ is called a *Markov kernel from* $(\Omega, \mathcal{F})$ *to* $(\mathcal{X}, \mathcal{B}(\mathcal{X}))$ if $K(\omega, \cdot) \in \mathcal{M}^1(\mathcal{X})$ for every $\omega \in \Omega$ and $K(\cdot, B)$ is $\mathcal{F}$-measurable for every $B \in \mathcal{B}(\mathcal{X})$. Let $\mathcal{K}^1 = \mathcal{K}^1(\mathcal{F}) = \mathcal{K}^1(\mathcal{F}, \mathcal{X})$ denote the set of all such Markov kernels. If $\mathcal{M}^1(\mathcal{X})$ is equipped with the σ-field $\Sigma\left(\mathcal{M}^1(\mathcal{X})\right) := \sigma(\nu \mapsto \nu(B), B \in \mathcal{B}(\mathcal{X}))$, then Markov kernels $K \in \mathcal{K}^1$ can be viewed as $\left(\mathcal{M}^1(\mathcal{X}), \Sigma\left(\mathcal{M}^1(\mathcal{X})\right)\right)$-valued random variables $\omega \mapsto K(\omega, \cdot)$. Furthermore, $\Sigma\left(\mathcal{M}^1(\mathcal{X})\right)$ coincides with the Borel σ-field of $\mathcal{M}^1(\mathcal{X})$ (see Lemma A.2).

For a Markov kernel $K \in \mathcal{K}^1$ and a probability distribution Q on $\mathcal{F}$ we define the *product measure* (which is a probability distribution again) on the product σ-field $\mathcal{F} \otimes \mathcal{B}(\mathcal{X})$ by

$$Q \otimes K(C) := \int \int 1_C(\omega, x) K(\omega, dx) \, dQ(\omega)$$

for $C \in \mathcal{F} \otimes \mathcal{B}(\mathcal{X})$ and its *marginal* on $\mathcal{B}(\mathcal{X})$ by

$$QK(B) := Q \otimes K(\Omega \times B) = \int K(\omega, B) \, dQ(\omega)$$

for $B \in \mathcal{B}(\mathcal{X})$. For functions $f : \Omega \to \mathbb{R}$ and $h : \mathcal{X} \to \mathbb{R}$ let $f \otimes h : \Omega \times \mathcal{X} \to \mathbb{R}$, $f \otimes h(\omega, x) := f(\omega)h(x)$, be the *tensor product*.

Lemma 2.1 (a) (Fubini's theorem for Markov kernels) *Let* $K \in \mathcal{K}^1$ *and* $g : (\Omega \times \mathcal{X}, \mathcal{F} \otimes \mathcal{B}(\mathcal{X})) \to \left(\overline{\mathbb{R}}, \mathcal{B}\left(\overline{\mathbb{R}}\right)\right)$ *be measurable such that* g^- *(or* g^+*)* $\in \mathcal{L}^1(P \otimes K)$. *Then*

$$\int g \, dP \otimes K = \int \int g(\omega, x) K(\omega, dx) \, dP(\omega).$$

(b) (Uniqueness) *For* $K_1, K_2 \in \mathcal{K}^1$, *we have* $\{\omega \in \Omega : K_1(\omega, \cdot) = K_2(\omega, \cdot)\} \in \mathcal{F}$, *and* $K_1(\cdot, B) = K_2(\cdot, B)$ *P-almost surely for every* $B \in \mathcal{B}(\mathcal{X})$ *implies* $P(\{\omega \in \Omega : K_1(\omega, \cdot) = K_2(\omega, \cdot)\}) = 1$, *that is,* $K_1 = K_2$ *P-almost surely.*

Proof (a) For $g = 1_C$ with $C \in \mathcal{F} \otimes \mathcal{B}(\mathcal{X})$ this is the definition of $P \otimes K$. The formula extends as usual by linearity, monotone convergence and the decomposition $g = g^+ - g^-$.

(b) Note that $\mathcal{B}(\mathcal{X})$ is countably generated. Let $\mathcal{C}$ be a countable generator of $\mathcal{B}(\mathcal{X})$ and let $\mathcal{C}_0$ be the (countable) system of all finite intersections of sets from $\mathcal{C}$. Then by measure uniqueness

$$\{\omega \in \Omega : K_1(\omega, \cdot) = K_2(\omega, \cdot)\} = \bigcap_{B \in \mathcal{C}_0} \{\omega \in \Omega : K_1(\omega, B) = K_2(\omega, B)\}.$$

Hence the assertion. $\square$

Exercise 2.1 Let $\mathcal{C} \subset \mathcal{B}(\mathcal{X})$ be closed under finite intersections with $\sigma(\mathcal{C}) = \mathcal{B}(\mathcal{X})$ and let $K : \Omega \times \mathcal{B}(\mathcal{X}) \to [0, 1]$ satisfy $K(\omega, \cdot) \in \mathcal{M}^1(\mathcal{X})$ for every $\omega \in \Omega$ and $K(\cdot, B)$ is $\mathcal{F}$-measurable for every $B \in \mathcal{C}$. Show that $K \in \mathcal{K}^1$.

Definition 2.2 The topology on $\mathcal{K}^1$ generated by the functions

$$K \mapsto \int f \otimes h \, dP \otimes K, \quad f \in \mathcal{L}^1(P), \, h \in C_b(\mathcal{X})$$

is called the *weak topology* and is denoted by $\tau = \tau(P) = \tau(\mathcal{F}, P)$. Accordingly, *weak convergence of a net* $(K_\alpha)_\alpha$ *in* $\mathcal{K}^1$ *to* $K \in \mathcal{K}^1$ means

$$\lim_\alpha \int f \otimes h \, dP \otimes K_\alpha = \int f \otimes h \, dP \otimes K$$

for every $f \in \mathcal{L}^1(P)$ and $h \in C_b(\mathcal{X})$.

The dependence of τ on P is usually not explicitly indicated. This topology is well known e.g. in statistical decision theory where $\mathcal{K}^1$ corresponds to all randomized decision rules and in areas such as dynamic programming, optimal control, game theory or random dynamical systems; see [7, 13, 18, 61, 62, 87].

Simple characterizations of weak convergence are as follows. For a sub-σ-field $\mathcal{G}$ of $\mathcal{F}$, let $\mathcal{K}^1(\mathcal{G}) = \mathcal{K}^1(\mathcal{G}, \mathcal{X})$ denote the subset of $\mathcal{K}^1$ consisting of all $\mathcal{G}$-measurable Markov kernels, that is of Markov kernels from $(\Omega, \mathcal{G})$ to $(\mathcal{X}, \mathcal{B}(\mathcal{X}))$. For $F \in \mathcal{F}$ with $P(F) > 0$ let $P_F := P(\cdot|F) = P(\cdot \cap F)/P(F)$ denote the *conditional probability measure given* F, and let E_F and Var_F denote expectation and variance, respectively, under P_F. Further recall that for a net $(y_\alpha)_\alpha$ in $\overline{\mathbb{R}} = \mathbb{R} \cup \{-\infty, \infty\}$

$$\limsup_\alpha y_\alpha := \inf_\alpha \sup_{\beta \geq \alpha} y_\beta \quad \text{and} \quad \liminf_\alpha y_\alpha := \sup_\alpha \inf_{\beta \geq \alpha} y_\beta.$$

Theorem 2.3 *Let* $\mathcal{G} \subset \mathcal{F}$ *be a sub-σ-field,* $(K_\alpha)_\alpha$ *a net in* $\mathcal{K}^1(\mathcal{G})$, $K \in \mathcal{K}^1(\mathcal{G})$ *and let* $\mathcal{E} \subset \mathcal{G}$ *be closed under finite intersections with* $\Omega \in \mathcal{E}$ *such that* $\sigma(\mathcal{E}) = \mathcal{G}$. *Then the following statements are equivalent:*

(i) $K_\alpha \to K$ *weakly,*
(ii) $\lim_\alpha \int f \otimes h \, dP \otimes K_\alpha = \int f \otimes h \, dP \otimes K$ *for every* $f \in \mathcal{L}^1(\mathcal{G}, P)$ *and* $h \in C_b(\mathcal{X})$,
(iii) $QK_\alpha \to QK$ *weakly (in* $\mathcal{M}^1(\mathcal{X})$) *for every probability distribution* Q *on* $\mathcal{F}$ *such that* $Q \ll P$.
(iv) $P_F K_\alpha \to P_F K$ *weakly for every* $F \in \mathcal{E}$ *with* $P(F) > 0$.

Proof (i) $\Rightarrow$ (iii). Let $Q \ll P$. Setting $f := dQ/dP$ and using Fubini's theorem for Markov kernels 2.1 (a), we obtain for $h \in C_b(\mathcal{X})$

$$\int h \, dQ K_\alpha = \int \int h(x) \, K_\alpha(\omega, dx) \, dQ(\omega) = \int f \otimes h \, dP \otimes K_\alpha$$
$$\rightarrow \int f \otimes h \, dP \otimes K = \int h \, dQ K \, .$$

(iii) $\Rightarrow$ (iv) is obvious because $P_F \ll P$.
(iv) $\Rightarrow$ (ii). Let

$$\mathcal{L} := \left\{ f \in \mathcal{L}^1(\mathcal{G}, P) : \lim_\alpha \int f \otimes h \, dP \otimes K_\alpha = \int f \otimes h \, dP \otimes K \right.$$
$$\text{for every } h \in C_b(\mathcal{X}) \left. \right\} .$$

Then $\mathcal{L}$ is a vector subspace of $\mathcal{L}^1(\mathcal{G}, P)$ with $\{1_G : G \in \mathcal{E}\} \subset \mathcal{L}$, in particular $1_\Omega \in \mathcal{L}$, and if $f_k \in \mathcal{L}$, $f \in \mathcal{L}^1(\mathcal{G}, P)$, $f_k \geq 0$, $f \geq 0$ such that $f_k \uparrow f$, then $f \in \mathcal{L}$. In fact,

$$\left| \int f \otimes h \, dP \otimes K_\alpha - \int f \otimes h \, dP \otimes K \right|$$
$$\leq \int |f \otimes h - f_k \otimes h| \, dP \otimes K_\alpha + \left| \int f_k \otimes h \, dP \otimes K_\alpha - \int f_k \otimes h \, dP \otimes K \right|$$
$$+ \int |f_k \otimes h - f \otimes h| \, dP \otimes K$$
$$\leq 2\|h\|_{\sup} \int (f - f_k) \, dP + \left| \int f_k \otimes h \, dP \otimes K_\alpha - \int f_k \otimes h \, dP \otimes K \right|$$

and hence

$$\limsup_\alpha \left| \int f \otimes h \, dP \otimes K_\alpha - \int f \otimes h \, dP \otimes K \right| \leq 2\|h\|_{\sup} \int (f - f_k) \, dP \, .$$

Letting $k \to \infty$ yields by monotone convergence

$$\lim_\alpha \int f \otimes h \, dP \otimes K_\alpha = \int f \otimes h \, dP \otimes K \, .$$

Thus $f \in \mathcal{L}$. One can conclude that $\mathcal{D} := \{G \in \mathcal{G} : 1_G \in \mathcal{L}\}$ is a Dynkin-system so that $\mathcal{D} = \sigma(\mathcal{E}) = \mathcal{G}$. This clearly yields $\mathcal{L} = \mathcal{L}^1(\mathcal{G}, P)$, hence (ii).

(ii) $\Rightarrow$ (i). For $f \in \mathcal{L}^1(P)$ we have $E(f|\mathcal{G}) \in \mathcal{L}^1(\mathcal{G}, P)$ and thus in view of the $\mathcal{G}$-measurability of K_α and K

$$\lim_\alpha \int f \otimes h \, dP \otimes K_\alpha = \lim_\alpha \int E\left(f|\mathcal{G}\right) \otimes h \, dP \otimes K_\alpha$$

$$= \int E\left(f|\mathcal{G}\right) \otimes h \, dP \otimes K = \int f \otimes h \, dP \otimes K$$

for every $h \in C_b(\mathcal{X})$. $\square$

Exercise 2.2 Prove that weak convergence $K_\alpha \to K$ is also equivalent to $QK_\alpha \to QK$ weakly for every probability distribution Q on $\mathcal{F}$ such that $Q \equiv P$, where $\equiv$ means mutual absolute continuity.

Exercise 2.3 Show that weak convergence is preserved under an absolutely continuous change of measure, that is, $\tau(Q) \subset \tau(P)$, if $Q \ll P$, and hence $\tau(Q) = \tau(P)$, if $Q \equiv P$.

Exercise 2.4 One may consider $\mathcal{M}^1(\mathcal{X})$ as a subset of $\mathcal{K}^1$. Show that $\tau \cap \mathcal{M}^1(\mathcal{X})$ is the weak topology on $\mathcal{M}^1(\mathcal{X})$.

The weak topology on $\mathcal{K}^1$ is not necessarily Hausdorff and the weak limit kernel is not unique, but it is P-almost surely unique. In fact, if $\int f \otimes h \, dP \otimes K_1 = \int f \otimes h \, dP \otimes K_2$ for $K_1, K_2 \in \mathcal{K}^1$ and every $f \in \mathcal{L}^1(P)$ and $h \in C_b(\mathcal{X})$, then $\int h \, dP_F K_1 = \int h \, dP_F K_2$ for every $h \in C_b(\mathcal{X})$ so that $P_F K_1 = P_F K_2$ for every $F \in \mathcal{F}$ with $P(F) > 0$. This implies $K_1(\cdot, B) = K_2(\cdot, B)$ P-almost surely for every $B \in \mathcal{B}(\mathcal{X})$ and thus $K_1 = K_2$ P-almost surely by Lemma 2.1 (b).

The following notion is sometimes useful.

Definition 2.4 Assume that $\mathcal{X}$ is polish. Let $K \in \mathcal{K}^1$ and $\mathcal{G} \subset \mathcal{F}$ be a sub-σ-field. Then by disintegration of measures there exists a (P-almost surely unique) kernel $H \in \mathcal{K}^1(\mathcal{G})$ such that

$$P \otimes H|\mathcal{G} \otimes \mathcal{B}(\mathcal{X}) = (P|\mathcal{G}) \otimes H = P \otimes K|\mathcal{G} \otimes \mathcal{B}(\mathcal{X})$$

(see Theorem A.6). The Markov kernel H is called the *conditional expectation of K w.r.t.* $\mathcal{G}$ and is denoted by $E(K|\mathcal{G})$.

For a sub-σ-field $\mathcal{G} \subset \mathcal{F}$, the weak topology on $\mathcal{K}^1(\mathcal{G})$ is denoted by $\tau(\mathcal{G}) = \tau(\mathcal{G}, P)$. We will see that the map $\mathcal{K}^1 \mapsto \mathcal{K}^1(\mathcal{G})$ or $\mathcal{K}^1, K \mapsto E(K|\mathcal{G})$, is weakly continuous.

Corollary 2.5 *Let $(K_\alpha)_\alpha$ be a net in $\mathcal{K}^1$, $K \in \mathcal{K}^1$ and $\mathcal{G} \subset \mathcal{F}$ a sub-σ-field.*
(a) $\tau(\mathcal{G})$ coincides with the topology induced by τ on $\mathcal{K}^1(\mathcal{G})$, that is $\tau(\mathcal{G}) = \tau \cap \mathcal{K}^1(\mathcal{G})$.
(b) Assume that $\mathcal{X}$ is polish. If $K_\alpha \to K$ weakly, then $E(K_\alpha|\mathcal{G}) \to E(K|\mathcal{G})$ weakly (in $\mathcal{K}^1$ and $\mathcal{K}^1(\mathcal{G})$).
(c) Assume that $\mathcal{X}$ is polish. If $\{N \in \mathcal{F} : P(N) = 0\} \subset \mathcal{G}$, then $\mathcal{K}^1(\mathcal{G})$ is τ-closed in $\mathcal{K}^1$.

Proof (a) is an immediate consequence of Theorem 2.3.

(b) is immediate from Theorem 2.3 and

$$\int f \otimes h \, dP \otimes E \, (K|\mathcal{G}) = \int f \otimes h \, dP \otimes K$$

for $K \in \mathcal{K}^1$, $f \in \mathcal{L}^1 (\mathcal{G}, P)$ and $h \in C_b (\mathcal{X})$.

(c) Let $(K_\alpha)_\alpha$ be a net in $\mathcal{K}^1 (\mathcal{G})$, $K \in \mathcal{K}^1$ and assume $K_\alpha \to K$ weakly in $\mathcal{K}^1$. Then by (b), $K_\alpha = E \, (K_\alpha|\mathcal{G}) \to E \, (K|\mathcal{G})$ weakly in $\mathcal{K}^1$ and hence, by almost sure uniqueness of limit kernels, we obtain $E \, (K|\mathcal{G}) = K$ P-almost surely. The assumption on $\mathcal{G}$ now implies $K \in \mathcal{K}^1 (\mathcal{G})$. Thus $\mathcal{K}^1 (\mathcal{G})$ is τ-closed. $\square$

We provide further characterizations of weak convergence. Recall that a function $h : \mathcal{Y} \to \overline{\mathbb{R}}$ on a topological space $\mathcal{Y}$ is said to be lower semicontinuous if $\{h \leq r\}$ is closed for every $r \in \overline{\mathbb{R}}$ or, what is the same, if $h \, (y) \leq \liminf_\alpha h \, (y_\alpha)$ for every net $(y_\alpha)_\alpha$ and y in $\mathcal{Y}$ with $y_\alpha \to y$. The function h is upper semicontinuous if $-h$ is lower semicontinuous. A function which is both upper and lower semicontinuous is continuous.

Theorem 2.6 *For a net $(K_\alpha)_\alpha$ and K in $\mathcal{K}^1$ the following statements are equivalent:*

(i) $K_\alpha \to K$ *weakly,*

(ii) $\lim_\alpha \int g \, dP \otimes K_\alpha = \int g \, dP \otimes K$ *for every measurable, bounded function* $g : (\Omega \times \mathcal{X}, \mathcal{F} \otimes \mathcal{B} (\mathcal{X})) \to (\mathbb{R}, \mathcal{B} (\mathbb{R}))$ *such that* $g \, (\omega, \cdot) \in C_b (\mathcal{X})$ *for every* $\omega \in \Omega$,

(iii) *(For $\mathcal{X}$ polish)* $\limsup_\alpha \int g \, dP \otimes K_\alpha \leq \int g \, dP \otimes K$ *for every measurable function* $g : (\Omega \times \mathcal{X}, \mathcal{F} \otimes \mathcal{B} (\mathcal{X})) \to \left(\overline{\mathbb{R}}, \mathcal{B} \left(\overline{\mathbb{R}}\right)\right)$ *which is bounded from above such that $g \, (\omega, \cdot)$ is upper semicontinuous for every $\omega \in \Omega$,*

(iv) *(For $\mathcal{X}$ polish)* $\liminf_\alpha \int g \, dP \otimes K_\alpha \geq \int g \, dP \otimes K$ *for every measurable function* $g : (\Omega \times \mathcal{X}, \mathcal{F} \otimes \mathcal{B} (\mathcal{X})) \to \left(\overline{\mathbb{R}}, \mathcal{B} \left(\overline{\mathbb{R}}\right)\right)$ *which is bounded from below such that $g \, (\omega, \cdot)$ is lower semicontinuous for every $\omega \in \Omega$.*

Note that statements (ii)–(iv) say that the function $\mathcal{K}^1 \to \overline{\mathbb{R}}, K \mapsto \int g \, dP \otimes K$, is weakly continuous, upper semicontinuous and lower semicontinuous, respectively. Moreover, it is interesting to note that the $\mathcal{F} \otimes \mathcal{B} (\mathcal{X})$-measurability of the function g in (ii) already follows from the $\mathcal{F}$-measurability of $g \, (\cdot, x)$ for every $x \in \mathcal{X}$; see [18], Lemma 1.1.

Proof (i) $\Rightarrow$ (ii) and (i) $\Rightarrow$ (iv). Let $g : \Omega \times \mathcal{X} \to \overline{\mathbb{R}}$ be as in (iv). Replacing g by $g - \inf g$, we may assume $g \geq 0$. There exists a totally bounded metric d inducing the topology of $\mathcal{X}$ so that the subspace $U_b (\mathcal{X}, d)$ of $C_b (\mathcal{X})$ consisting of all d-uniformly continuous, bounded functions is separable; see [26], Theorem 2.8.2, [69], Lemma II.6.3. Let $\{h_n : n \in \mathbb{N}\}$ be a countable dense subset of $U_b (\mathcal{X}, d)$. We obtain the representation

$$g \, (\omega, x) = \sup \left\{h_n^+ \, (x) : h_n \leq g \, (\omega, \cdot), n \in \mathbb{N}\right\}$$

for every $\omega \in \Omega$ and $x \in \mathcal{X}$. To see this, let $\varepsilon > 0$, fix $\omega \in \Omega$ and $x \in \mathcal{X}$ and consider the functions

$$g_k : \mathcal{X} \to \mathbb{R}, \quad g_k(y) := \inf_{z \in \mathcal{X}} \{k \wedge g(\omega, z) + kd(y, z)\} - \varepsilon$$

for $k \in \mathbb{N}$. One easily checks that g_k is d-Lipschitz and thus $g_k \in U_b(\mathcal{X}, d)$, $g_k \leq g(\omega, \cdot) - \varepsilon$ and $g_k(y) \uparrow g(\omega, y) - \varepsilon$ for every $y \in \mathcal{X}$. If $g(\omega, x) < \infty$, choose $k \in \mathbb{N}$ such that $g_k(x) \geq g(\omega, x) - 2\varepsilon$ and then $m \in \mathbb{N}$ such that $\|g_k - h_m\|_{\sup} \leq \varepsilon$. This implies $h_m \leq g(\omega, \cdot)$ and $h_m(x) \geq g(\omega, x) - 3\varepsilon$, hence

$$\sup \{h_n^+(x) : h_n \leq g(\omega, \cdot)\} \geq \sup \{h_n(x) : h_n \leq g(\omega, \cdot)\} \geq g(\omega, x) - 3\varepsilon.$$

Since ε was arbitrary, we get the above representation. If $g(\omega, x) = \infty$, for $t > 0$, choose $k \in \mathbb{N}$ such that $g_k(x) > t + \varepsilon$ and $m \in \mathbb{N}$ such that $\|g_k - h_m\|_{\sup} \leq \varepsilon$. Then $h_m \leq g(\omega, \cdot)$ and $h_m(x) > t$ which yields $\sup \{h_n^+(x) : h_n \leq g(\omega, \cdot)\} = \infty$.

Setting $F_n := \{\omega \in \Omega : h_n \leq g(\omega, \cdot)\}$ for $n \in \mathbb{N}$ we obtain $g(\omega, x) = \sup \{1_{F_n} \otimes h_n^+(\omega, x) : n \in \mathbb{N}\}$ for every $\omega \in \Omega$ and $x \in \mathcal{X}$.

Now assume that g is bounded and $g(\omega, \cdot) \in C_b(\mathcal{X})$ for every $\omega \in \Omega$. Then

$$F_n = \bigcap_{x \in \mathcal{X}_0} \{h_n(x) \leq g(\cdot, x)\}$$

for some countable dense subset $\mathcal{X}_0$ of $\mathcal{X}$ and hence $F_n \in \mathcal{F}$. In view of the rather obvious fact that

$$V := \left\{ \sum_{i=1}^n 1_{H_i} \otimes k_i : H_i \in \mathcal{F} \text{ pairwise disjoint}, k_i \in C_b(\mathcal{X})_+, n \in \mathbb{N} \right\}$$

is a lattice in the pointwise ordering there exists a nondecreasing sequence $(v_n)_{n \geq 1}$ in V such that $g(\omega, x) = \sup_{n \in \mathbb{N}} v_n(\omega, x)$ for every $\omega \in \Omega$ and $x \in \mathcal{X}$. Using monotone convergence we obtain that the map $K \mapsto \int g \, dP \otimes K = \sup_{n \in \mathbb{N}} \int v_n \, dP \otimes K$ is lower τ-semicontinuous on $\mathcal{K}^1$. This can be applied to the function $-g + \sup g$ and yields that the map $K \mapsto \int g \, dP \otimes K$ is τ-continuous, hence (ii).

In the setting of (iv) the proof is a bit more involved because F_n is not necessarily in $\mathcal{F}$. However,

$$F_n^c = \bigcup_{x \in \mathcal{X}} \{\omega \in \Omega : h_n(x) > g(\omega, x)\}$$

is the image of $A_n := \{(\omega, x) \in \Omega \times \mathcal{X} : h_n(x) > g(\omega, x)\} \in \mathcal{F} \otimes \mathcal{B}(\mathcal{X})$ under the projection $\pi_\Omega : \Omega \times \mathcal{X} \to \Omega$ onto Ω, that is

$$\pi_\Omega(A_n) = \bigcup_{x \in \mathcal{X}} A_{n,x} = F_n^c,$$

and hence, using that $\mathcal{X}$ is polish, it follows from a projection theorem that F_n belongs to the P-completion of $\mathcal{F}$; see [83], Theorem 4. Therefore, for every $n \in \mathbb{N}$ there is a set $G_n \in \mathcal{F}$ and a P-null set $N_n \in \mathcal{F}$ such that $G_n \subset F_n$ and $F_n \backslash G_n \subset N_n$. Defining $N := \bigcup_{n \in \mathbb{N}} N_n$ we obtain $g(\omega, x) = \sup\{1_{G_n} \otimes h_n^+(\omega, x) : n \in \mathbb{N}\}$ for every $\omega \in N^c$ and $x \in \mathcal{X}$. As above, this yields the lower τ-semicontinuity of $K \mapsto \int g \, dP \otimes K$, hence (iv).

(ii) $\Rightarrow$ (i) is obvious, as is (iv) $\Leftrightarrow$ (iii) $\Rightarrow$ (ii). $\qquad\qquad\qquad\qquad\qquad\qquad$ $\square$

Finally we mention a characterization of compactness in $\mathcal{K}^1$. For this, it is convenient to identify Markov kernels in $\mathcal{K}^1$ that agree P-almost surely. One arrives at the space $K^1(P) = K^1(\mathcal{F}, P) = K^1(\mathcal{F}, P, \mathcal{X})$ of P-equivalence classes. The weak topology on $K^1(P)$, still denoted by $\tau(P)$, is now Hausdorff. For a sub-σ-field $\mathcal{G} \subset \mathcal{F}$, let $K^1(\mathcal{G}, P)$ denote the subspace of $K^1(P)$ consisting of equivalence classes which contain at least one representative from $\mathcal{K}^1(\mathcal{G})$. By Corollary 2.5 (c), the set $K^1(\mathcal{G}, P)$ is weakly closed in $K^1(P)$ provided $\mathcal{X}$ is polish.

A net in $\mathcal{M}^1(\mathcal{X})$ is called *tight* if the corresponding subset is tight. A weakly convergent sequence in $\mathcal{M}^1(\mathcal{X})$ is tight provided $\mathcal{X}$ is polish. In fact, weak convergence $\nu_n \to \nu$ in $\mathcal{M}^1(\mathcal{X})$ obviously implies weak compactness of $\{\nu_n : n \in \mathbb{N}\} \cup \{\nu\}$, hence $\{\nu_n : n \in \mathbb{N}\}$ is relatively weakly compact and thus tight.

Theorem 2.7 *Assume that $\mathcal{X}$ is polish. For a subset $\Gamma \subset K^1(P)$,*

(i) Γ *is relatively $\tau(P)$-compact*
if and only if
(ii) $P\Gamma := \{PK : K \in \Gamma\}$ *is relatively compact in $\mathcal{M}^1(\mathcal{X})$,*
and then
(iii) Γ *is relatively sequentially $\tau(P)$-compact.*
In particular, if $(K_\alpha)_\alpha$ is a net (sequence) in $\mathcal{K}^1$ such that $(PK_\alpha)_\alpha$ is tight, then $(K_\alpha)_\alpha$ has a weakly convergent subnet (subsequence).

Proof (i) $\Rightarrow$ (ii) is an immediate consequence of the continuity of the map $K \mapsto PK$.

(ii) $\Rightarrow$ (i). Choose as in the proof of Theorem 2.6 a totally bounded metrization of $\mathcal{X}$. Then the completion $\mathcal{Y}$ of $\mathcal{X}$ is compact and $\mathcal{X} \in \mathcal{B}(\mathcal{Y})$ because $\mathcal{X}$ is, as a polish subspace of the polish space $\mathcal{Y}$, a G_δ-set, i.e. a countable intersection of open subsets of $\mathcal{Y}$. Hence $\mathcal{B}(\mathcal{X}) \subset \mathcal{B}(\mathcal{Y})$. Because $U_b(\mathcal{X})$ and $C_b(\mathcal{Y})$ are isometrically isomorphic, it follows from the Portmanteau theorem that $(K^1(P, \mathcal{X}), \tau(P, \mathcal{X}))$ is homeomorphic to the subspace $\{K \in K^1(P, \mathcal{Y}) : PK(\mathcal{X}) = 1\}$ of $(K^1(P, \mathcal{Y}), \tau(P, \mathcal{Y}))$. Identifying these spaces and because $K^1(P, \mathcal{Y})$ is $\tau(P, \mathcal{Y})$-compact, see [29], [65], [33], Theorem 3.58, the $\tau(P, \mathcal{Y})$-closure $\overline{\Gamma}$ of Γ in $K^1(P, \mathcal{Y})$ is compact. Let $K \in \overline{\Gamma}$ and let $(K_\alpha)_\alpha$ be a net in Γ such that $K_\alpha \to K$ weakly in $K^1(P, \mathcal{Y})$. Because $P\Gamma$ is tight in $\mathcal{M}^1(\mathcal{X})$, for every $\varepsilon > 0$ we find a compact set $A \subset \mathcal{X}$ such that $PK_\alpha(A) \geq 1 - \varepsilon$ for every α. By Theorem 2.3 and the Portmanteau theorem we obtain

$$1 - \varepsilon \leq \limsup_\alpha PK_\alpha(A) \leq PK(A) \leq PK(\mathcal{X}).$$

This implies $PK(\mathcal{X}) = 1$ and hence $K \in K^1(P, \mathcal{X})$.

(i) $\Rightarrow$ (iii). Let $(K_n)_{n \geq 1}$ be a sequence in Γ and $\mathcal{G} := \sigma(K_n, n \in \mathbb{N})$. If $\mathcal{A}$ denotes a countable generator of $\mathcal{B}(\mathcal{X})$ which is stable under finite intersections, then $\mathcal{G} = \sigma(K_n(\cdot, B), B \in \mathcal{A}, n \in \mathbb{N})$ so that $\mathcal{G}$ is a countably generated sub-σ-field of $\mathcal{F}$. In view of Corollary 2.5 (a) the set $\{K_n : n \in \mathbb{N}\}$ is relatively $\tau(\mathcal{G}, P)$-compact and because $(K^1(\mathcal{G}, P), \tau(\mathcal{G}, P))$ is metrizable, see [33], Proposition 3.25, [18], Theorem 4.16, $(K_n)_{n \geq 1}$ has a $\tau(\mathcal{G}, P)$-convergent subsequence which is again by Corollary 2.5 (a) also $\tau(P)$-convergent. $\qquad\square$

Exercise 2.5 Show that one can replace in the last part of Theorem 2.7 the tightness of the net $(PK_\alpha)_\alpha$ by its weak convergence in $\mathcal{M}^1(\mathcal{X})$.

Exercise 2.6 Assume that $\mathcal{X}$ is polish and let $\Gamma \subset \mathcal{K}^1$. Regarding each $K \in \mathcal{K}^1$ as an $(\mathcal{M}^1(\mathcal{X}), \mathcal{B}(\mathcal{M}^1(\mathcal{X})))$-valued random variable, prove that $P\Gamma$ is tight in $\mathcal{M}^1(\mathcal{X})$ if and only if $\{P^K : K \in \Gamma\}$ is tight in $\mathcal{M}^1(\mathcal{M}^1(\mathcal{X}))$. Here P^K denotes the image measure.

Exercise 2.7 Let $\mathcal{Y}$ be a further separable metrizable topological space. Show that the weak topology on $\mathcal{M}^1(\mathcal{X} \times \mathcal{Y})$ is generated by the functions

$$\mu \mapsto \int h \otimes k \, d\mu, \quad h \in C_b(\mathcal{X}), \ k \in C_b(\mathcal{Y})$$

and the weak topology on $\mathcal{K}^1(\mathcal{F}, \mathcal{X} \times \mathcal{Y})$ is generated by the functions

$$H \mapsto \int 1_F \otimes h \otimes k \, dP \otimes H, \quad F \in \mathcal{F}, \ h \in C_b(\mathcal{X}), \ k \in C_b(\mathcal{Y}).$$

Exercise 2.8 Let $\mathcal{Y}$ be a further separable metrizable space. Let $(H_\alpha)_\alpha$ be a net in $\mathcal{K}^1(\mathcal{F}, \mathcal{X})$, $H \in \mathcal{K}^1(\mathcal{F}, \mathcal{X})$ and let $(K_\alpha)_\alpha$ be a net in $\mathcal{K}^1(\mathcal{F}, \mathcal{Y})$, $K \in \mathcal{K}^1(\mathcal{F}, \mathcal{Y})$. Assume that $H_\alpha \to H$ weakly and

$$\int k(y) \, K_\alpha(\cdot, dy) \to \int k(y) \, K(\cdot, dy) \quad \text{in } \mathcal{L}^1(P) \text{ for every } k \in C_b(\mathcal{Y}).$$

Show that $H_\alpha \otimes K_\alpha \to H \otimes K$ weakly in $\mathcal{K}^1(\mathcal{F}, \mathcal{X} \times \mathcal{Y})$.

Chapter 3
Stable Convergence of Random Variables

Based on the notions and results of Chap. 2 we may now introduce and deeply investigate the mode of stable convergence of random variables. Starting from the papers [76–78] expositions can be found in [4, 13, 48, 50, 57].

Let $\mathcal{X}$ still be a separable metrizable topological space and fix a metric d that induces the topology on $\mathcal{X}$. For an $(\mathcal{X}, \mathcal{B}(\mathcal{X}))$-valued random variable X and a sub-σ-field $\mathcal{G} \subset \mathcal{F}$ let $P^{X|\mathcal{G}}$ denote the conditional distribution which exists, for example, provided that $\mathcal{X}$ is polish. It is a Markov kernel from $(\Omega, \mathcal{G})$ to $(\mathcal{X}, \mathcal{B}(\mathcal{X}))$ such that $P^{X|\mathcal{G}}(\cdot, B) = P(X \in B|\mathcal{G})$ almost surely for all $B \in \mathcal{B}(\mathcal{X})$. The conditional distribution is P-almost surely unique by Lemma 2.1 (b) and characterized by the Radon-Nikodym equations

$$\int_G P^{X|\mathcal{G}}(\omega, B) \, dP(\omega) = P\left(X^{-1}(B) \cap G\right) \quad \text{for every } G \in \mathcal{G}, B \in \mathcal{B}(\mathcal{X}) \,,$$

or, what is the same,

$$P \otimes P^{X|\mathcal{G}} = P \otimes \delta_X \text{ on } \mathcal{G} \otimes \mathcal{B}(\mathcal{X}) \,,$$

where δ_X is the *Dirac-kernel* associated with X given by $\delta_X(\omega) := \delta_{X(\omega)}$. If, for example, X is $\mathcal{G}$-measurable, then $P^{X|\mathcal{G}} = \delta_X$. The distribution of X (under P) is denoted by P^X. In the sequel we restrict our attention to sequences of random variables, all defined on the same probability space $(\Omega, \mathcal{F}, P)$.

3.1 First Approach

Definition 3.1 Let $\mathcal{G} \subset \mathcal{F}$ be a sub-σ-field. A sequence $(X_n)_{n \geq 1}$ of $(\mathcal{X}, \mathcal{B}(\mathcal{X}))$-valued random variables is said to *converge $\mathcal{G}$-stably* to $K \in \mathcal{K}^1(\mathcal{G})$, written as $X_n \to K$ $\mathcal{G}$-stably, if $P^{X_n|\mathcal{G}} \to K$ weakly as $n \to \infty$. In case K does not depend on

© Springer International Publishing Switzerland 2015
E. Häusler and H. Luschgy, *Stable Convergence and Stable Limit Theorems*,
Probability Theory and Stochastic Modelling 74,
DOI 10.1007/978-3-319-18329-9_3

$\omega \in \Omega$ in the sense that $K = \nu$ P-almost surely for some $\nu \in \mathcal{M}^1 (\mathcal{X})$, then $(X_n)_{n \geq 1}$ is said to *converge $\mathcal{G}$-mixing to ν*, and we write $X_n \rightarrow \nu$ $\mathcal{G}$-mixing. *Stable* and *mixing convergence* are short for $\mathcal{F}$-stable and $\mathcal{F}$-mixing convergence, respectively.

In Definition 3.1 and in the sequel we always assume that the conditional distributions involved exist. (Existence is not part of the subsequent assertions.)

Using Fubini's theorem for Markov kernels (see Lemma 2.1 (a)) and the fact that $\int h (x) \, P^{X_n | \mathcal{G}} (dx) = E (h (X_n) | \mathcal{G})$, $\mathcal{G}$-stable convergence $X_n \rightarrow K$ reads

$$\lim_{n \rightarrow \infty} E (f E (h (X_n) | \mathcal{G})) = \int f \int h (x) \, K (\cdot, dx) \, dP$$

for every $f \in \mathcal{L}^1 (P)$ and $h \in C_b (\mathcal{X})$. The choice $f = 1$ implies $X_n \overset{d}{\rightarrow} PK$, that is, $P^{X_n} \rightarrow PK$ weakly. Here and elsewhere the reference measure for distributional convergence is always P. The $\mathcal{G}$-mixing convergence $X_n \rightarrow \nu$ means

$$\lim_{n \rightarrow \infty} E (f E (h (X_n) | \mathcal{G})) = \int f \, dP \int h \, d\nu$$

for every $f \in \mathcal{L}^1 (P)$ and $h \in C_b (\mathcal{X})$, which implies $X_n \overset{d}{\rightarrow} \nu$. Because $P^{X_n | \mathcal{G}} = E (\delta_{X_n} | \mathcal{G})$ in the sense of Definition 2.4, $\mathcal{G}$-stable convergence $X_n \rightarrow K$ can also be read as $E (\delta_{X_n} | \mathcal{G}) \rightarrow K$ weakly. In the extreme case $\mathcal{G} = \{\emptyset, \Omega\}$, $\mathcal{G}$-stable convergence $X_n \rightarrow K$ coincides with distributional convergence $X_n \overset{d}{\rightarrow} \nu$, because $K = \nu$ for some $\nu \in \mathcal{M}^1 (\mathcal{X})$ by $\mathcal{G}$-measurability of K.

Typical limit kernels for $\mathcal{G}$-stable convergence $X_n \rightarrow K$ are of the type $K (\omega, \cdot) = \mu^{\varphi(\omega, \cdot)}$, where $\mu \in \mathcal{M}^1 (\mathcal{Y})$, $\mathcal{Y}$ is a separable metrizable space and $\varphi : (\Omega \times \mathcal{Y}, \mathcal{G} \otimes \mathcal{B} (\mathcal{Y})) \rightarrow (\mathcal{X}, \mathcal{B} (\mathcal{X}))$ is some "concrete" measurable map. Here $\mu^{\varphi(\omega, \cdot)}$ is the image measure of μ under the map $\varphi (\omega, \cdot)$ so that $K (\omega, B) = \mu (\{y \in \mathcal{Y} : \varphi (\omega, y) \in B\})$. In fact, every kernel has such a representation provided $\mathcal{X}$ is polish; see [51], Lemma 3.22)). In particular, if $\mathcal{X} = \mathcal{Y} = \mathbb{R}$, $\mu = N (0, 1)$ and $\varphi (\omega, x) := \eta (\omega) x$ for some $\mathcal{G}$-measurable and nonnegative real random variable η, we obtain the *Gauss-kernel* $K (\omega, \cdot) = N (0, 1)^{\varphi(\omega, \cdot)} = N (0, \eta^2 (\omega))$.

The results of Chap. 2 provide the following characterizations of $\mathcal{G}$-stable convergence.

Theorem 3.2 *Let X_n be $(\mathcal{X}, \mathcal{B} (\mathcal{X}))$-valued random variables, $K \in \mathcal{K}^1 (\mathcal{G})$ and let $\mathcal{E} \subset \mathcal{G}$ be closed under finite intersections with $\Omega \in \mathcal{E}$ and $\sigma (\mathcal{E}) = \mathcal{G}$. Then the following statements are equivalent:*

(i) $X_n \rightarrow K$ $\mathcal{G}$-stably,
(ii) $\lim_{n \rightarrow \infty} Efh (X_n) = \int f \otimes h \, dP \otimes K$ *for every $f \in \mathcal{L}^1 (\mathcal{G}, P)$ and $h \in C_b (\mathcal{X})$,*
(iii) $Q^{X_n} \rightarrow QK$ *weakly (in $\mathcal{M}^1 (\mathcal{X})$) for every probability distribution Q on $\mathcal{F}$ such that $Q \ll P$ and dQ/dP is $\mathcal{G}$-measurable,*
(iv) $P_F^{X_n} \rightarrow P_F K$ *weakly for every $F \in \mathcal{E}$ with $P (F) > 0$,*

(v) $\lim_{n \to \infty} \int g(\omega, X_n(\omega)) \, dP(\omega) = \int g \, dP \otimes K$ *for every measurable,*
    bounded function $g : (\Omega \times \mathcal{X}, \mathcal{G} \otimes \mathcal{B}(\mathcal{X})) \to (\mathbb{R}, \mathcal{B}(\mathbb{R}))$ *such that* $g(\omega, \cdot) \in$
    $C_b(\mathcal{X})$ *for every* $\omega \in \Omega$,

(vi) *(For $\mathcal{X}$ polish)* $\limsup_{n \to \infty} \int g(\omega, X_n(\omega)) \, dP(\omega) \leq \int g \, dP \otimes K$ *for every*
    measurable function $g : (\Omega \times \mathcal{X}, \mathcal{G} \otimes \mathcal{B}(\mathcal{X})) \to \left(\overline{\mathbb{R}}, \mathcal{B}\left(\overline{\mathbb{R}}\right)\right)$ *which is*
    bounded from above such that $g(\omega, \cdot)$ *is upper semicontinuous for every*
    $\omega \in \Omega$,

(vii) $(X_n, Y) \to K \otimes \delta_Y$ $\mathcal{G}$-*stably for every separable metrizable space* $\mathcal{Y}$ *and*
    every $\mathcal{G}$-*measurable* $(\mathcal{Y}, \mathcal{B}(\mathcal{Y}))$-*valued random variable* Y, *where* $K \otimes \delta_Y \in$
    $\mathcal{K}^1(\mathcal{G}, \mathcal{X} \times \mathcal{Y})$, $K \otimes \delta_Y(\omega, \cdot) = K(\omega, \cdot) \otimes \delta_{Y(\omega)}$,

(viii) $(X_n, 1_F) \xrightarrow{d} P\left(K \otimes \delta_{1_F}\right)$ *for every* $F \in \mathcal{E}$.

Proof The equivalences (i)–(vi) follow from Theorems 2.3 and 2.6. Here are some comments. First, observe that for $Q \ll P$ such that dQ/dP is $\mathcal{G}$-measurable we have $Q \otimes P^{X_n|\mathcal{G}} = Q \otimes \delta_{X_n}$ on $\mathcal{G} \otimes \mathcal{B}(\mathcal{X})$ and hence $QP^{X_n|\mathcal{G}} = Q\delta_{X_n} = Q^{X_n}$ for the marginals on $\mathcal{B}(\mathcal{X})$ (see Lemma A.4 (d)).

(i) $\Leftrightarrow$ (ii). For $f \in \mathcal{L}^1(\mathcal{G}, P)$ and $h \in C_b(\mathcal{X})$ we have $E(fE(h(X_n)|\mathcal{G})) = Efh(X_n)$.

(i) $\Rightarrow$ (iii) $\Rightarrow$ (iv) $\Rightarrow$ (i) are clear from the above formulas and Theorem 2.3.

(ii) $\Leftrightarrow$ (v) $\Leftrightarrow$ (vi). For a measurable function $g : (\Omega \times \mathcal{X}, \mathcal{G} \otimes \mathcal{B}(\mathcal{X})) \to \left(\overline{\mathbb{R}}, \mathcal{B}\left(\overline{\mathbb{R}}\right)\right)$ which is bounded from above,

$$\int g \, dP \otimes P^{X_n|\mathcal{G}} = \int g \, dP \otimes \delta_{X_n} = \int g(\omega, X_n(\omega)) \, dP(\omega).$$

Therefore the equivalences follow from Theorem 2.6 applied to the weak topology $\tau(\mathcal{G})$ on $\mathcal{K}^1(\mathcal{G})$ instead of τ.

(v) $\Rightarrow$ (vii). For $F \in \mathcal{G}$ and $h \in C_b(\mathcal{X} \times \mathcal{Y})$ define $g : \Omega \times \mathcal{X} \to \mathbb{R}$ by $g(\omega, x) := 1_F(\omega) h(x, Y(\omega))$. Using $\mathcal{B}(\mathcal{X} \times \mathcal{Y}) = \mathcal{B}(\mathcal{X}) \otimes \mathcal{B}(\mathcal{Y})$ we see that g is $\mathcal{G} \otimes \mathcal{B}(\mathcal{X})$-measurable and $g(\omega, \cdot) \in C_b(\mathcal{X})$ for every $\omega \in \Omega$, so that

$$\lim_{n \to \infty} E 1_F h(X_n, Y) = \lim_{n \to \infty} \int g(\omega, X_n(\omega)) \, dP(\omega) = \int g \, dP \otimes K$$

$$= \int \int \int 1_F(\omega) h(x, y) \, d\delta_{Y(\omega)}(y) \, K(\omega, dx) \, dP(\omega)$$

$$= \int 1_F \otimes h \, dP \otimes (K \otimes \delta_Y).$$

Now $\mathcal{G}$-stable convergence (vii) follows in view of (iv) $\Leftrightarrow$ (i). Note that no further assumption on $\mathcal{Y}$ is needed to assure the existence of conditional distributions because $P^{(X_n, Y)|\mathcal{G}} = P^{X_n|\mathcal{G}} \otimes \delta_Y$ (see Lemma A.5 (a)).

(vii) $\Rightarrow$ (viii) is clear.

(viii) $\Rightarrow$ (iv). For $F \in \mathcal{E}$, $h \in C_b(\mathcal{X})$ and $k \in C_b(\mathbb{R})$ satisfying $k(x) = x$ for $x \in [0, 1]$ and $Y = 1_F$, we have $h \otimes k \in C_b(\mathcal{X} \times \mathbb{R})$ and thus

$$\lim_{n \to \infty} E 1_F h(X_n) = \lim_{n \to \infty} Eh \otimes k(X_n, Y) = \int h \otimes k \, dP(K \otimes \delta_Y)$$

$$= \int \int \int h(x) k(y) \, d\delta_{Y(\omega)}(y) \, K(\omega, dx) \, dP(\omega)$$

$$= \int 1_F \otimes h \, dP \otimes K. \qquad \qquad \square$$

Some of the above equivalent conditions are more useful in a given situation than the others. So, for proving a particular stable limit theorem, Theorem 3.2 (iv) is usually used. In order to obtain theoretical consequences of stability, the other conditions are more interesting.

Unlike convergence in distribution, stable convergence $X_n \to K$ is a property of the random variables X_n rather than of their distributions. Consider, for example, a $U(0, 1)$-distributed random variable U and set $X_n := U$ if n is even, $X_n := 1 - U$ if n is odd and $Y_n := U$ for every n. Then $P^{X_n} = P^{Y_n}$ for every n and $Y_n \to \delta_U$ stably, but X_n does not converge stably, because otherwise $\delta_U = \delta_{1-U}$ by uniqueness of limit kernels so that $U = 1 - U$ or $U = 1/2$ almost surely.

Exercise 3.1 Let $(F_n)_{n \geq 1}$ be a nonincreasing (nondecreasing) sequence in $\mathcal{F}$, $F = \bigcap_{n=1}^{\infty} F_n$ $(F = \bigcup_{n=1}^{\infty} F_n)$ and $P(F) > 0$. Show that if $P_F^{X_n} \to \nu$ weakly for some $\nu \in \mathcal{M}^1(\mathcal{X})$, then $P_{F_n}^{X_n} \to \nu$ weakly as $n \to \infty$.

Exercise 3.2 Let $F_n \in \mathcal{F}, \alpha : \Omega \to [0, 1]$ $\mathcal{G}$-measurable and $K(\omega, \cdot) := \alpha(\omega) \delta_1 + (1 - \alpha(\omega)) \delta_0$. Show that $1_{F_n} \to K$ $\mathcal{G}$-stably if and only if $\lim_{n \to \infty} P(F_n \cap G) = \int_G \alpha \, dP$ for every $G \in \mathcal{G}$.

Exercise 3.3 Let $(\Omega, \mathcal{F}, P) = ([0, 1], \mathcal{B}([0, 1]), \lambda_{[0,1]})$, $a_n \in [0, 1/2]$, $X_n := 1_{[a_n, a_n + 1/2]}$ and $\mathcal{G} := \sigma(Y)$, where $Y : \Omega \to \mathbb{R}$, $Y(\omega) = \omega$. Show that $P^{X_n} = (\delta_0 + \delta_1)/2$ for every n but, if $(a_n)_{n \geq 1}$ is not convergent, $(X_n)_{n \geq 1}$ does not converge $\mathcal{G}$-stably.

Corollary 3.3 (Mixing convergence) *In the situation of Theorem 3.2 let $K = \nu$ almost surely for some $\nu \in \mathcal{M}^1(\mathcal{X})$. Then the following assertions are equivalent:*

(i) $X_n \to \nu$ $\mathcal{G}$-mixing,
(ii) $\lim_{n \to \infty} Efh(X_n) = \int f \, dP \int h \, d\nu$ for every $f \in \mathcal{L}^1(\mathcal{G}, P)$ and $h \in C_b(\mathcal{X})$,
(iii) $Q^{X_n} \to \nu$ weakly for every probability distribution Q on $\mathcal{F}$ such that $Q \ll P$ and dQ/dP is $\mathcal{G}$-measurable,
(iv) $P_F^{X_n} \to \nu$ weakly for every $F \in \mathcal{E}$ with $P(F) > 0$,
(v) $(X_n, Y) \xrightarrow{d} \nu \otimes P^Y$ for every separable metrizable space $\mathcal{Y}$ and every $\mathcal{G}$-measurable $(\mathcal{Y}, \mathcal{B}(\mathcal{Y}))$-valued random variable Y.

Proof The equivalences (i)–(iv) are obvious consequences of Theorem 3.2.

(i) $\Rightarrow$ (v). By Theorem 3.2, we have $(X_n, Y) \to \nu \otimes \delta_Y$ $\mathcal{G}$-stably so that $(X_n, Y) \overset{d}{\to} P(\nu \otimes \delta_Y) = \nu \otimes P^Y$.

(v) $\Rightarrow$ (i) is again immediate from Theorem 3.2. $\qquad\square$

Exercise 3.4 Assume that $\sigma(X_n)$ and $\mathcal{G}$ are independent for every $n \in \mathbb{N}$. Prove that

(i) (X_n) converges $\mathcal{G}$-stably,
(ii) (X_n) converges $\mathcal{G}$-mixing,
(iii) (X_n) converges in distribution

are equivalent assertions.

Next we state various further features of stable convergence.

Proposition 3.4 (a) (*For $\mathcal{X}$ polish*) *If $\left(P^{X_n}\right)_{n \geq 1}$ is tight in $\mathcal{M}^1(\mathcal{X})$, then $(X_n)_{n \geq 1}$ has a stably convergent subsequence.*
(b) (*For $\mathcal{X}$ polish*) *Let $\mathcal{G}_1 \subset \mathcal{G}_2 \subset \mathcal{F}$ be sub-σ-fields and $K \in \mathcal{K}^1(\mathcal{G}_2)$. If $X_n \to K$ $\mathcal{G}_2$-stably, then $X_n \to E(K|\mathcal{G}_1)$ $\mathcal{G}_1$-stably.*
(c) *Let $\mathcal{Y}$ be a separable metrizable space, Y a $(\mathcal{Y}, \mathcal{B}(\mathcal{Y}))$-valued random variable, $\mathcal{G} = \sigma(Y)$ and $K \in \mathcal{K}^1(\mathcal{G})$. Then $X_n \to K$ $\mathcal{G}$-stably if and only if $(X_n, Y) \overset{d}{\to} P(K \otimes \delta_Y)$.*

Proof (a) By Theorem 2.7, there exists a subsequence (X_k) of (X_n) with $\delta_{X_k} \to K$ weakly for some $K \in \mathcal{K}^1$ because $\left(P\delta_{X_n}\right) = \left(P^{X_n}\right)$ is tight. Using $P^{X_k|\mathcal{F}} = \delta_{X_k}$, this means $X_k \to K$ stably.

(b) It follows from Lemma A.7 (b) and Corollary 2.5 (b) that $P^{X_n|\mathcal{G}_1} = E(P^{X_n|\mathcal{G}_2}|\mathcal{G}_1) \to E(K|\mathcal{G}_1)$ weakly in $\mathcal{K}^1$, that is, $X_n \to E(K|\mathcal{G}_1)$ $\mathcal{G}_1$-stably.

(c) The "if" part. One checks that

$$\mathcal{L} := \left\{ f \in \mathcal{L}^1(\mathcal{G}, P) : \lim_{n \to \infty} Efh(X_n) = \int f \otimes h\, dP \otimes K \text{ for every } h \in C_b(\mathcal{X}) \right\}$$

is a closed vector subspace of $\mathcal{L}^1(\mathcal{G}, P)$. Moreover, functions f of the type $f = k(Y)$ with $k \in C_b(\mathcal{Y})$ belong to $\mathcal{L}$ because

$$Efh(X_n) = Eh \otimes k(X_n, Y) \to \int h \otimes k\, dP(K \otimes \delta_Y) = \int f \otimes h\, dP \otimes K.$$

Since $C_b(\mathcal{Y})$ is dense in $\mathcal{L}^1(P^Y)$, the vector space $\{k(Y) : k \in C_b(\mathcal{Y})\}$ is dense in $\mathcal{L}^1(\mathcal{G}, P)$ so that $\mathcal{L} = \mathcal{L}^1(\mathcal{G}, P)$. Theorem 3.2 yields $X_n \to K$ $\mathcal{G}$-stably.

The "only if" part follows from Theorem 3.2. $\qquad\square$

The most powerful case concerns $\mathcal{G}$-stability when X_n is $\mathcal{G}$-measurable for every n.

Proposition 3.5 *Let X_n be $\mathcal{G}$-measurable, $(\mathcal{X}, \mathcal{B}(\mathcal{X}))$-valued random variables and let $K \in \mathcal{K}^1(\mathcal{G})$. Then the following assertions are equivalent:*

(i) $X_n \to K$ $\mathcal{G}$-stably,

(ii) $X_n \to K$ stably,

(iii) $\delta_{X_n} \to K$ weakly.

Proof The result is an immediate consequence of Definition 3.1 and $P^{X_n | \mathcal{G}} = P^{X_n | \mathcal{F}} = \delta_{X_n}$ P-almost surely. □

Exercise 3.5 Show that the following assertions are equivalent:

(i) $X_n \to \nu$ mixing,

(ii) $P_F^{X_n} \to \nu$ weakly for every $F \in \mathcal{E} := \bigcup_{k=1}^{\infty} \sigma(X_k)$ with $P(F) > 0$,

(iii) $X_n \xrightarrow{d} \nu$ and $\lim_{n \to \infty} P^{(X_n, X_k)}(B \times B) = \nu(B) P^{X_k}(B)$ for every $k \in \mathbb{N}$ and $B \in \mathcal{B}(\mathcal{X})$ with $\nu(\partial B) = 0$,

(iv) $(X_n, X_k) \xrightarrow{d} \nu \otimes P^{X_k}$ as $n \to \infty$ for every $k \in \mathbb{N}$.

(Note that $\mathcal{E}$ is a generator of $\sigma(X_n, n \geq 1)$ which is generally not closed under finite intersections.)

In case of a Dirac kernel as limit kernel, stable convergence turns into convergence in probability. Recall that a sequence $(X_n)_{n \geq 1}$ of $(\mathcal{X}, \mathcal{B}(\mathcal{X}))$-valued random variables is said to *converge in probability* to an $(\mathcal{X}, \mathcal{B}(\mathcal{X}))$-valued random variable X if $\lim_{n \to \infty} P(d(X_n, X) > \varepsilon) = 0$ for every $\varepsilon > 0$, where d is any metric which metrizes $\mathcal{X}$ and $d(X_n, X)$ is $\mathcal{F}$-measurable because $\mathcal{B}(\mathcal{X} \times \mathcal{X}) = \mathcal{B}(\mathcal{X}) \otimes \mathcal{B}(\mathcal{X})$. This feature does not depend on the choice of the metric, see e.g. [35], p. 335, and is equivalent to $\lim_{n \to \infty} E(d(X_n, X) \wedge 1) = 0$.

Corollary 3.6 (Convergence in probability) *For $(\mathcal{X}, \mathcal{B}(\mathcal{X}))$-valued random variables X_n and X, where X is $\mathcal{G}$-measurable for some sub-σ-field $\mathcal{G}$ of $\mathcal{F}$, the following assertions are equivalent:*

(i) $X_n \to X$ in probability,

(ii) $X_n \to \delta_X$ $\mathcal{G}$-stably,

(iii) $Q^{X_n} \to Q^X$ weakly for every probability distribution Q on $\mathcal{F}$ such that $Q \ll P$ and dQ/dP is $\mathcal{G}$-measurable.

This corollary may of course be applied with $\mathcal{G} = \mathcal{F}$.

Proof (i) ⇒ (iii). For Q with $Q \ll P$ it follows from (i) that $X_n \to X$ in Q-probability and hence (iii).

(ii) ⇔ (iii) is an immediate consequence of Theorem 3.2 because $Q^X = Q\delta_X$.

(ii) ⇒ (i). Define $g : \Omega \times \mathcal{X} \to \mathbb{R}$ by $g(\omega, x) := d(x, X(\omega)) \wedge 1$. Since g is $\mathcal{G} \otimes \mathcal{B}(\mathcal{X})$-measurable and $g(\omega, \cdot) \in C_b(\mathcal{X})$ for every $\omega \in \Omega$, Theorem 3.2 yields

$$\lim_{n \to \infty} E(d(X_n, X) \wedge 1) = \lim_{n \to \infty} \int g(\omega, X_n(\omega)) \, dP(\omega) = \int g \, dP \otimes \delta_X$$

$$= \int g(\omega, X(\omega)) \, dP(\omega) = 0,$$

hence (i). □

Exercise 3.6 Assume $X_n \rightarrow \nu$ mixing, where $\nu \in \mathcal{M}^1(\mathcal{X})$ is no Dirac-measure. Show that X_n cannot converge in probability.

Exercise 3.7 (a) Assume that $\mathcal{X}$ is polish. Find a direct proof of the implication (ii) $\Rightarrow$ (i) in Corollary 3.6 based only on the definition of $\mathcal{G}$-stable convergence (that is, on Theorem 3.2, (i) $\Rightarrow$ (ii)).
(b) Find a proof of the same implication based on Theorem 3.2, (i) $\Rightarrow$ (vii).

The main advantage of stable convergence when compared with distributional convergence is contained in part (b) of the next result.

Theorem 3.7 *Assume $X_n \rightarrow K$ $\mathcal{G}$-stably for $(\mathcal{X}, \mathcal{B}(\mathcal{X}))$-valued random variables X_n and $K \in \mathcal{K}^1(\mathcal{G})$. Let $\mathcal{Y}$ be a separable metrizable space and Y_n, Y random variables with values in $(\mathcal{Y}, \mathcal{B}(\mathcal{Y}))$.*

(a) *Let $\mathcal{X} = \mathcal{Y}$. If $d(X_n, Y_n) \rightarrow 0$ in probability, then $Y_n \rightarrow K$ $\mathcal{G}$-stably.*
(b) *If $Y_n \rightarrow Y$ in probability and Y is $\mathcal{G}$-measurable, then $(X_n, Y_n) \rightarrow K \otimes \delta_Y$ $\mathcal{G}$-stably.*
(c) *If $g : \mathcal{X} \rightarrow \mathcal{Y}$ is Borel-measurable and PK-almost surely continuous, then $g(X_n) \rightarrow K^g$ $\mathcal{G}$-stably with $K^g(\omega, \cdot) := K(\omega, \cdot)^g$. The PK-almost sure continuity of g means that the Borel set $\{x \in \mathcal{X} : g$ is not continuous at $x\}$ has PK-measure zero.*

Proof (a) For $F \in \mathcal{G}$ with $P(F) > 0$ we have $d(X_n, Y_n) \rightarrow 0$ in P_F-probability and, by Theorem 3.2, $P_F^{X_n} \rightarrow P_F K$ weakly. This implies $P_F^{Y_n} \rightarrow P_F K$ weakly by Theorem 4.1 in [9]. Hence $Y_n \rightarrow K$ $\mathcal{G}$-stably again by Theorem 3.2.

(b) Since $(X_n, Y) \rightarrow K \otimes \delta_Y$ $\mathcal{G}$-stably by Theorem 3.2, (b) follows from (a).

(c) For any distribution Q on $\mathcal{F}$ such that $Q \ll P$ and dQ/dP is $\mathcal{G}$-measurable we have weak convergence $Q^{X_n} \rightarrow QK$ by Theorem 3.2. Since $QK \ll PK$, the function g is QK-almost surely continuous so that $(Q^{X_n})^g \rightarrow (QK)^g$ weakly (in $\mathcal{M}^1(\mathcal{Y})$) by [9], Theorem 5.1. In view of $(Q^{X_n})^g = Q^{g(X_n)}$ and $(QK)^g = QK^g$ the assertion follows from Theorem 3.2. $\qquad\square$

We now consider special spaces $\mathcal{X}$. In case $\mathcal{X} = \mathbb{R}^d$, let $\langle \cdot, \cdot \rangle$ denote the usual scalar product.

Corollary 3.8 *Let $\mathcal{X} = \mathbb{R}^d$. Let X_n be $\mathbb{R}^d$-valued random variables, $K \in \mathcal{K}^1(\mathcal{G}, \mathbb{R}^d)$ and let $\mathcal{E} \subset \mathcal{G}$ be closed under finite intersections with $\Omega \in \mathcal{E}$ and $\sigma(\mathcal{E}) = \mathcal{G}$. Then the following assertions are equivalent:*

(i) $X_n \rightarrow K$ $\mathcal{G}$-stably,
(ii) $\lim_{n \to \infty} E 1_F \exp(i \langle u, X_n \rangle) = E 1_F \int \exp(i \langle u, x \rangle) K(\cdot, dx)$ *for every $F \in \mathcal{E}$ and $u \in \mathbb{R}^d$,*
(iii) *(Cramér-Wold device) $\langle u, X_n \rangle \rightarrow K^u$ $\mathcal{G}$-stably for every $u \in \mathbb{R}^d$, where $K^u \in \mathcal{K}^1(\mathcal{G}, \mathbb{R})$ is given by $K^u(\omega, \cdot) := K(\omega, \cdot)^{\langle u, \cdot \rangle}$.*

Proof This follows from Theorem 3.2 and Lévy's continuity theorem. $\qquad\square$

Now let $\mathcal{X} = C([0, T]) = C_b([0, T])$, for $0 < T < \infty$ and equipped with the sup-norm, or $\mathcal{X} = C(\mathbb{R}_+)$. Then $C([0, T])$ is polish. The space $C(\mathbb{R}_+)$ of all continuous functions $x : \mathbb{R}_+ \to \mathbb{R}$ is equipped with the local uniform topology induced by the metric $d(x, y) = \sum_{n=1}^{\infty} 2^{-n} \left(\max_{t \in [0, n]} |x(t) - y(t)| \wedge 1 \right)$. This metric is complete, $C(\mathbb{R}_+)$ is a polish space and $\mathcal{B}(C(I)) = \sigma(\pi_t, t \in I)$, $I = [0, T]$ or $I = \mathbb{R}_+$, where $\pi_t : C(I) \to \mathbb{R}$, $\pi_t(x) = x(t)$ denotes the projection (see [53], Theorems 21.30 and 21.31). Consequently, any path-continuous stochastic process $X = (X_t)_{t \in I}$ may be viewed as a $(C(I), \mathcal{B}(C(I)))$-valued random variable. For $t_j \in I$ let $\pi_{t_1, \ldots, t_k} : C(I) \to \mathbb{R}^k$, $\pi_{t_1, \ldots, t_k}(x) = (x(t_1), \ldots, x(t_k))$.

Proposition 3.9 *Let $\mathcal{X} = C(I)$ with $I = [0, T]$ or $\mathbb{R}_+$, and let $X^n = \left(X_t^n\right)_{t \in I}$ be path-continuous processes and $K \in \mathcal{K}^1(\mathcal{G})$. Then the following assertions are equivalent:*

(i) $X^n \to K$ $\mathcal{G}$-stably,
(ii) $\left(P^{X^n}\right)_{n \geq 1}$ *is tight and* $\left(X_{t_1}^n, \ldots, X_{t_k}^n\right) \to K^{\pi_{t_1, \ldots, t_k}}$ $\mathcal{G}$-stably *for every* $k \geq 1$ *and* $0 \leq t_1 < \cdots < t_k$, $t_j \in I$.

Proof (i) $\Rightarrow$ (ii). Since $P^{X^n} \to PK$ weakly, the sequence $\left(P^{X^n}\right)_{n \geq 1}$ is tight. The second assertion follows from Theorem 3.7 (c).

(ii) $\Rightarrow$ (i). If $X^n \nrightarrow K$ $\mathcal{G}$-stably, we may choose functions $f \in \mathcal{L}^1(\mathcal{G}, P)$ and $h \in C_b(\mathcal{X})$ and some $\varepsilon > 0$ such that $\left| Efh(X^r) - \int f \otimes h\, dP \otimes K \right| \geq \varepsilon$ along a subsequence (r) of the sequence (n) of all positive integers. By Proposition 3.4 (a), (b) there exists a further subsequence (m) of (r) and an $H \in \mathcal{K}^1(\mathcal{G})$ such that $X^m \to H$ $\mathcal{G}$-stably. But then by Theorem 3.7 (c) and Theorem 3.2

$$P_F^{\left(X_{t_1}^m, \ldots, X_{t_k}^m\right)} \to P_F H^{\pi_{t_1, \ldots, t_k}} = (P_F H)^{\pi_{t_1, \ldots, t_k}} \quad \text{weakly,}$$

and because also by (ii)

$$P_F^{\left(X_{t_1}^m, \ldots, X_{t_k}^m\right)} \to (P_F K)^{\pi_{t_1, \ldots, t_k}} \quad \text{weakly,}$$

for every $\mathcal{F} \in \mathcal{G}$ with $P(F) > 0$ and every $k \geq 1, 0 \leq t_1 < \cdots < t_k, t_j \in I$, we obtain $P_F H = P_F K$ for every $F \in \mathcal{G}$ with $P(F) > 0$, which yields $H = K$ almost surely. Thus $X^m \to K$ $\mathcal{G}$-stably, and so $Efh(X^m) \to \int f \otimes h\, dP \otimes K$, which is a contradiction. $\square$

Note that characterizations of stable convergence similar to Proposition 3.9 may by given for spaces of càdlàg functions, e.g. $\mathcal{X} = D([0, T])$, $D(\mathbb{R}_+)$, $D\left(\mathbb{R}_+, \mathbb{R}^k\right)$ etc.

The following approximation result provides a useful tool for proving stable convergence.

Theorem 3.10 (Approximation) *Let $X_{n,r}$ and Y_n be $(\mathcal{X}, \mathcal{B}(\mathcal{X}))$-valued random variables and $K_r, K \in \mathcal{K}^1(\mathcal{G})$ for $n, r \in \mathbb{N}$. Assume that*

(i) $X_{n,r} \to K_r$ $\mathcal{G}$-stably for $n \to \infty$ and all $r \in \mathbb{N}$,
(ii) $K_r \to K$ weakly for $r \to \infty$,
(iii) $\lim_{r \to \infty} \limsup_{n \to \infty} P\left(d\left(X_{n,r}, Y_n\right) > \varepsilon\right) = 0$ for every $\varepsilon > 0$.

Then $Y_n \to K$ $\mathcal{G}$-stably.

Proof For $F \in \mathcal{G}$ with $P(F) > 0$ we have $P_F^{X_{n,r}} \to P_F K_r$ weakly for $n \to \infty$ by (i) and Theorem 3.2, and $P_F K_r \to P_F K$ weakly for $r \to \infty$ by (ii) and Theorem 2.3. It remains to show that this combined with (iii) implies $P_F^{Y_n} \to P_F K$ weakly. Then Theorem 3.2 yields $\mathcal{G}$-stable convergence $Y_n \to K$.

For a closed set $B \subset \mathcal{X}$ and $\varepsilon > 0$ let $B_\varepsilon := \{y \in \mathcal{X} : \inf_{x \in B} d(y, x) \leq \varepsilon\}$. Since $\{Y_n \in B\} \subset \{X_{n,r} \in B_\varepsilon\} \cup \{d(X_{n,r}, Y_n) > \varepsilon\}$, we obtain $P_F^{Y_n}(B) \leq P_F^{X_{n,r}}(B_\varepsilon) + P_F(d(X_{n,r}, Y_n) > \varepsilon)$. Since B_ε is closed, the subadditivity of limsup and the Portmanteau theorem yield

$$\limsup_{n \to \infty} P_F^{Y_n}(B) \leq P_F K_r(B_\varepsilon) + \limsup_{n \to \infty} P_F\left(d\left(X_{n,r}, Y_n\right) > \varepsilon\right)$$

and furthermore $\limsup_{r \to \infty} P_F K_r(B_\varepsilon) \leq P_F K(B_\varepsilon)$. By (iii) and since $B_\varepsilon \downarrow B$ as $\varepsilon \downarrow 0$ we get $\limsup_{n \to \infty} P_F^{Y_n}(B) \leq P_F K(B)$ so that, B being arbitrary closed, again by the Portmanteau theorem $P_F^{Y_n} \to P_F K$ weakly. $\qquad \square$

Exercise 3.8 Show that condition (iii) of Theorem 3.10 is equivalent to the condition $\lim_{r \to \infty} \limsup_{n \to \infty} E\left(d\left(X_{n,r}, Y_n\right) \wedge 1\right) = 0$.

The following observation is sometimes useful.

Proposition 3.11 *Let $P = \sum_{i=1}^\infty s_i Q_i$ for probability distributions Q_i on $\mathcal{F}$ and $s_i \in [0, 1]$ satisfying $\sum_{i=1}^\infty s_i = 1$. If $X_n \to K$ $\mathcal{G}$-stably under Q_i for every $i \in \mathbb{N}$ with $s_i > 0$ for $(\mathcal{X}, \mathcal{B}(\mathcal{X}))$-valued random variables X_n and $K \in \mathcal{K}^1(\mathcal{G})$, then $X_n \to K$ $\mathcal{G}$-stably (under P).*

Proof This is an immediate consequence of Theorem 3.2. In fact, let $I = \{i \in \mathbb{N} : s_i > 0\}$, $F \in \mathcal{G}$ and $h \in C_b(\mathcal{X})$. Then

$$\int 1_F h(X_n) \, dP = \sum_{i \in I} s_i \int 1_F h(X_n) \, dQ_i$$

$$\to \sum_{i \in I} s_i \int 1_F \otimes h \, dQ_i \otimes K = \int 1_F \otimes h \, dP \otimes K. \qquad \square$$

Finally, we state an unspecified limit version of (parts of) Theorem 3.2. Typically, unspecified limit results are not of great interest. However, the subsequent condition (iii) with $\mathcal{E} = \mathcal{G} = \mathcal{F}$ was the original definition of stable convergence.

Proposition 3.12 (Unspecified limit) *Assume that $\mathcal{X}$ is polish. Let $\mathcal{E} \subset \mathcal{G}$ be closed under finite intersections with $\Omega \in \mathcal{E}$ and $\sigma(\mathcal{E}) = \mathcal{G}$. Then the following assertions are equivalent:*

(i) *(X_n) converges $\mathcal{G}$-stably,*
(ii) *$\left(P^{X_n}\right)$ is tight and the sequence $(E1_F h(X_n))$ converges in $\mathbb{R}$ for every $F \in \mathcal{E}$ and $h \in C_b(\mathcal{X})$,*
(iii) *$\left(P_F^{X_n}\right)$ converges weakly for every $F \in \mathcal{E}$ with $P(F) > 0$,*
(iv) *$((X_n, Y))$ converges in distribution for every separable metrizable space $\mathcal{Y}$ and every $\mathcal{G}$-measurable $(\mathcal{Y}, \mathcal{B}(\mathcal{Y}))$-valued random variable Y.*

Proof The implications (i) $\Rightarrow$ (iii) $\Rightarrow$ (ii) are obvious in view of Theorem 3.2.

(ii) $\Rightarrow$ (i). For $F \in \mathcal{E}$ and $h \in C_b(\mathcal{X})$, let $c_{F,h} := \lim_{n\to\infty} E1_F h(X_n)$. By Proposition 3.4 (a) and (b), there is a subsequence (X_k) of (X_n) with $X_k \to K$ $\mathcal{G}$-stably for some $K \in \mathcal{K}^1(\mathcal{G})$. Hence, $\lim_{k\to\infty} E1_F h(X_k) = \int 1_F \otimes h \, dP \otimes K$ so that $c_{F,h} = \int 1_F \otimes h \, dP \otimes K$ for every $F \in \mathcal{E}, h \in C_b(\mathcal{X})$. Again Theorem 3.2 yields $X_n \to K$ $\mathcal{G}$-stably.

(i) $\Rightarrow$ (iv) follows from Theorem 3.2.

(iv) $\Rightarrow$ (ii). Clearly, $\left(P^{X_n}\right)$ is tight. For $F \in \mathcal{E}$, let $(X_n, 1_F) \overset{d}{\to} \mu_F$ for some $\mu_F \in \mathcal{M}^1(\mathcal{X} \times \mathbb{R})$. Then for $h \in C_b(\mathcal{X})$ and $k \in C_b(\mathbb{R})$ satisfying $k(x) = x$ for $x \in [0, 1]$, we obtain

$$\lim_{n\to\infty} E1_F h(X_n) = \lim_{n\to\infty} Eh \otimes k(X_n, 1_F) = \int h \otimes k \, d\mu_F. \qquad \square$$

Exercise 3.9 Assume that $\mathcal{X}$ is polish. Show that for (general) stable convergence an unspecified limit version of most parts of Exercise 3.5 is true, that is,

(i) *$(X_n)_{n \geq 1}$ converges stably,*
(ii) *$\left(P_F^{X_n}\right)_{n \geq 1}$ converges weakly for every $F \in \mathcal{E} := \bigcup_{k=1}^{\infty} \sigma(X_k)$ with $P(F) > 0$,*
(iii) *$((X_n, X_k))_{n \geq 1}$ converges in distribution for every $k \in \mathbb{N}$*

are equivalent assertions.

Here is a first example.

Example 3.13 (Classical stable central limit theorem; Takahashi, Rényi)
(a) We observe automatic stability in the following setting. Let $(Z_n)_{n \geq 1}$ be an independent sequence of real random variables, $b_n \in \mathbb{R}$, $a_n > 0$, $a_n \to \infty$ and $\nu \in \mathcal{M}^1(\mathbb{R})$. If

$$X_n := \frac{1}{a_n} \left(\sum_{j=1}^{n} Z_j - b_n \right) \overset{d}{\to} \nu,$$

then $X_n \to \nu$ mixing as $n \to \infty$. To see this, let $\mathcal{G} := \sigma(Z_n, n \geq 1)$ and $\mathcal{E} := \bigcup_{k=1}^{\infty} \sigma(Z_1, \ldots, Z_k)$. Then $\mathcal{E}$ is a field with $\sigma(\mathcal{E}) = \mathcal{G}$ and the X_n are $\mathcal{G}$-measurable. If $F \in \sigma(Z_1, \ldots, Z_k)$ for some $k \in \mathbb{N}$ with $P(F) > 0$ and

$$Y_n := \frac{1}{a_n} \left(\sum_{j=k+1}^{n} Z_j - b_n \right), \quad n > k,$$

then

$$|X_n - Y_n| = \left| \frac{1}{a_n} \sum_{j=1}^{k} Z_j \right| \to 0 \quad \text{everywhere on } \Omega \text{ as } n \to \infty$$

so that $Y_n \overset{d}{\to} \nu$. Since $\sigma(Z_1, \ldots, Z_k)$ and $\sigma(Z_n, n \geq k+1)$ are independent, we have $P_F^{Y_n} = P^{Y_n} \to \nu$ weakly (in $\mathcal{M}^1(\mathbb{R})$) and hence $P_F^{X_n} \to \nu$ weakly. The assertion follows from Corollary 3.3 and Proposition 3.5.

(b) Now let $(Z_n)_{n \geq 1}$ be an independent and identically distributed sequence of real random variables with $Z_1 \in \mathcal{L}^2(P)$ and $\sigma^2 := \text{Var } Z_1$. Then by the classical central limit theorem and (a),

$$X_n := \frac{1}{\sqrt{n}} \sum_{j=1}^{n} (Z_j - EZ_1) \to N\left(0, \sigma^2\right) \quad \text{mixing as } n \to \infty.$$

Consequences of the mixing feature are, for example, statements such as

$$\lim_{n \to \infty} P(X_n \leq Y) = \int N\left(0, \sigma^2\right) ((-\infty, y]) \, dP^Y(y)$$

for any real random variable Y, which is out of scope under mere distributional convergence. In fact, by Corollary 3.3, $(X_n, Y) \overset{d}{\to} N(0, \sigma^2) \otimes P^Y$ so that for the closed set $D := \{(x, y) \in \mathbb{R}^2 : x \leq y\}$, by the Portmanteau theorem,

$$P(X_n \leq Y) = P((X_n, Y) \in D)$$

$$\to N\left(0, \sigma^2\right) \otimes P^Y(D) = \int N\left(0, \sigma^2\right) ((-\infty, y]) \, dP^Y(y)$$

because $N(0, \sigma^2) \otimes P^Y(\partial D) = 0$ provided $\sigma^2 > 0$.

We can also easily derive a multivariate version of the above stable central limit theorem using the Cramér-Wold device from Corollary 3.8 (iii). $\qquad \square$

Example 3.14 (Classical stable functional central limit theorem, cf. [9], Theorem 16.3) Let $(Z_n)_{n \geq 1}$ be an independent and identically distributed sequence of real random variables with $Z_1 \in \mathcal{L}^2(P)$, $E Z_1 = 0$ and $\sigma^2 := \mathrm{Var}\, Z_1 > 0$. For $n \in \mathbb{N}$, consider the path-continuous process $X^n = (X_t^n)_{t \geq 0}$ defined by

$$X_t^n := \frac{1}{\sigma \sqrt{n}} \left(\sum_{j=1}^{[nt]} Z_j + (nt - [nt]) Z_{[nt]+1} \right), \quad t \geq 0$$

$\left(\sum_{j=1}^0 Z_j := 0 \right)$, where $[nt]$ denotes the integer part. By Donsker's theorem, $X^n \xrightarrow{d} \nu$ in $C(\mathbb{R}_+)$, where $\nu \in \mathcal{M}^1(C(\mathbb{R}_+))$ denotes the Wiener measure ([53], Theorem 21.43). We show that $X^n \to \nu$ mixing. Arguing as in Example 3.13 (a), it is enough to show that $P_F^{X^n} \to \nu$ weakly for every $F \in \bigcup_{k=1}^\infty \sigma(Z_1, \ldots, Z_k)$ with $P(F) > 0$. If $F \in \sigma(Z_1, \ldots, Z_k)$ for some $k \in \mathbb{N}$ with $P(F) > 0$ and

$$Y_t^n := \begin{cases} 0 & , \ 0 \leq t \leq \frac{k}{n} \\ \dfrac{1}{\sigma \sqrt{n}} \left(\displaystyle\sum_{j=k+1}^{[nt]} Z_j + (nt - [nt]) Z_{[nt]+1} \right) & , \ t > \dfrac{k}{n} \end{cases}$$

for $n \in \mathbb{N}$, then

$$d(X^n, Y^n) \leq \frac{2}{\sigma \sqrt{n}} \sum_{i=1}^k |Z_i| \to 0 \quad \text{everywhere on } \Omega \text{ as } n \to \infty$$

so that $Y^n \xrightarrow{d} \nu$. Since $\sigma(Z_1, \ldots, Z_k)$ and $\sigma(Y^n)$ are independent, we have $P_F^{Y_n} = P^{Y_n} \to \nu$ weakly and hence $P_F^{X_n} \to \nu$ weakly. For a martingale approach to the mixing Donsker theorem, see Chap. 7.   □

Exercise 3.10 Show in the situation of Example 3.14 that

$$\frac{1}{\sigma \sqrt{n}} \max_{0 \leq j \leq n} \sum_{i=1}^j Z_i \to \mu \quad \text{mixing},$$

where

$$\frac{d\mu}{d\lambda}(t) = \frac{2}{\sqrt{2\pi}} \exp\left(-\frac{t^2}{2} \right) 1_{\mathbb{R}_+}(t).$$

Hint: μ is the distribution of $\max_{t \in [0,1]} W_t$ for a Brownian motion W.

3.2 Second Approach

The limit kernel for $\mathcal{G}$-stable convergence $X_n \to K$ can always be represented as a $\mathcal{G}$-conditional distribution of a further random variable X defined on a suitable extension of the underlying probability space $(\Omega, \mathcal{F}, P)$: Take $\overline{\Omega} = \Omega \times \mathcal{X}, \overline{\mathcal{F}} = \mathcal{F} \otimes \mathcal{B}(\mathcal{X})$, $\overline{P} = P \otimes K$ and $X(\omega, x) = x$. So, for instance, the Gauss-kernel $N(0, \eta^2)$, where η is a $\mathcal{G}$-measurable, nonnegative real random variable, satisfies $N(0, \eta^2) = P^{\eta Z | \mathcal{G}}$ assuming the existence of a $N(0, 1)$-distributed random variable Z on $(\Omega, \mathcal{F}, P)$ which is independent of $\mathcal{G}$. This motivates the following approach.

Definition 3.15 Let $\mathcal{G} \subset \mathcal{F}$ be a sub-σ-field. A sequence $(X_n)_{n \geq 1}$ of $(\mathcal{X}, \mathcal{B}(\mathcal{X}))$-valued random variables is said to *converge $\mathcal{G}$-stably* to an $(\mathcal{X}, \mathcal{B}(\mathcal{X}))$-valued random variable X if $X_n \to P^{X | \mathcal{G}}$ $\mathcal{G}$-stably for $n \to \infty$. Then we write $X_n \to X$ $\mathcal{G}$-stably.

As before, we assume the existence of conditional distributions. By Definition 3.1 $\mathcal{G}$-stable convergence $X_n \to X$ reads

$$\lim_{n \to \infty} E\left(f E\left(h\left(X_n\right) | \mathcal{G}\right)\right) = E\left(f E\left(h\left(X\right) | \mathcal{G}\right)\right)$$

for every $f \in \mathcal{L}^1(P)$ and $h \in C_b(\mathcal{X})$ and implies $X_n \xrightarrow{d} X$. The $\mathcal{G}$-mixing convergence $X_n \to X$ corresponds to $P^{X | \mathcal{G}} = P^X$ P-almost surely which is equivalent to the independence of $\sigma(X)$ and $\mathcal{G}$. Thus $X_n \to X$ $\mathcal{G}$-mixing means $X_n \to X$ $\mathcal{G}$-stably and $\sigma(X)$ and $\mathcal{G}$ are independent which is also equivalent to $X_n \to P^X$ $\mathcal{G}$-mixing and independence of $\sigma(X)$ and $\mathcal{G}$.

For the formulation of stable limit theorems in subsequent chapters we sometimes use the "K-approach", sometimes the "X-approach", and sometimes both.

Example 3.16 In the situation of Example 3.13 (b) with $\mathcal{G} = \sigma(Z_n, n \geq 1)$ let X be $N(0, \sigma^2)$-distributed and independent of $\mathcal{G}$. Such an X exists at least after a suitable extension of $(\Omega, \mathcal{F}, P)$. Then Example 3.13 (b) yields

$$\frac{1}{\sqrt{n}} \sum_{j=1}^{n} \left(Z_j - E Z_1\right) \to X \quad \mathcal{G}\text{-mixing.}$$

However, there is nothing special about this $\mathcal{G}$. The above statement holds for any pair $(\mathcal{G}, X)$, where $P^X = N(0, \sigma^2)$ and $\sigma(X)$, $\mathcal{G}$ are independent. The random variable X is merely an "artificial" construct to describe the limit kernel. In practice, $\mathcal{G}$ can and will be chosen so large that all random variables of interest are measurable w.r.t. $\mathcal{G}$. □

The previous characterizations of $\mathcal{G}$-stable convergence now read as follows.

Theorem 3.17 *Let X_n and X be $(\mathcal{X}, \mathcal{B}(\mathcal{X}))$-valued random variables and let $\mathcal{E} \subset \mathcal{G}$ be closed under finite intersections with $\Omega \in \mathcal{E}$ and $\sigma(\mathcal{E}) = \mathcal{G}$. Then the following assertions are equivalent:*

(i) $X_n \to X$ $\mathcal{G}$-stably,

(ii) $\lim_{n\to\infty} Efh(X_n) = Efh(X)$ for every $f \in \mathcal{L}^1(\mathcal{G}, P)$ and $h \in C_b(\mathcal{X})$,

(iii) $Q^{X_n} \to Q^X$ weakly for every probability distribution Q on $\mathcal{F}$ such that $Q \ll P$ and dQ/dP is $\mathcal{G}$-measurable,

(iv) $P_F^{X_n} \to P_F^X$ weakly for every $F \in \mathcal{E}$ with $P(F) > 0$,

(v) $\lim_{n\to\infty} \int g(\omega, X_n(\omega))\, dP(\omega) = \int g(\omega, X(\omega))\, dP(\omega)$ for every measurable, bounded function $g : (\Omega \times \mathcal{X}, \mathcal{G} \otimes \mathcal{B}(\mathcal{X})) \to (\mathbb{R}, \mathcal{B}(\mathbb{R}))$ such that $g(\omega, \cdot) \in C_b(\mathcal{X})$ for every $\omega \in \Omega$,

(vi) (For X polish) $\limsup_{n\to\infty} \int g(\omega, X_n(\omega))\, dP(\omega) \le \int g(\omega, X(\omega))\, dP(\omega)$ for every measurable function $g : (\Omega \times \mathcal{X}, \mathcal{G} \otimes \mathcal{B}(\mathcal{X})) \to \left(\overline{\mathbb{R}}, \mathcal{B}\left(\overline{\mathbb{R}}\right)\right)$ which is bounded from above such that $g(\omega, \cdot)$ is upper semicontinuous for every $\omega \in \Omega$,

(vii) $(X_n, Y) \to (X, Y)$ $\mathcal{G}$-stably for every separable metrizable space $\mathcal{Y}$ and every $\mathcal{G}$-measurable $(\mathcal{Y}, \mathcal{B}(\mathcal{Y}))$-valued random variable Y,

(viii) $(X_n, 1_F) \xrightarrow{d} (X, 1_F)$ for every $F \in \mathcal{E}$.

Proof Just apply Theorem 3.2. As for (vii) and (viii) one has to recall that $P^{X|\mathcal{G}} \otimes \delta_Y = P^{(X,Y)|\mathcal{G}}$ by $\mathcal{G}$-measurability of Y so that $P\left(P^{X|\mathcal{G}} \otimes \delta_Y\right) = P^{(X,Y)}$. □

Exercise 3.11 Let $\mathcal{Y}$ be a separable metrizable space, Y a $(\mathcal{Y}, \mathcal{B}(\mathcal{Y}))$-valued random variable and $\mathcal{G} = \sigma(Y)$. Show that $X_n \to X$ $\mathcal{G}$-stably if and only if $(X_n, Y) \xrightarrow{d} (X, Y)$ as $n \to \infty$.

Exercise 3.12 Let $\mathcal{G} = \sigma(X_n, n \ge 1)$. Prove that $X_n \to X$ $\mathcal{G}$-stably if and only if $(X_n, X_1, \ldots, X_k) \xrightarrow{d} (X, X_1, \ldots, X_k)$ as $n \to \infty$ for every $k \ge 1$.

In case $\mathcal{G}_1 \subset \mathcal{G}_2 \subset \mathcal{F}$ it is clear from Theorem 3.17 that $\mathcal{G}_2$-stable convergence $X_n \to X$ implies $\mathcal{G}_1$-stable convergence $X_n \to X$.

The $\mathcal{G}$-measurability of all X_n in $\mathcal{G}$-stable convergence $X_n \to X$ has no specific impact (in contrast to Proposition 3.5) while the $\mathcal{G}$-measurability of X has a very strong impact. In fact, if $\sigma(X) \subset \mathcal{G}$, then $X_n \to X$ $\mathcal{G}$-stably if and only if $X_n \to X$ in probability. This is a reformulation of Corollary 3.6 because $P^{X|\mathcal{G}} = \delta_X$. In particular, still under $\mathcal{G}$-measurability of X, we have $X_n \to X$ $\mathcal{G}$-mixing if and only if $X = c$ almost surely for some constant $c \in \mathcal{X}$ and $X_n \to c$ in probability.

Since $\mathcal{G} = \{\emptyset, \Omega\}$ reduces $\mathcal{G}$-stable convergence $X_n \to X$ to distributional convergence and $\sigma(X) \subset \mathcal{G}$ gives convergence in probability, $\mathcal{G}$-stability provides a type of convergence in between.

In the "X-approach" Theorem 3.7 reads as follows.

Theorem 3.18 Assume $X_n \to X$ $\mathcal{G}$-stably and let Y_n and Y be random variables with values in $(\mathcal{Y}, \mathcal{B}(\mathcal{Y}))$ for some separable metrizable space.

(a) *Let $\mathcal{X} = \mathcal{Y}$. If $d\,(X_n, Y_n) \to 0$ in probability, then $Y_n \to X$ $\mathcal{G}$-stably.*
(b) *If $Y_n \to Y$ in probability and Y is $\mathcal{G}$-measurable, then $(X_n, Y_n) \to (X, Y)$ $\mathcal{G}$-stably.*
(c) *If $g : \mathcal{X} \to \mathcal{Y}$ is Borel-measurable and P^X-almost surely continuous, then $g\,(X_n) \to g\,(X)$ $\mathcal{G}$-stably.*

Proof Recall that $P^{X|\mathcal{G}} \otimes \delta_Y = P^{(X,Y)|\mathcal{G}}$, note that $(P^{X|\mathcal{G}})^g = P^{g(X)|\mathcal{G}}$ and use Theorem 3.7. $\square$

Corollary 3.8 reads as follows.

Corollary 3.19 *Let $\mathcal{X} = \mathbb{R}^d$. Let X_n and X be $\mathbb{R}^d$-valued random variables and let $\mathcal{E} \subset \mathcal{G}$ be closed under finite intersections with $\Omega \in \mathcal{E}$ and $\sigma\,(\mathcal{E}) = \mathcal{G}$. Then the following assertions are equivalent:*

(i) *$X_n \to X$ $\mathcal{G}$-stably,*
(ii) *$\lim_{n\to\infty} E1_F \exp\,(i\,\langle u, X_n\rangle) = E1_F \exp\,(i\,\langle u, X\rangle)$ for every $F \in \mathcal{E}$ and $u \in \mathbb{R}^d$,*
(iii) *$\langle u, X_n\rangle \to \langle u, X\rangle$ $\mathcal{G}$-stably for every $u \in \mathbb{R}^d$.*

Proposition 3.9 reads as follows.

Proposition 3.20 *Let $\mathcal{X} = C\,(I)$ with $I = [0, T]$ or $\mathbb{R}_+$. For path-continuous processes $X^n = \left(X_t^n\right)_{t\in I}$ and $X = (X_t)_{t\in I}$ the following assertions are equivalent:*

(i) *$X_n \to X$ $\mathcal{G}$-stably,*
(ii) *$\left(P^{X^n}\right)_{n\geq 1}$ is tight and $\left(X_{t_1}^n, \ldots, X_{t_k}^n\right) \to \left(X_{t_1}, \ldots, X_{t_k}\right)$ $\mathcal{G}$-stably for every $k \geq 1$ and $0 \leq t_1 < \cdots < t_k$, $t_j \in I$.*

Theorem 3.10 reads as follows.

Theorem 3.21 (Approximation) *Let $X_{n,r}$, X_r, X and Y_n be $(\mathcal{X}, \mathcal{B}\,(\mathcal{X}))$-valued random variables. Assume that*

(i) *$X_{n,r} \to X_r$ $\mathcal{G}$-stably for $n \to \infty$ and all $r \in \mathbb{N}$,*
(ii) *$X_r \to X$ $\mathcal{G}$-stably for $r \to \infty$,*
(iii) *$\lim_{r\to\infty} \limsup_{n\to\infty} P\left(d\left(X_{n,r}, Y_n\right) > \varepsilon\right) = 0$ for every $\varepsilon > 0$.*

Then $Y_n \to X$ $\mathcal{G}$-stably.

Using Theorem 3.21 we can treat a further special case quite easily.

Proposition 3.22 *Let $\mathcal{X} = \prod_{j\in\mathbb{N}} \mathcal{Y}_j$ for separable metrizable spaces $\mathcal{Y}_j$. For $(\mathcal{X}, \mathcal{B}\,(\mathcal{X}))$-valued random variables $X^n = \left(X_k^n\right)_{k\geq 1}$ and $X = (X_k)_{k\geq 1}$ are equivalent:*

(i) $X^n \to X$ $\mathcal{G}$-stably,
(ii) $\left(X_1^n, \ldots, X_k^n\right) \to (X_1, \ldots, X_k)$ $\mathcal{G}$-stably for every $k \geq 1$.

Proof (i) $\Rightarrow$ (ii) follows from the continuity of $\pi_{1,\ldots,k} : \mathcal{X} \to \prod_{j=1}^{k} \mathcal{Y}_j$, $\pi_{1,\ldots,k}$ $((x_n)) := (x_1, \ldots, x_k)$ for every $k \in \mathbb{N}$ and Theorem 3.18 (c).

(ii) $\Rightarrow$ (i). Fix any $(c_n) \in \mathcal{X}$. For $k \in \mathbb{N}$, the map $\varphi_k : \prod_{j=1}^{k} \mathcal{Y}_j \to \mathcal{X}$, $\varphi_k ((x_1, \ldots, x_k)) := (x_1, \ldots, x_k, c_{k+1}, c_{k+2}, \ldots)$ is continuous so that by Theorem 3.18 (c)

$$Z^{n,k} := \varphi_k \left(\left(X_1^n, \ldots, X_k^n \right) \right) \to Z^k := \varphi_k ((X_1, \ldots, X_k)) \text{ } \mathcal{G}\text{-stably as } n \to \infty$$

for every $k \in \mathbb{N}$. Note that if d_j denotes a metric inducing the topology of $\mathcal{Y}_j$, then the metric $d \left((x_j), (y_j) \right) := \sum_{j=1}^{\infty} 2^{-j} \left(d_j (x_j, y_j) \wedge 1 \right)$ provides a metrization of the product topology of $\mathcal{X}$, and note also that $d \left(Z^{n,k}, X^n \right) \leq \sum_{j=k+1}^{\infty} 2^{-j}$ and $d \left(Z^k, X \right) \leq \sum_{j=k+1}^{\infty} 2^{-j}$ for all $k, n \in \mathbb{N}$. The $\mathcal{G}$-stable convergence $X^n \to X$ now follows from Theorem 3.21.   □

One can deduce a characterization of stable convergence of continuous processes.

Corollary 3.23 *Let $\mathcal{X} = C (\mathbb{R}_+)$. For path-continuous processes $X^n = \left(X_t^n \right)_{t \geq 0}$ and $X = (X_t)_{t \geq 0}$ are equivalent:*

(i) $X^n \to X$ $\mathcal{G}$-stably,
(ii) $\left(X_t^n \right)_{t \in [0,k]} \to (X_t)_{t \in [0,k]}$ $\mathcal{G}$-stably in $C ([0, k])$ for every $k \in \mathbb{N}$.

Proof (i) $\Rightarrow$ (ii) follows from the continuity of the restriction maps $\varphi_k : C (\mathbb{R}_+) \to C ([0, k])$ and Theorem 3.18 (c).

(ii) $\Rightarrow$ (i). By hypothesis $\varphi_k (X^n) \to \varphi_k (X)$ $\mathcal{G}$-stably in $C ([0, k])$ for every $k \in \mathbb{N}$. Since the restriction map $C ([0, k]) \to \prod_{m=1}^{k} C ([0, m])$, $y \mapsto (y| [0, 1], \ldots, y| [0, k])$ is continuous, Theorem 3.18 (c) implies

$$\left(\varphi_1 (X^n), \ldots, \varphi_k (X^n) \right) \to (\varphi_1 (X), \ldots, \varphi_k (X)) \text{ } \mathcal{G}\text{-stably in } \prod_{m=1}^{k} C ([0, m])$$

for every $k \in \mathbb{N}$ so that Proposition 3.22 yields

$$\left(\varphi_m (X^n) \right)_{m \in \mathbb{N}} \to (\varphi_m (X))_{m \in \mathbb{N}} \text{ } \mathcal{G}\text{-stably in } \prod_{m \in \mathbb{N}} C ([0, m])$$

as $n \to \infty$. Now $(\varphi_m)_{m \in \mathbb{N}}$ is a homeomorphism from $C (\mathbb{R}_+)$ onto its range $\mathcal{Z}$, say, in $\prod_{m \in \mathbb{N}} C ([0, m])$. ($\mathcal{Z}$ is a Borel subset of $\prod_{m \in \mathbb{N}} C ([0, m])$; see [69], Theorem I.3.9.) Using the Portmanteau theorem one checks that

$$\left(\varphi_m\left(X^n\right)\right)_{m\in\mathbb{N}} \to \left(\varphi_m\left(X\right)\right)_{m\in\mathbb{N}} \quad \mathcal{G}\text{-stably in } \mathcal{Z}.$$

Assertion (i) follows again from Theorem 3.18 (c).   □

Proposition 3.11 reads as follows.

Proposition 3.24 *Let* $P = \sum_{i=1}^{\infty} s_i Q_i$ *for probability distributions* Q_i *on* $\mathcal{F}$ *and* $s_i \in [0, 1]$ *satisfying* $\sum_{i=1}^{\infty} s_i = 1$. *If* $X_n \to X$ $\mathcal{G}$-*stably under* Q_i *for every* i *with* $s_i > 0$, *then* $X_n \to X$ $\mathcal{G}$-*stably (under* P).

Chapter 4
Applications

The goal of this chapter is to establish consequences of stable convergence of random variables. We thus demonstrate the importance of this notion simply because many distributional limit theorems can be shown to be stable. Stable convergence implies convergence in distribution. But it implies much more. Stable convergence is useful, for example, in connection with random normalization and random index limit theorems and can be used to prove results on the fluctuations of sample paths of stochastic processes. Also the δ-method with random centering works under stable convergence, and stable convergence $X_n \rightarrow K$ implies the existence of a subsequence (X_m) such that the associated empirical measures of every further subsequence of (X_m) converge weakly to $K(\omega, \cdot)$, almost surely. Thus the stable limit kernel specifies the almost sure limit of empirical measures.

As before, let $\mathcal{X}$ be a separable metrizable space and let d be a metric inducing the topology on $\mathcal{X}$.

4.1 Limit Points

In order to describe the fluctuation behavior of stably convergent random variables recall that $x \in \mathcal{X}$ is said to be a *limit point* of a sequence $(x_n)_{n \geq 1}$ in $\mathcal{X}$ if it has a subsequence converging to x. We denote by $L((x_n))$ the set of all limit points of $(x_n)_{n \geq 1}$. Since $\mathcal{X}$ is first countable (each point has a countable neighborhood basis) the limit points of a sequence are precisely the cluster (or accumulation) points of the sequence, so that $L((x_n)) = \bigcap_{n \in \mathbb{N}} \overline{\{x_k : k \geq n\}}$, where $\overline{B}$ denotes the closure of $B \subset \mathcal{X}$. Furthermore, the set $L := \{((x_n), x) \in \mathcal{X}^{\mathbb{N}} \times \mathcal{X} : x \in L((x_n))\}$ can be written as $L = \bigcap_{n \in \mathbb{N}} L_n$, where

© Springer International Publishing Switzerland 2015
E. Häusler and H. Luschgy, *Stable Convergence and Stable Limit Theorems*,
Probability Theory and Stochastic Modelling 74,
DOI 10.1007/978-3-319-18329-9_4

$$L_n := \left\{ ((x_j), x) \in \mathcal{X}^{\mathbb{N}} \times \mathcal{X} : x \in \overline{\{x_k : k \geq n\}} \right\}$$

$$= \bigcap_{i=1}^{\infty} \bigcup_{k=n}^{\infty} \left\{ ((x_j), x) \in \mathcal{X}^{\mathbb{N}} \times \mathcal{X} : d(x_k, x) < \frac{1}{i} \right\},$$

hence $L_n, L \in \mathcal{B}(\mathcal{X})^{\mathbb{N}} \otimes \mathcal{B}(\mathcal{X})$. For $\nu \in \mathcal{M}^1(\mathcal{X})$, let supp($\nu$) denote the support of ν (i.e. the smallest closed set B such that $\nu(B) = 1$), which exists in our setting ([69], Theorem II.2.1).

Theorem 4.1 (Limit points) *Assume that $\mathcal{X}$ is polish. If $X_n \to K$ stably for $(\mathcal{X}, \mathcal{B}(\mathcal{X}))$-valued random variables X_n and $K \in \mathcal{K}^1$, then $L((X_n(\omega))) \supset$ supp($K(\omega, \cdot)$) almost surely.*

Proof The map $\varphi : (\Omega \times \mathcal{X}, \mathcal{F} \otimes \mathcal{B}(\mathcal{X})) \to \left(\mathcal{X}^{\mathbb{N}} \times \mathcal{X}, \mathcal{B}(\mathcal{X})^{\mathbb{N}} \otimes \mathcal{B}(\mathcal{X}) \right), \varphi(\omega, x) :=$ $((X_n(\omega)), x)$, is measurable. Hence the sets

$$C_n := \left\{ (\omega, x) \in \Omega \times \mathcal{X} : x \in \overline{\{X_k(\omega) : k \geq n\}} \right\} = \{\varphi \in L_n\}$$

and

$$C := \{(\omega, x) \in \Omega \times \mathcal{X} : x \in L((X_n(\omega)))\} = \{\varphi \in L\} = \bigcap_{n=1}^{\infty} C_n$$

satisfy $C_n, C \in \mathcal{F} \otimes \mathcal{B}(\mathcal{X})$, and the ω-sections $C_{n,\omega}$ are closed so that $1_{C_n}(\omega, \cdot)$ is upper semicontinuous for every $\omega \in \Omega$. Since obviously

$$\int 1_{C_n}(\omega, X_k(\omega)) \, dP(\omega) = 1$$

for every $k \geq n$, Theorem 3.2 yields

$$1 = \limsup_{k \to \infty} \int 1_{C_n}(\omega, X_k(\omega)) \, dP(\omega) \leq P \otimes K(C_n)$$

for every $n \in \mathbb{N}$. This implies $P \otimes K(C) = 1$ and thus $K(\omega, C_\omega) = 1$ for almost all $\omega \in \Omega$, where $C_\omega = L((X_n(\omega)))$. □

In the mixing case the above theorem first appeared in [80] and for the general case see [7], Corollary 3.18. A sharper "subsequence principle" may be found in [48].

Example 4.2 In the situation of Example 3.13 (b) with $\sigma^2 \in (0, \infty)$ we obtain from Theorem 4.1 that

$$L\left(\left(n^{-1/2} \sum_{j=1}^{n} (Z_j - EZ_1) \right) \right) = \mathbb{R} \quad \text{a.s.}$$

This may be compared with Strassen's law of the iterated logarithm

$$L\left(\left((2n \log \log n)^{-1/2} \sum_{j=1}^{n} (Z_j - E Z_1)\right)\right) = [-\sigma, \sigma] \quad \text{a.s.}$$

which, of course, is much better and implies the above statement as well as the strong law of large numbers

$$L\left(\left(n^{-\alpha} \sum_{j=1}^{n} (Z_j - E Z_1)\right)\right) = \{0\} \quad \text{a.s.}$$

for all $\alpha > 1/2$. □

Example 4.3 (Occupation time of Brownian motion) Let $W = (W_t)_{t \geq 0}$ be an (everywhere path-continuous) Brownian motion and η its *occupation measure*, defined by

$$\eta_t (A) := \int_0^t 1_A (W_s) \, ds = \lambda (s \leq t : W_s \in A)$$

for $t \geq 0$ and $A \in \mathcal{B}(\mathbb{R})$. Using Theorem 4.1 we show for $A = (0, \infty)$ that the limit points of the sequence $(n^{-1}\eta_n ((0, \infty)))_{n \geq 1}$ coincide almost surely with $[0, 1]$ and, in particular,

$$\limsup_{n \to \infty} \frac{1}{n} \lambda (t \leq n : W_t > 0) = 1 \quad \text{a.s.}$$

and

$$\liminf_{n \to \infty} \frac{1}{n} \lambda (t \leq n : W_t > 0) = 0 \quad \text{a.s.}$$

We proceed as follows. Let $\mathcal{X} = C ([0, 1])$, $\nu := P^{(W_t)_{t \in [0,1]}} \in \mathcal{M}^1 (\mathcal{X})$ and for $n \in \mathbb{N}$ let $X_t^n := n^{-1/2} W_{nt}, t \in [0, 1]$. By the scaling invariance of Brownian motion we obtain $P^{X^n} = \nu$ for every n (and obviously $X^n \xrightarrow{d} \nu$ and $(P^{X^n})_{n \geq 1}$ is tight). We first observe that $X^n \to \nu$ mixing as $n \to \infty$. By Proposition 3.9 and Corollary 3.8 it is enough to show that

$$\sum_{j=1}^{k} u_j X_{t_j}^n \to P^{\sum_{j=1}^{k} u_j W_{t_j}} \quad \text{mixing as } n \to \infty$$

for every $k \in \mathbb{N}, 0 < t_1 < \cdots < t_k \leq 1$ and $u_1, \ldots, u_k \in \mathbb{R}$. (The case $t_1 = 0$ can be excluded since $X_0^n = W_0 = 0$.) Choose $a_n \in (0, \infty)$ such that $a_n < n$, $a_n \uparrow \infty$ and $a_n/n \to 0$ as $n \to \infty$ and define $Y_n := n^{-1/2} \sum_{j=1}^{k} u_j \left(W_{nt_j} - W_{a_n t_j} \right)$ for $n \in \mathbb{N}$. Since $E \left(W_{a_n t}/\sqrt{n} \right)^2 = a_n t/n \to 0$ for every $t \geq 0$, we obtain

$$\sum_{j=1}^{k} u_j X_{t_j}^n - Y_n = n^{-1/2} \sum_{j=1}^{k} u_j W_{a_n t_j} \to 0 \quad \text{in } \mathcal{L}^2 (P)$$

and thus in probability as $n \to \infty$. Hence by Theorem 3.7 (a) it is enough to show that $Y_n \to P^{\sum_{j=1}^{k} u_j W_{t_j}}$ mixing. Let $\mathcal{G} := \sigma (Y_n, n \in \mathbb{N})$ and $\mathcal{E} := \bigcup_{m=1}^{\infty} \sigma (Y_1, \ldots, Y_m)$, satisfying $\sigma (\mathcal{E}) = \mathcal{G}$. For all $m \in \mathbb{N}$, we have $\sigma (Y_1, \ldots, Y_m) \subset \sigma (W_t, t \leq m)$ and for all $n \in \mathbb{N}$ such that $a_n t_1 \geq m$, we have $\sigma (Y_n) \subset \sigma (W_t - W_m, t \geq m)$. Also, the σ-fields $\sigma (W_t, t \leq m)$ and $\sigma (W_t - W_m, t \geq m)$ are independent by the independence of the increments of W. Thus, if $F \in \sigma (Y_1, \ldots, Y_m)$ with $P (F) > 0$, then for $n \in \mathbb{N}$ with $a_n t_1 \geq m$

$$P_F^{Y_n} = P^{Y_n} \to P^{\sum_{j=1}^{k} u_j W_{t_j}} \quad \text{weakly}.$$

The desired mixing convergence of Y_n follows from Corollary 3.3 and Proposition 3.5.

We can mention, as a first consequence of Proposition 4.1, that

$$L \left((X^n) \right) = \text{supp}(\nu) = \{x \in C([0, 1]) : x(0) = 0\} \ P\text{-a.s.}$$

and compare this with Strassen's law of the iterated logarithm for Brownian motion, saying that the processes $Z_t^n := (2n \log \log n)^{-1/2} W_{nt}, t \in [0, 1]$, satisfy

$$L \left((Z^n) \right) = \text{unit ball of the reproducing kernel Hilbert space of } \nu$$

$$= \left\{ x \in C ([0, 1]) : x (0) = 0, x \text{ absolutely continuous and} \right.$$

$$\left. \int_0^1 \dot{x} (t)^2 \, dt \leq 1 \right\} \quad \text{a.s.;}$$

see [91], Theorem 1.17.

Now consider the occupation time functional $g : \mathcal{X} \to [0, 1]$ defined by

$$g (x) := \int_0^1 1_{(0,\infty)} (x (t)) \, dt = \lambda (t \leq 1 : x (t) > 0) .$$

Since $\mathcal{X} \times [0, 1] \to [0, 1]$, $(x, t) \mapsto x(t)$, is obviously continuous, hence Borel-measurable, and $\mathcal{B}(\mathcal{X} \times [0, 1]) = \mathcal{B}(\mathcal{X}) \otimes \mathcal{B}([0, 1])$, the functional g is also Borel-measurable. Furthermore, g is ν-almost surely continuous. In fact, for the t-sections of

$$D := \{(x, t) \in \mathcal{X} \times [0, 1] : x(t) = 0\} \in \mathcal{B}(\mathcal{X}) \otimes \mathcal{B}([0, 1])$$

we have $\nu(D_t) = P^{W_t}(\{0\}) = N(0, t)(\{0\}) = 0$ for every $t > 0$ and by Fubini's theorem

$$0 = \int_0^1 \nu(D_t)\, dt = \nu \otimes \lambda(D) = \int_{\mathcal{X}} \lambda(D_x)\, d\nu(x) .$$

Hence there exists a set $N \in \mathcal{B}(\mathcal{X})$ with $\nu(N) = 0$ such that $\lambda(D_x) = 0$ for every $x \in N^c$. For $x \in N^c$ and $x_n \in \mathcal{X}$ such that $x_n \to x$ we obtain $1_{(0,\infty)}(x_n(t)) \to 1_{(0,\infty)}(x(t))$ for every $t \in D_x^c$, hence λ-almost surely, so that by dominated convergence $g(x_n) \to g(x)$. This gives continuity of g at every point $x \in N^c$. By Theorem 3.7 (c) we can conclude $g(X^n) \to \nu^g$ mixing. Since

$$g(X^n) = \int_0^1 1_{(0,\infty)}(W_{nt})\, dt = \frac{1}{n} \int_0^n 1_{[0,\infty)}(W_s)\, ds = \frac{1}{n}\eta_n((0, \infty)),$$

and $\nu^g = P^{g((W_t)_{t \in [0,1]})} = P^{\eta_1((0,\infty))}$ as well as $\operatorname{supp}(P^{\eta_1((0,\infty))}) = [0, 1]$ simply because $\eta_1((0, \infty))$ has an arcsine distribution with strictly positive λ-density on $(0, 1)$, see e.g. [51], Theorem 13.16, the assertion about the limit points of $\left(n^{-1}\eta_n((0, \infty))\right)_{n \geq 1}$ stated at the beginning follows from Theorem 4.1. $\quad\square$

Example 4.4 (Borel-Cantelli-type features; [30]) Let $F_n \in \mathcal{F}$ and $\alpha \in (0, 1)$. Assume that $(F_n)_{n \geq 1}$ is *mixing with density* α in the sense of

$$\lim_{n \to \infty} P(F_n \cap G) = \alpha P(G) \quad \text{for every } G \in \mathcal{F}$$

(cf. [76]). Then $1_{F_n} \to \alpha \delta_1 + (1 - \alpha) \delta_0$ mixing so that by Theorem 4.1, the limit points of $\left(1_{F_n}\right)_{n \geq 1}$ coincide almost surely with $\{0, 1\}$. In particular,

$$1_{\limsup\limits_{n \to \infty} F_n} = \limsup_{n \to \infty} 1_{F_n} = 1 \quad \text{a.s.}$$

and

$$1_{\liminf\limits_{n \to \infty} F_n} = \liminf_{n \to \infty} 1_{F_n} = 0 \quad \text{a.s.}$$

which implies

$$P\left(\limsup_{n\to\infty} F_n\right) = 1 \quad \text{and} \quad P\left(\liminf_{n\to\infty} F_n\right) = 0.$$

For instance, if $X_n \to \nu$ mixing, $B \in \mathcal{B}(\mathcal{X})$ with $\nu(\partial B) = 0$, $\nu(B) \in (0, 1)$ and $F_n := \{X_n \in B\}$, then by Corollary 3.3 and the Portmanteau theorem, $P_G(F_n) = P_G^{X_n}(B) \to \nu(B)$ for every $G \in \mathcal{F}$ with $P(G) > 0$, so that the sequence $(F_n)_{n\geq 1}$ satisfies the above mixing condition with $\alpha = \nu(B)$.

More generally, let $\alpha : (\Omega, \mathcal{F}) \to ([0, 1], \mathcal{B}([0, 1]))$ be measurable and assume that $(F_n)_{n\geq 1}$ is *stable with density* α in the sense of

$$\lim_{n\to\infty} P(F_n \cap G) = \int_G \alpha \, dP \quad \text{for every } G \in \mathcal{F}$$

(cf. [77]). If $K(\omega, \cdot) := \alpha(\omega)\delta_1 + (1 - \alpha(\omega))\delta_0$, then $1_{F_n} \to K$ stably. Since $1 \in \mathrm{supp}(K(\omega, \cdot))$ for $\omega \in \{\alpha > 0\}$, Theorem 4.1 yields $\limsup_{n\to\infty} 1_{F_n} = 1$ almost surely on $\{\alpha > 0\}$ so that

$$P\left(\limsup_{n\to\infty} F_n\right) \geq P(\alpha > 0).$$

Analogously, one obtains

$$P\left(\liminf_{n\to\infty} F_n\right) \leq 1 - P(\alpha < 1).$$

If $X_n \to H$ stably, $B \in \mathcal{B}(\mathcal{X})$ with $PH(\partial B) = 0$ and $F_n := \{X_n \in B\}$, then by Theorem 3.2 and the Portmanteau theorem,

$$P_G(F_n) = P_G^{X_n}(B) \to P_G H(B) = \frac{1}{P(G)} \int_G H(\omega, B) \, dP(\omega)$$

for every $G \in \mathcal{F}$ with $P(G) > 0$. Consequently, the sequence $(F_n)_{n\geq 1}$ satisfies the above stability condition with $\alpha = H(\cdot, B)$. $\square$

4.2 Random Indices

Let τ_n be an $\mathbb{N}$-valued random variable for every $n \in \mathbb{N}$. We are interested in the convergence of $\left(X_{\tau_n}\right)_{n\geq 1}$ for $(\mathcal{X}, \mathcal{B}(\mathcal{X}))$-valued random variables X_n provided $\tau_n \to \infty$ in probability as $n \to \infty$, that is $\lim_{n\to\infty} P(\tau_n \geq C) = 1$ for every $C \in (0, \infty)$.

We start with the simple independent setting where $(\tau_n)_{n\geq 1}$ and $(X_n)_{n\geq 1}$ are independent. Here we observe that stable convergence is preserved by such a random time change with the same limit.

Proposition 4.5 *Let X_n be $(\mathcal{X}, \mathcal{B}(\mathcal{X}))$-valued random variables, $K \in \mathcal{K}^1$ and $\tau_n \to \infty$ in probability as $n \to \infty$. Assume that*

(i) $\mathcal{H}_1 := \sigma(\tau_n, n \geq 1)$ *and* $\mathcal{H}_2 := \sigma(K, X_n, n \geq 1)$ *are independent.*
Let $\mathcal{H}'_i \subset \mathcal{H}_i$ *be sub-σ-fields and* $\mathcal{G} := \sigma\left(\mathcal{H}'_1 \cup \mathcal{H}'_2\right)$. *If* $K \in \mathcal{K}^1(\mathcal{G})$ *and*
(ii) $X_n \to K$ $\mathcal{G}$-*stably,*
then $X_{\tau_n} \to K$ $\mathcal{G}$-*stably as* $n \to \infty$.

Proof The system $\mathcal{E} := \left\{F_1 \cap F_2 : F_1 \in \mathcal{H}'_1, F_2 \in \mathcal{H}'_2\right\}$ is closed under finite intersections, $\Omega \in \mathcal{E}$ and $\sigma(\mathcal{E}) = \mathcal{G}$. Thus by Theorem 3.2 it is enough to show that

$$\lim_{n\to\infty} E 1_{F_1 \cap F_2} h\left(X_{\tau_n}\right) = \int 1_{F_1 \cap F_2} \otimes h \, dP \otimes K$$

for every $F_i \in \mathcal{H}'_i$ and $h \in C_b(\mathcal{X})$. For this, let $F_i \in \mathcal{H}'_i$ and $h \in C_b(\mathcal{X})$ be fixed. The independence of $\mathcal{H}'_1$ and $\mathcal{H}_2$ yields

$$\int 1_{F_1 \cap F_2} \otimes h \, dP \otimes K = P(F_1) \int 1_{F_2} \otimes h \, dP \otimes K.$$

Let $\varepsilon > 0$. By (ii), there exists an $N \in \mathbb{N}$ such that for every $n \geq N$,

$$\left| E 1_{F_2} h(X_n) - \int 1_{F_2} \otimes h \, dP \otimes K \right| \leq \varepsilon.$$

Furthermore, there exists an $N_1 \in \mathbb{N}$ such that $P(\tau_n < N) \leq \varepsilon$ for every $n \geq N_1$. We obtain for $n \geq N_1$, using (i),

$$\left| E 1_{F_1 \cap F_2} h\left(X_{\tau_n}\right) - \int 1_{F_1 \cap F_2} \otimes h \, dP \otimes K \right|$$

$$= \left| \sum_{k=1}^{\infty} \int_{\{\tau_n = k\}} 1_{F_1 \cap F_2} h(X_k) \, dP - P(F_1) \int 1_{F_2} \otimes h \, dP \otimes K \right|$$

$$\leq P(\tau_n < N) \|h\|_{\sup}$$

$$+ \left| \sum_{k \geq N} \int_{\{\tau_n = k\}} 1_{F_1 \cap F_2} h(X_k) \, dP - P(F_1) \int 1_{F_2} \otimes h \, dP \otimes K \right|$$

$$= P(\tau_n < N) \|h\|_{\sup}$$

$$+ \left| \sum_{k \geq N} P(F_1 \cap \{\tau_n = k\}) E 1_{F_2} h(X_k) - P(F_1) \int 1_{F_2} \otimes h \, dP \otimes K \right|$$

$$\leq P\left(\tau_n < N\right) \|h\|_{\sup}$$

$$+ \left| \sum_{k \geq N} \left(P\left(F_1 \cap \{\tau_n = k\}\right) - P\left(F_1\right)\right) \int 1_{F_2} \otimes h \, dP \otimes K \right|$$

$$+ \sum_{k \geq N} P\left(F_1 \cap \{\tau_n = k\}\right) \left| E 1_{F_2} h\left(X_k\right) - \int 1_{F_2} \otimes h \, dP \otimes K \right|$$

$$\leq 2P\left(\tau_n < N\right) \|h\|_{\sup}$$

$$+ \sum_{k \geq N} P\left(F_1 \cap \{\tau_n = k\}\right) \left| E 1_{F_2} h\left(X_k\right) - \int 1_{F_2} \otimes h \, dP \otimes K \right|$$

$$\leq \varepsilon \left(1 + 2 \|h\|_{\sup}\right),$$

which completes the proof. □

Exercise 4.1 Show that in Proposition 4.5 it is enough to assume $K \in \mathcal{K}^1\left(\mathcal{H}_2'\right)$ and $X_n \to K \; \mathcal{H}_2'$-stably instead of condition (ii).

Exercise 4.2 Assume $X_n \to X$ almost surely and $\tau_n \to \infty$ in probability. Show that $X_{\tau_n} \to X$ in probability.

In case $\mathcal{H}_i' = \mathcal{H}_i$, the $\mathcal{G}$-stable convergence $X_n \to K$ and $X_{\tau_n} \to K$ is, by Proposition 3.5, the same as stable convergence, while in case $\mathcal{H}_1' = \mathcal{H}_2' = \{\emptyset, \Omega\}$, $\mathcal{G}$-stable convergence means distributional convergence. So for mere distributional convergence of X_{τ_n} there is no need of stable convergence of X_n. This is different in the general (dependent) case, where stable convergence plays an essential role. Now we need the condition $\tau_n / a_n \to \eta$ in probability as $n \to \infty$ for some random variable $\eta > 0$, where $a_n \in (0, \infty)$, $a_n \to \infty$. For simplicity, we specialize from general sequences of $(\mathcal{X}, \mathcal{B}\left(\mathcal{X}\right))$-valued random variables to normalized real processes (thus avoiding explicit use of Anscombe-type conditions); see [9], Theorem 17.1, [2], [13], Theorem 9.4.3, [32].

Theorem 4.6 *Let $\mathcal{G} \subset \mathcal{F}$ be a sub-σ-field, $(X_t)_{t \geq 0}$ a path-continuous real process, $X_t^n := n^{-\alpha} X_{nt}, t \geq 0, n \in \mathbb{N}$, with $\alpha \in (0, \infty)$ and $\mathcal{X} = C\left(\mathbb{R}_+\right)$. Assume*

(i) $\tau_n / a_n \to \eta$ *in probability as $n \to \infty$*
for some $\mathbb{R}_+$-valued, $\mathcal{G}$-measurable random variable η with $P\left(\eta > 0\right) > 0$
and $a_n \in (0, \infty)$ satisfying $a_n \to \infty$. If
(ii) $X^n \to K \; \mathcal{G}$-stably
for some $K \in \mathcal{K}^1\left(\mathcal{G}, C\left(\mathbb{R}_+\right)\right)$, then $X^{\tau_n} \to K \; \mathcal{G}$-stably under $P_{\{\eta > 0\}}$ as $n \to \infty$.

Proof Choose $k_n \in \mathbb{N}$ such that $\lim_{n \to \infty} k_n / a_n = 1$. Clearly, by (i), $\tau_n / k_n \to \eta$ in probability. Let $\varphi : C\left(\mathbb{R}_+\right) \times \mathbb{R}_+ \to C\left(\mathbb{R}_+\right)$ be defined by

$$\varphi\left(x, b\right)\left(t\right) := \begin{cases} b^{-\alpha} x\left(bt\right), & \text{if } b > 0 \\ 0, & \text{if } b = 0. \end{cases}$$

Then φ is Borel-measurable, $\varphi \left(X^{k_n}, \tau_n / k_n \right) = X^{\tau_n}$ and one checks that φ is continuous on $C \left(\mathbb{R}_+ \right) \times (0, \infty)$. Condition (ii) yields $\mathcal{G}$-stable convergence $X^{k_n} \to K$ and hence by Theorem 3.7 (b),

$$\left(X^{k_n}, \frac{\tau_n}{k_n} \right) \to K_\eta := K \otimes \delta_\eta \quad \mathcal{G}\text{-stably}.$$

In particular, we have $\mathcal{G}$-stable convergence under $P_{\{\eta > 0\}}$ because $\{\eta > 0\} \in \mathcal{G}$. Since $P_{\{\eta > 0\}} K_\eta \left(C \left(\mathbb{R}_+ \right) \times \{0\} \right) = P_{\{\eta > 0\}} (\eta = 0) = 0$, φ is $P_{\{\eta > 0\}} K_\eta$-almost surely continuous so that we derive from Theorem 3.7 (c) that

$$X^{\tau_n} = \varphi \left(X^{k_n}, \frac{\tau_n}{k_n} \right) \to K_\eta^\varphi \quad \mathcal{G}\text{-stably under } P_{\{\eta > 0\}} \text{ as } n \to \infty.$$

It remains to show that $K_\eta^\varphi = K \, P_{\{\eta > 0\}}$-almost surely. Setting $\varphi_b (x) := \varphi (x, b)$ for $b > 0$, the limiting kernel reads

$$K_\eta^\varphi (\omega, \cdot) = \left(K (\omega, \cdot) \otimes \delta_{\eta(\omega)} \right)^\varphi = K (\omega, \cdot)^{\varphi_{\eta(\omega)}}, \quad \omega \in \{\eta > 0\}.$$

Since $\varphi_b \circ \varphi_c = \varphi_{bc}$ and $X^n = \varphi_n (X)$, we have for $b = N \in \mathbb{N}$, $\varphi_N (X^n) = X^{nN} \to K \, \mathcal{G}$-stably as $n \to \infty$ while continuity of φ_b yields $\varphi_N (X^n) \to K^{\varphi_N} \, \mathcal{G}$-stably as $n \to \infty$ (see Theorem 3.7 (c)). Hence, by almost sure uniqueness of stable limits, $K^{\varphi_N} = K \, P$-almost surely. Moreover, $\varphi_{1/N} \left(X^{nN} \right) = X^n \to K \, \mathcal{G}$-stably while continuity of φ_b yields $\varphi_{1/N} \left(X^{nN} \right) \to K^{\varphi_{1/N}} \, \mathcal{G}$-stably so that $K^{\varphi_{1/N}} = K \, P$-almost surely. We obtain $K^{\varphi_b} = K \, P$-almost surely for every $b \in \mathbb{Q}$, $b > 0$. Consequently, there exists a $\Omega_0 \in \mathcal{G}$ with $P (\Omega_0) = 1$ such that $K (\omega, \cdot)^{\varphi_b} = K (\omega, \cdot)$ for every $\omega \in \Omega_0$, $b \in \mathbb{Q}$, $b > 0$. Since the map $(0, \infty) \to \mathcal{M}^1 (C (\mathbb{R}_+))$, $b \mapsto K (\omega, \cdot)^{\varphi_b}$, is continuous for every $\omega \in \Omega$, the above equality holds for all $b \in (0, \infty)$. This implies $K (\omega, \cdot)^{\varphi_{\eta(\omega)}} = K (\omega, \cdot)$ for every $\omega \in \Omega_0 \cap \{\eta > 0\}$ and thus $K_\eta^\varphi = K$ $P_{\{\eta > 0\}}$-almost surely. □

One obtains the same result for càdlàg processes X and $\mathcal{X} = D (\mathbb{R}_+)$.

Remark 4.7 (a) Literally the same proof shows that Theorem 4.6 is still true for $(0, \infty)$-valued random variables τ_n, where $X_t^{\tau_n} = X_{\tau_n t} / \tau_n^\alpha$.
(b) Condition (i) may be weakened. For instance, Theorem 4.6 still holds if (i) is replaced by

(i)' $\displaystyle \sum_{i=1}^m \frac{\tau_n}{a_{n,i}} 1_{G_i} \to \eta := \sum_{i=1}^m \eta_i 1_{G_i}$ in probability,

where $m \in \mathbb{N}$, $\{G_1, \ldots, G_m\}$ is a $\mathcal{G}$-measurable partition of Ω, η_i are $\mathbb{R}_+$-valued, $\mathcal{G}$-measurable random variables with $P (\eta > 0) > 0$ and $a_{n,i} \in (0, \infty)$ satisfying $a_{n,i} \to \infty$ as $n \to \infty$.

In fact, for $i \in I := \left\{ j \in \{1, \ldots, m\} : P \left(G_j \cap \{\eta > 0\} \right) > 0 \right\}$ we have $\tau_n / a_{n,i} \to \eta_i$ in P_{G_i}-probability and $X^n \to K \, \mathcal{G}$-stably under P_{G_i} so that by Theorem 4.6 (with

P replaced by P_{G_i}) $X^{\tau_n} \to K$ $\mathcal{G}$-stably under $P_{G_i \cap \{\eta_i > 0\}}$. Since $P_{G_i \cap \{\eta_i > 0\}} = P_{G_i \cap \{\eta > 0\}}$ and

$$P_{\{\eta > 0\}} = \sum_{i \in I} P_{\{\eta > 0\}}(G_i) P_{G_i \cap \{\eta > 0\}},$$

the assertion follows from Proposition 3.11.

(c) ([2]) The role of condition (ii) can be further clarified as follows. Assume that $\mathcal{X}$ is polish. If X_n are $(\mathcal{X}, \mathcal{B}(\mathcal{X}))$-valued random variables (like X^n in Theorem 4.6) such that $X_{\tau_n} \xrightarrow{d} \nu$ for some $\nu \in \mathcal{M}^1(\mathcal{X})$ and every sequence $(\tau_n)_{n \geq 1}$ satisfying condition (i)' from (b) with $m \leq 2$ and $P(\eta > 0) = 1$, then $(X_n)_{n \geq 1}$ must converge $\mathcal{G}$-stably. Otherwise, by Proposition 3.12, there exists $G \in \mathcal{G}$ with $P(G) \in (0, 1)$ such that $\left(P_G^{X_n}\right)_{n \geq 1}$ does not converge weakly in $\mathcal{M}^1(\mathcal{X})$. Thus we can find $h \in C_b(\mathcal{X})$ and subsequences (k_n) and (m_n) of (n) such that

$$\int_G h\left(X_{k_n}\right) dP \to c_1 \quad \text{and} \quad \int_G h\left(X_{m_n}\right) dP \to c_2,$$

where $c_1, c_2 \in \mathbb{R}$, $c_1 \neq c_2$. The $\mathbb{N}$-valued random variables $\tau_n := m_n$ and $\sigma_n := k_n 1_G + m_n 1_{G^c}$ satisfy (i)' with $\eta = 1$ and

$$Eh\left(X_{\sigma_n}\right) - Eh\left(X_{\tau_n}\right) = \int_G h\left(X_{k_n}\right) dP - \int_G h\left(X_{m_n}\right) dP \to c_1 - c_2 \neq 0,$$

a contradiction.

Exercise 4.3 Show that $(X_n)_{n \geq 1}$ converges stably if and only if $X_{\tau_n} \xrightarrow{d} \nu$ for some $\nu \in \mathcal{M}^1(\mathcal{X})$ and all sequences $(\tau_n)_{n \geq 1}$ of $\mathbb{N}$-valued random variables such that $\tau_n \to \infty$ in probability and $P|\sigma(\tau_n, n \geq 1)$ is purely atomic.

Example 4.8 (Classical stable functional random-sum central limit theorem) In the situation of Example 3.14 let

$$X_t := \frac{1}{\sigma}\left(\sum_{j=1}^{[t]} Z_j + (t - [t]) Z_{[t]+1}\right), \quad t \geq 0$$

and $X_t^n := n^{-1/2} X_{nt}$, $t \geq 0$, $n \in \mathbb{N}$. Since $X^n \to \nu$ mixing in $C(\mathbb{R}_+)$ for the Wiener measure $\nu \in \mathcal{M}^1(C(\mathbb{R}_+))$, it follows from Theorem 4.6 (with $\mathcal{G} = \mathcal{F}$) that $X^{\tau_n} \to \nu$ mixing under $P_{\{\eta > 0\}}$ provided condition (i) of Theorem 4.6 is satisfied. In particular, using Theorem 3.7 (c),

$$\frac{1}{\sigma \tau_n^{1/2}} \sum_{j=1}^{\tau_n} Z_j = X_1^{\tau_n} \to N(0, 1) \quad \text{mixing under } P_{\{\eta > 0\}}.$$

It is not enough to assume $\tau_n \to \infty$ in probability as $n \to \infty$ in Theorem 4.6. For instance, if $P(Z_1 = \pm 1) = 1/2$, $\tau_1 := \inf\left\{k \geq 1 : \sum_{i=1}^{k} Z_i = 0\right\}$ and $\tau_n := \inf\left\{k > \tau_{n-1} : \sum_{i=1}^{k} Z_i = 0\right\}$ for $n \geq 2$, then $P(\tau_n \in \mathbb{N}) = 1$ and $\tau_n \geq n$ so that $\tau_n \to \infty$ almost surely but $\sum_{i=1}^{\tau_n} Z_i = 0$ for every $n \in \mathbb{N}$. $\qquad\square$

Exercise 4.4 Let $W = (W_t)_{t \geq 0}$ be a (path-continuous) Brownian motion, $X_t^b := b^{-1/2} W_{bt}$, $t \geq 0$, $b > 0$, and let τ_n be $(0, \infty)$-valued random variables satisfying condition (i) of Theorem 4.6 with $\mathcal{G} = \mathcal{F}$. Show that $X^{\tau_n} \to \nu$ mixing under $P_{\{\eta > 0\}}$ in $C(\mathbb{R}_+)$, where $\nu = P^W \in \mathcal{M}^1(C(\mathbb{R}_+))$.

Exercise 4.5 In the situation of Example 4.8 let

$$\tau_n := n1_{\left\{\sum_{j=1}^{n} Z_j > 0\right\}} + 2n1_{\left\{\sum_{j=1}^{n} Z_j \leq 0\right\}}.$$

Show that $\tau_n/n \xrightarrow{d} (\delta_1 + \delta_2)/2$, but $\sigma^{-1}\tau_n^{-1/2}\sum_{j=1}^{\tau_n} Z_j$ does not even converge in distribution to $N(0, 1)$ as $n \to \infty$. Thus in condition (i) of Theorem 4.6 convergence in probability cannot be replaced by convergence in distribution.

4.3 The Empirical Measure Theorem and the δ-Method

The following result (see [7], Corollary 3.16, Theorem 4.7, [31]) allows us to pass from stable convergence to almost sure convergence and has the Komlós theorem as its point of departure.

Theorem 4.9 (Empirical measure theorem) *If $X_n \to K$ stably for $(\mathcal{X}, \mathcal{B}(\mathcal{X}))$-valued random variables X_n and $K \in \mathcal{K}^1$, then there exists a subsequence (X_m) of (X_n) such that for every further subsequence (X_k) of (X_m), almost surely*

$$\frac{1}{r}\sum_{k=1}^{r} \delta_{X_k(\omega)} \to K(\omega, \cdot) \quad \text{weakly} \left(\text{in } \mathcal{M}^1(\mathcal{X})\right) \text{ as } r \to \infty.$$

The above assertion simply means almost sure convergence of $\frac{1}{r}\sum_{k=1}^{r} \delta_{X_k}$ to K when the Markov kernels are seen as $(\mathcal{M}^1(\mathcal{X}), \mathcal{B}(\mathcal{M}^1(\mathcal{X})))$-valued random variables. Note that the exceptional null set may vary with the subsequence. In general, the assertion is not true for (X_n) itself (see [7], Example 3.17). However, in the classical case of an independent and identically distributed sequence (X_n) it is well known that $(X_n \to P^{X_1}$ mixing and) almost surely

$$\frac{1}{r}\sum_{n=1}^{r} \delta_{X_n}(\omega) \to P^{X_1} \quad \text{weakly as } r \to \infty.$$

Proof of Theorem 4.9. *Step 1.* We rely on the Komlós theorem: If $(f_n)_{n \geq 1}$ is a sequence in $\mathcal{L}^1(P)$ satisfying $\sup_{n \geq 1} E|f_n| < \infty$, then there exists a subsequence (f_m) of (f_n) and a function $f \in \mathcal{L}^1(P)$ such that for every further subsequence (f_k) of (f_m)

$$\frac{1}{r} \sum_{k=1}^{r} f_k \to f \quad \text{a.s. as } r \to \infty$$

(see [14], Théorème IX.1).

Step 2. Let $\{h_i : i \in \mathbb{N}\}$ be a countable *convergence-determining* subset of $C_b(\mathcal{X})$ for $\mathcal{M}^1(\mathcal{X})$, that is, the weak topology on $\mathcal{M}^1(\mathcal{X})$ is generated by the functions $\nu \mapsto \int h_i \, d\nu, i \in \mathbb{N}$. For instance, any countable dense subset of $U_b(\mathcal{X}, \widetilde{d})$ for some totally bounded metric $\widetilde{d}$ inducing the topology of $\mathcal{X}$ has the desired property (see the proof of Theorem 2.6). If $f_{i,n} := h_i(X_n)$, then $\sup_{n \geq 1} E|f_{i,n}| \leq \|h_i\|_{\sup} < \infty$. Applying Step 1 (to $(f_{i,n})_{n \geq 1}$) in a diagonal procedure, we obtain a subsequence (X_m) of (X_n) and functions $f_i \in \mathcal{L}^1(P)$ such that for every further subsequence (X_k) of (X_m)

$$\frac{1}{r} \sum_{k=1}^{r} f_{i,k} \to f_i \quad \text{a.s. as } r \to \infty$$

for every $i \in \mathbb{N}$. Setting $K_r := \sum_{k=1}^{r} \delta_{X_k}/r$ this reads $\int h_i(x) \, K_r(\cdot, dx) \to f_i$ almost surely. The exceptional null set is denoted by N_1. Dominated convergence yields

$$\lim_{r \to \infty} \int_F \int h_i(x) \, K_r(\cdot, dx) \, dP = \int_F f_i \, dP$$

for every $F \in \mathcal{F}, i \in \mathbb{N}$. On the other hand, by stable convergence $X_k \to K$,

$$\lim_{k \to \infty} \int_F h_i(X_k) \, dP = \int_F \int h_i(x) \, K(\cdot, dx) \, dP$$

and hence

$$\lim_{r \to \infty} \int_F \int h_i(x) \, K_r(\cdot, dx) \, dP = \int_F \int h_i(x) \, K(\cdot, dx) \, dP$$

for every $F \in \mathcal{F}, i \in \mathbb{N}$. Consequently, $f_i = \int h_i(x) \, K(\cdot, dx)$ almost surely for every $i \in \mathbb{N}$. The exceptional null set is denoted by N_2. We obtain for every $\omega \in N_1^c \cap N_2^c$ and $i \in \mathbb{N}$

$$\lim_{r \to \infty} \int h_i(x) \, K_r(\omega, dx) = \int h_i(x) \, K(\omega, dx)$$

and thus

$$K_r(\omega, \cdot) = \frac{1}{r} \sum_{k=1}^{r} \delta_{X_k(\omega)} \to K(\omega, \cdot) \quad \text{weakly}$$

which completes the proof.   □

Exercise 4.6 Let A be countable and dense in $\mathcal{X}$ and $\mathcal{B}$ the collection of all finite unions of open balls with centers in A and radius in $\mathbb{Q} \cap (0, \infty)$. Associate to each $B \in \mathcal{B}$ and $n \in \mathbb{N}$ the function $h_{B,n} \in C_b(\mathcal{X})$, where $h_{B,n}(x) := 1 \wedge n \inf_{y \in B^c} d(x, y)$. The resulting collection of all such functions is countable. Show that it is convergence-determining for $\mathcal{M}^1(\mathcal{X})$.

Exercise 4.7 Assume that $\mathcal{X}$ is polish. Let $K \in \mathcal{K}^1$ and let X_n be $(\mathcal{X}, \mathcal{B}(\mathcal{X}))$-valued random variables such that $\left(P^{X_n}\right)_{n \geq 1}$ is tight and for every subsequence (X_m) of (X_n), almost surely

$$\frac{1}{r} \sum_{m=1}^{r} \delta_{X_m(\omega)} \to K(\omega, \cdot) \quad \text{weakly as } r \to \infty.$$

Show that $X_n \to K$ stably.

The δ-method with random centering needs stable convergence.

Proposition 4.10 (*δ-method*) *Let* $\mathcal{G} \subset \mathcal{F}$ *be a sub-σ-field,* $g : \mathbb{R}^d \to \mathbb{R}$ *continuously differentiable and* $a_n \in (0, \infty)$ *with* $a_n \to \infty$ *as* $n \to \infty$. *If*

$$a_n(Y_n - Y) \to X \quad \mathcal{G}\text{-stably}$$

for $\mathbb{R}^d$-*valued random variables* X, Y_n *and* Y, *where* Y *is* $\mathcal{G}$-*measurable, then*

$$a_n(g(Y_n) - g(Y)) \to \langle \nabla g(Y), X \rangle \quad \mathcal{G}\text{-stably as } n \to \infty.$$

Proof The mean value theorem implies that

$$a_n(g(Y_n) - g(Y)) = \langle \nabla g(\xi_n), a_n(Y_n - Y) \rangle = \langle \nabla g(Y), a_n(Y_n - Y) \rangle + R_n$$

for some map $\xi_n : \Omega \to \mathbb{R}^d$ (not necessarily measurable) with $\|\xi_n - Y\| \leq \|Y_n - Y\|$ everywhere on Ω, where $R_n := \langle \nabla g(\xi_n) - \nabla g(Y), a_n(Y_n - Y) \rangle$ (which is measurable) and $\|\cdot\|$ denotes the euclidean norm on $\mathbb{R}^d$. By Theorems 3.17 and 3.18 (b), (c),

$$\langle \nabla g(Y), a_n(Y_n - Y) \rangle \to \langle \nabla g(Y), X \rangle \quad \mathcal{G}\text{-stably}.$$

In view of Theorem 3.18 (a) it remains to show that $R_n \to 0$ in probability. Since $a_n (Y_n - Y) \overset{d}{\to} X$, we have $\|Y_n - Y\| \to 0$ in probability. Let $\varepsilon > 0$ and $0 < N < \infty$ be arbitrary. The map ∇g is uniformly continuous on the compact subset $B_N := \{x \in \mathbb{R}^d : \|x\| \le N + 1\}$ of $\mathbb{R}^d$ so that there exists a $\delta > 0$ such that $\|\nabla g(x) - \nabla g(y)\| \le \varepsilon/N$ for all $x, y \in B_N$ with $\|x - y\| \le \delta$. On the event $\{|R_n| > \varepsilon\} \cap \{\|Y\| \le N\} \cap \{a_n \|Y_n - Y\| \le N\}$ we have, by the Cauchy-Schwarz inequality,

$$\varepsilon < |R_n| \le \|\nabla g(\xi_n) - \nabla g(Y)\| a_n \|Y_n - Y\| \le \|\nabla g(\xi_n) - \nabla g(Y)\| N$$

so that $\|\nabla g(\xi_n) - \nabla g(Y)\| > \varepsilon/N$. Moreover, on this event we have $\|\xi_n - Y\| \le \|Y_n - Y\| \le N/a_n \le 1$ for all large $n \in \mathbb{N}$, which implies $\xi_n, Y \in B_N$. Consequently, $\delta < \|\xi_n - Y\| \le \|Y_n - Y\|$, yielding, for all large $n \in \mathbb{N}$,

$$\{|R_n| > \varepsilon\} \subset (\{|R_n| > \varepsilon\} \cap \{\|Y\| \le N\} \cap \{a_n \|Y_n - Y\| \le N\})$$
$$\cup \{\|Y\| > N\} \cup \{a_n \|Y_n - Y\| > N\}$$
$$\subset \{\|Y_n - Y\| > \delta\} \cup \{\|Y\| > N\} \cup \{a_n \|Y_n - Y\| > N\}.$$

Therefore, for all large $n \in \mathbb{N}$,

$$P(|R_n| > \varepsilon) \le P(\|Y_n - Y\| > \delta) + P(\|Y\| > N) + P(a_n \|Y_n - Y\| > N).$$

From $\|Y_n - Y\| \to 0$ in probability and $a_n \|Y_n - Y\| \overset{d}{\to} \|X\|$ we obtain, by the Portmanteau theorem,

$$\limsup_{n \to \infty} P(|R_n| > \varepsilon) \le P(\|Y\| > N) + P(\|X\| \ge N)$$

for every $\varepsilon > 0$ and $0 < N < \infty$. Letting $N \to \infty$ yields the assertion. $\square$

Remark 4.11 (a) For $\mathcal{G} = \{\emptyset, \Omega\}$ Proposition 4.10 reduces to the usual δ-method for convergence in distribution in which Y is almost surely constant.
(b) To see that the δ-method for convergence in distribution does not in general work with random centering we consider the probability space and sequence $(X_n)_{n \ge 1}$ from Example 1.2. For any sequence $b_n \in (0, \infty)$ with $b_n \to \infty$ as $n \to \infty$ we set $Y_n := b_n^{-1} X_n + Y$, where Y is as in Example 1.2. Then

$$b_n (Y_n - Y) = X_n \overset{d}{\to} X_1 \quad \text{as } n \to \infty.$$

For the continuously differentiable function $g(x) = x^2$, $x \in \mathbb{R}$, we have

$$b_n (g(Y_n) - g(Y)) = b_n^{-1} X_n^2 + 2 X_n Y \quad \text{for all } n \ge 1.$$

Now $\left(b_n^{-1} X_n^2\right)_{n \geq 1}$ converges almost surely to zero because $b_n \to \infty$ as $n \to \infty$ and $|X_n| \leq 1$, and the sequence $(Y X_n)_{n \geq 1}$ does not converge in distribution as seen in Example 1.2, if the sequence $(a_n)_{n \geq 1}$ used to define the random variables X_n is not convergent. Thus, the random variables $b_n \left(g\left(Y_n\right) - g\left(Y\right)\right)$ do not converge in distribution.

Random centering occurs, for example, in connection with exchangeable processes; see Corollary 6.27. Stable convergence in connection with random normalization occurs in various subsequent chapters.

Chapter 5
Stability of Limit Theorems

In this chapter we present some first results on the stability of limit theorems taken from [28] (see also [79, 100]). More precisely, we derive simple sufficient conditions for distributional limit theorems to be mixing.

To this end, let Z_n be $(\mathcal{Z}, \mathcal{C})$-valued random variables for some measurable space $(\mathcal{Z}, \mathcal{C})$ and $f_n : (\mathcal{Z}^n, \mathcal{C}^n) \to (\mathcal{X}, \mathcal{B}(\mathcal{X}))$ measurable maps for every $n \in \mathbb{N}$, where we need a vector space structure for $\mathcal{X}$. So, let $\mathcal{X}$ be a polish topological vector space (like $\mathbb{R}^d$, $C([0, T])$ for $0 < T < \infty$ or $C(\mathbb{R}_+)$). Then there exists a translation invariant metric d on $\mathcal{X}$ inducing the topology ([86], Theorem 1.6.1) so that $U_n - V_n \to 0$ in probability for $(\mathcal{X}, \mathcal{B}(\mathcal{X}))$-valued random variables U_n and V_n means $d(U_n, V_n) = d(U_n - V_n, 0) \to 0$ in probability or, what is the same, $E(d(U_n, V_n) \wedge 1) \to 0$.

Furthermore, let $b_n \in \mathcal{X}$ and $a_n \in (0, \infty)$. We consider the $(\mathcal{X}, \mathcal{B}(\mathcal{X}))$-valued random variables

$$X_n := \frac{1}{a_n}(f_n(Z_1, \ldots, Z_n) - b_n)$$

for $n \in \mathbb{N}$ and assume $X_n \xrightarrow{d} \nu$ for some $\nu \in \mathcal{M}^1(\mathcal{X})$. The *tail σ-field* of $Z = (Z_n)$ is given by

$$\mathcal{T}_Z = \bigcap_{n=1}^{\infty} \sigma(Z_k, k \geq n).$$

Proposition 5.1 *Assume* $X_n \xrightarrow{d} \nu$ *and*

(i) *for every* $k \in \mathbb{N}$,

$$\frac{1}{a_n}(f_n(Z_1, \ldots, Z_n) - f_{n-k}(Z_{k+1}, \ldots, Z_n)) \to 0 \quad \text{in probability as } n \to \infty,$$

© Springer International Publishing Switzerland 2015
E. Häusler and H. Luschgy, *Stable Convergence and Stable Limit Theorems*,
Probability Theory and Stochastic Modelling 74,
DOI 10.1007/978-3-319-18329-9_5

(ii) T_Z *is trivial, i.e.* $P(T_Z) = \{0, 1\}$.

Then $X_n \rightarrow v$ *mixing as* $n \rightarrow \infty$.

Proof Since $\left(P^{X_n}\right)_{n\geq 1}$ is tight in $\mathcal{M}^1(\mathcal{X})$, $(X_n)_{n\geq 1}$ has a stably convergent subsequence by Proposition 3.4 (a). Let (X_m) be any subsequence of (X_n) with $X_m \rightarrow K$ stably for some $K \in \mathcal{K}^1$ and for $k \in \mathbb{N}$, let $X_m(k) := (f_{m-k}(Z_{k+1}, \ldots, Z_m) - b_m) / a_m$, $m > k$. Distributional convergence $X_m \overset{d}{\rightarrow} v$ yields $PK = v$. By (i), we have $X_m - X_m(k) \rightarrow 0$ in probability as $m \rightarrow \infty$. Consequently, by Theorem 3.7 (a), $X_m(k) \rightarrow K$ stably as $m \rightarrow \infty$. Now, $X_m(k)$ is $\mathcal{H}_{k+1}$-measurable, where $\mathcal{H}_k := \sigma(Z_j, j \geq k)$, so that by Propositions 3.4 (b) and 3.5, $X_m(k) \rightarrow E(K|\mathcal{H}_{k+1})$ stably as $m \rightarrow \infty$. The P-almost sure uniqueness of stable limits yields $K = E(K|\mathcal{H}_{k+1})$ P-almost surely for every $k \in \mathbb{N}$. Letting $H := E(K|T_Z)$, the martingale convergence theorem and Lemma A.7 (c) imply for every $B \in \mathcal{B}(\mathcal{X})$

$$K(\cdot, B) = E(K(\cdot, B)|\mathcal{H}_{k+1}) \rightarrow E(K(\cdot, B)|T_Z) = H(\cdot, B) \quad P\text{-a.s.}$$

as $k \rightarrow \infty$ and hence, $K = H$ P-almost surely by Lemma 2.1 (b). Therefore, by (ii), K is P-almost surely constant and thus $K = PK = v$ P-almost surely. Thus all subsequences of (X_n) which converge stably, converge mixing to v and so the original sequence must converge mixing to v. $\qquad\square$

Condition (ii) in Proposition 5.1 is met for independent sequences $(Z_n)_{n\geq 1}$ by the Kolmogorov zero-one law. In this case, for instance, the choice $(\mathcal{Z}, \mathcal{C}) = (\mathbb{R}, \mathcal{B}(\mathbb{R}))$ and $f_n(z_1, \ldots, z_n) = \sum_{i=1}^n z_i$ yields Example 3.13 (a).

Triviality of the tail σ-field may be characterized by asymptotic independence in the following sense (see [11]).

Lemma 5.2 *Let* $\mathcal{F}_k := \sigma(Z_1, \ldots, Z_k)$ *and* $\mathcal{H}_k := \sigma(Z_j, j \geq k)$. *Then the assertions*

(i) $P(T_Z) = \{0, 1\}$,

(ii) *for every* $G \in \bigcup_{k=1}^{\infty} \mathcal{F}_k$,

$$\lim_{n \to \infty} \sup_{F \in \mathcal{H}_n} |P(F \cap G) - P(F)P(G)| = 0,$$

(iii) *for every* $G \in \mathcal{F}$,

$$\lim_{n \to \infty} \sup_{F \in \mathcal{H}_n} |P(F \cap G) - P(F)P(G)| = 0$$

are equivalent.

Proof (i) $\Rightarrow$ (iii). Let $G \in \mathcal{F}$. The martingale convergence theorem and (i) yield $P(G|\mathcal{H}_n) \rightarrow P(G|T_Z) = P(G)$ in $\mathcal{L}^1(P)$. Consequently, for every $F \in \mathcal{H}_n$,

$$|P(F \cap G) - P(F) P(G)| = \left| \int_F (1_G - P(G)) \, dP \right|$$

$$= \left| \int_F (P(G|\mathcal{H}_n) - P(G)) \, dP \right| \le \int |P(G|\mathcal{H}_n) - P(G)| \, dP \to 0 ,$$

hence (iii).

The implication (iii) $\Rightarrow$ (ii) is obvious.

(ii) $\Rightarrow$ (i). Let $F \in \mathcal{T}_Z = \bigcap_{j=1}^{\infty} \mathcal{H}_j$ with $P(F) > 0$ and $G \in \mathcal{E} := \bigcup_{k=1}^{\infty} \mathcal{F}_k$ with $P(G) > 0$. Then for every $n \in \mathbb{N}$

$$|P_G(F) - P(F)| \le \frac{1}{P(G)} \sup_{D \in \mathcal{H}_n} |P(D \cap G) - P(D) P(G)| .$$

Condition (ii) yields $P_G(F) = P(F)$ or, what is the same, $P_F(G) = P(G)$. Clearly, this also holds if $P(G) = 0$. We obtain $P_F = P$ on the field $\mathcal{E}$ and thus on $\sigma(\mathcal{E}) = \mathcal{H}_1$. Consequently, $P(F) = P_F(F) = 1$ because $F \in \sigma(\mathcal{E})$. $\square$

Second proof of Proposition 5.1. Let $G \in \mathcal{F}$ with $P(G) > 0$ and $\varepsilon > 0$. By (ii) and Lemma 5.2, there exists a $k \in \mathbb{N}$ such that $\sup_{F \in \mathcal{H}_k} |P_G(F) - P(F)| \le \varepsilon$, where $\mathcal{H}_k = \sigma(Z_j, j \ge k)$. For $n > k$, let $Y_n := (f_{n-k}(Z_{k+1}, \ldots, Z_n) - b_n) / a_n$. By (i), we have $X_n - Y_n \to 0$ in probability so that $Y_n \overset{d}{\to} \nu$ as $n \to \infty$. Now for all closed sets $B \subset \mathcal{X}$ we have $\{Y_n \in B\} \in \mathcal{H}_k$ and hence $P_G(Y_n \in B) \le P(Y_n \in B) + \varepsilon$ for every $n > k$. The Portmanteau theorem yields $\limsup_{n \to \infty} P_G(Y_n \in B) \le \nu(B) + \varepsilon$ and letting ε tend to zero gives $\limsup_{n \to \infty} P_G(Y_n \in B) \le \nu(B)$. Using again the Portmanteau theorem, this implies $P_G^{Y_n} \to \nu$ weakly and thus $P_G^{X_n} \to \nu$ weakly. The assertion follows from Corollary 3.3. $\square$

Exercise 5.1 ([92]) Assume $Z_n \overset{d}{\to} \nu$, where $\mathcal{Z}$ is a polish space and $\mathcal{C} = \mathcal{B}(\mathcal{Z})$, and condition (ii) of Proposition 5.1. Show that $Z_n \to \nu$ mixing.

The process $Z = (Z_n)_{n \ge 1}$ is called *stationary* if $P^{S(Z)} = P^Z$ on the σ-field $\mathcal{C}^{\mathbb{N}}$, where $S : \mathcal{Z}^{\mathbb{N}} \to \mathcal{Z}^{\mathbb{N}}$, $S((z_n)_{n \in \mathbb{N}}) = (z_{n+1})_{n \in \mathbb{N}}$ denotes the shift operator. Clearly, S is $(\mathcal{C}^{\mathbb{N}}, \mathcal{C}^{\mathbb{N}})$-measurable. Let $\mathcal{C}^{\mathbb{N}}(S) := \{D \in \mathcal{C}^{\mathbb{N}} : D = S^{-1}(D)\}$ denote the σ-field of invariant measurable subsets of $\mathcal{Z}^{\mathbb{N}}$ and for $Q \in \mathcal{M}^1(\mathcal{Z}^{\mathbb{N}})$, $\mathcal{C}^{\mathbb{N}}(S, Q) := \{D \in \mathcal{C}^{\mathbb{N}} : Q(D \Delta S^{-1}(D)) = 0\}$ is the σ-field of Q-almost invariant measurable sets. If $Q^S \ll Q$, we have $\mathcal{C}^{\mathbb{N}}(S) = \mathcal{C}^{\mathbb{N}}(S, Q)$ Q-almost surely, that is, for every $D \in \mathcal{C}^{\mathbb{N}}(S, Q)$ there exists a set $C \in \mathcal{C}^{\mathbb{N}}(S)$ such that $Q(D \Delta C) = 0$. In fact, if $D \in \mathcal{C}^{\mathbb{N}}(S, Q)$ and $S^n = S^{n-1} \circ S$, then $Q\left(D \Delta (S^n)^{-1}(D)\right) = 0$ for every $n \in \mathbb{N}$ because $Q^S \ll Q$. Defining $C := \limsup_{n \to \infty} (S^n)^{-1}(D)$ yields $C \in \mathcal{C}^{\mathbb{N}}(S)$ and $D \Delta C \subset \bigcup_{n=1}^{\infty} D \Delta (S^n)^{-1}(D)$, hence $Q(D \Delta C) = 0$.

A stationary process Z is said to be *ergodic* if $P^Z(\mathcal{C}^{\mathbb{N}}(S, P^Z)) = \{0, 1\}$ which is equivalent to $P^Z(\mathcal{C}^{\mathbb{N}}(S)) = \{0, 1\}$. Since $Z^{-1}(\mathcal{C}^{\mathbb{N}}(S)) \subset \mathcal{T}_Z$, asymptotic independence of Z in the sense of Lemma 5.2 implies ergodicity.

We only need *quasi-stationarity* of Z, that is $P^{S(Z)} \ll P^Z$.

Proposition 5.3 *Assume that* $Z = (Z_n)_{n \in \mathbb{N}}$ *is quasi-stationary. Assume further* $X_n \overset{d}{\to} v$ *and*

(i) $\dfrac{1}{a_n} (f_n (Z_1, \ldots, Z_n) - f_n (Z_2, \ldots, Z_{n+1})) \to 0$ *in probability as* $n \to \infty$,

(ii) $P^Z (\mathcal{C}^{\mathbb{N}} (S)) = \{0, 1\}$.

Then $X_n \to v$ *mixing as* $n \to \infty$.

Since $Z^{-1} (\mathcal{C}^{\mathbb{N}} (S)) \subset \mathcal{T}_Z$, condition (ii) in Proposition 5.3 is weaker than condition (ii) in Proposition 5.1.

Proof Step 1. First, we consider the canonical model $(\mathcal{Z}^{\mathbb{N}}, \mathcal{C}^{\mathbb{N}}, P^Z)$ with projections $\pi_n : \mathcal{Z}^{\mathbb{N}} \to \mathcal{Z}$. Letting $Y_n := (f_n (\pi_1, \ldots, \pi_n) - b_n) / a_n$ we will show that $Y_n \to v$ mixing under P^Z as $n \to \infty$. For this, let (Y_m) be any subsequence of (Y_n) with $Y_m \to K$ stably under P^Z for some $K \in \mathcal{K}^1 (\mathcal{C}^{\mathbb{N}}, \mathcal{X})$. As in the proof of Proposition 5.1, it is enough to show $K = v$ P^Z-almost surely. Since $X_n \overset{d}{\to} v$ and $(P^Z)^{Y_n} = P^{X_n}$, we have $Y_n \overset{d}{\to} v$ under P^Z and thus $P^Z K = v$. Condition (i) implies

$$\frac{1}{a_n} (f_n (\pi_1, \ldots, \pi_n) - f_n (\pi_2, \ldots, \pi_{n+1})) \to 0 \quad \text{in } P^Z\text{-probability}.$$

Hence, $Y_m - Y_m \circ S \to 0$ in P^Z-probability so that by Theorem 3.7 (a), $Y_m \circ S \to K$ stably under P^Z as $m \to \infty$. On the other hand, we have $Y_m \circ S \to K \circ S$ stably under P^Z, where $K \circ S (z, B) := K (S (z), B), z \in \mathcal{Z}^{\mathbb{N}}, B \in \mathcal{B} (\mathcal{X})$. In fact, by Theorem 3.2 and quasi-stationarity $(P^Z)^S = P^{S(Z)} \ll P^Z, Y_m \to K$ stably under $(P^Z)^S$. This implies, for every $C \in S^{-1} (\mathcal{C}^{\mathbb{N}}), C = S^{-1} (D)$ with $D \in \mathcal{C}^{\mathbb{N}}$ and $h \in C_b (\mathcal{X})$,

$$\int_C h (Y_m \circ S) \, dP^Z = \int_D h (Y_m) \, d (P^Z)^S$$

$$\to \int 1_D \otimes h \, d (P^Z)^S \otimes K = \int 1_C \otimes h \, dP^Z \otimes K \circ S.$$

Hence, again by Theorem 3.2, $Y_m \circ S \to K \circ S$ $S^{-1} (\mathcal{C}^{\mathbb{N}})$-stably under P^Z. Since the maps $Y_m \circ S$ are $S^{-1} (\mathcal{C}^{\mathbb{N}})$-measurable, it follows from Proposition 3.5 that $Y_m \circ S \to K \circ S$ stably under P^Z.

Now, almost sure uniqueness of stable limits yields $K \circ S = K$ P^Z-almost surely. Therefore, K is $\mathcal{C}^{\mathbb{N}} (S, P^Z)$-measurable because for all $A \in \mathcal{B} (\mathbb{R}), B \in \mathcal{B} (\mathcal{X})$

$$\{K (\cdot, B) \in A\} \, \Delta S^{-1} (\{K (\cdot, B) \in A\}) \subset \{K \circ S \neq K\}.$$

Consequently, by (ii) (which is the same as $P^Z (\mathcal{C}^{\mathbb{N}} (S, P^Z)) = \{0, 1\}$), K must be P^Z-almost surely constant and thus $K = v$ P^Z-almost surely.

Step 2. Let $\mathcal{G} := \sigma(Z_n, n \geq 1) = Z^{-1}(\mathcal{C}^{\mathbb{N}})$. Then it follows from Step 1 and Theorem 3.2 that $X_n \to \nu$ $\mathcal{G}$-mixing because for every $G \in \mathcal{G}$, $P(G) > 0$, $G = Z^{-1}(D)$ with $D \in \mathcal{C}^{\mathbb{N}}$, we have $P_G^{X_n} = (P^Z)_D^{Y_n}$. Since the maps X_n are $\mathcal{G}$-measurable, Proposition 3.5 yields $X_n \to \nu$ mixing.   □

Remark 5.4 One may consider even more general maps f_n in Proposition 5.3. In fact, Proposition 5.3 still holds for $f_n : (\mathcal{Z}^{\mathbb{N}}, \mathcal{C}^{\mathbb{N}}) \to (\mathcal{X}, \mathcal{B}(\mathcal{X}))$ and condition 5.3 (i) replaced by

$$\frac{1}{a_n}\left(f_n\left((Z_j)_{j\geq 1}\right) - f_n\left((Z_j)_{j\geq 2}\right)\right) \to 0 \quad \text{in probability as } n \to \infty.$$

This is obvious from the proof of Proposition 5.3.

Most applications are for stationary ergodic processes.

Example 5.5 Let $(Z_n)_{n\geq 1}$ be a stationary and ergodic real process and $\mathcal{X} = \mathcal{Z} = \mathbb{R}$.
(a) If $X_n := \left(\sum_{j=1}^n Z_j - b_n\right)/a_n \overset{d}{\to} \nu$ and $a_n \to \infty$, then $X_n \to \nu$ mixing. This follows from Proposition 5.3 because

$$\frac{1}{a_n}\left(\sum_{j=1}^n Z_j - \sum_{j=2}^{n+1} Z_j\right) = \frac{1}{a_n}(Z_1 - Z_{n+1}) \to 0 \quad \text{in probability}$$

by stationarity. (As for $X_n \overset{d}{\to} \nu$ see e.g. [41], Chap. 5.)
(b) If $X_n := \left(\max_{0\leq j\leq n} \sum_{i=1}^j Z_i - b_n\right)/a_n \overset{d}{\to} \nu$ and $a_n \to \infty$, then $X_n \to \nu$ mixing. In fact, one checks that

$$\frac{1}{a_n}\left|\max_{0\leq j\leq n}\sum_{i=1}^j Z_i - \max_{1\leq j\leq n+1}\sum_{i=2}^j Z_i\right| \leq \frac{1}{a_n}(|Z_1| + |Z_{n+1}|) \to 0$$

in probability so that the assertion follows from Proposition 5.3.
(c) If $X_n := \left(\max_{1\leq i\leq n} Z_i - b_n\right)/a_n \overset{d}{\to} \nu$ and $a_n \to \infty$, then $X_n \to \nu$ mixing. This follows again from Proposition 5.3 because

$$\frac{1}{a_n}\left|\max_{1\leq i\leq n} Z_i - \max_{2\leq i\leq n+1} Z_i\right| \leq \frac{1}{a_n}|Z_1 - Z_{n+1}| \to 0 \quad \text{in probability.} \quad □$$

The condition $a_n \to \infty$ in Example 5.5 (c) excludes most extreme value distributions ν. So let us explore this situation further.

Example 5.6 Let $(Z_j)_{j \in \mathbb{N}}$ be a sequence of real random variables. In order to establish mixing convergence of the normalized maxima $a_n^{-1} \left(\max_{1 \le j \le n} Z_j - b_n \right)$ with real constants $a_n > 0$ and b_n for all $n \in \mathbb{N}$ via Proposition 5.1, we set

$$f_n(z_1, \ldots, z_n) := \max_{1 \le j \le n} z_j \quad \text{for all } z_1, \ldots, z_n \in \mathbb{R}$$

so that for all $k \in \mathbb{N}$ and $n \ge k + 1$

$$\frac{1}{a_n} \left(f_n \left(Z_1, \ldots, Z_n \right) - f_{n-k} \left(Z_{k+1}, \ldots, Z_n \right) \right) = \frac{1}{a_n} \left(\max_{1 \le j \le n} Z_j - \max_{k+1 \le j \le n} Z_j \right) .$$

(a) If $a_n \to \infty$ as $n \to \infty$, then condition (i) of Proposition 5.1 is satisfied without further assumptions on the sequence $(Z_j)_{j \in \mathbb{N}}$. For a proof observe that $\left(\max_{1 \le j \le n} Z_j \right)_{n \in \mathbb{N}}$ is a nondecreasing sequence of real random variables so that for every $\omega \in \Omega$ the limit

$$Z_\infty(\omega) = \lim_{n \to \infty} \max_{1 \le j \le n} Z_j(\omega)$$

exists with $Z_\infty(\omega) \in (-\infty, \infty]$. Let $k \in \mathbb{N}$ be arbitrary. We consider two cases.

Case 1. $Z_\infty(\omega) = \infty$. Since $\max_{1 \le j \le k} Z_j(\omega) < \infty$, there exists an $n_0(\omega, k) \in \mathbb{N}$ with $n_0(\omega, k) \ge k + 1$ and $\max_{1 \le j \le k} Z_j(\omega) < \max_{1 \le j \le n} Z_j(\omega)$ for all $n \ge n_0(\omega, k)$. Hence for all $n \ge n_0(\omega, k)$

$$\max_{1 \le j \le k} Z_j(\omega) < \max_{1 \le j \le n} Z_j(\omega) = \max_{1 \le j \le k} Z_j(\omega) \vee \max_{k+1 \le j \le n} Z_j(\omega) = \max_{k+1 \le j \le n} Z_j(\omega)$$

so that

$$\frac{1}{a_n} \left(\max_{1 \le j \le n} Z_j(\omega) - \max_{k+1 \le j \le n} Z_j(\omega) \right) = 0 .$$

Case 2. $Z_\infty(\omega) < \infty$. By monotonicity, the limit

$$Z_{\infty,k}(\omega) = \lim_{n \to \infty} \max_{k+1 \le j \le n} Z_j(\omega)$$

exists with $-\infty < Z_{k+1}(\omega) \le Z_{\infty,k}(\omega) \le Z_\infty(\omega) < \infty$ so that, because $-\infty < Z_1(\omega) \le Z_\infty(\omega) < \infty$ and $a_n \to \infty$,

$$\frac{1}{a_n} \left(\max_{1 \le j \le n} Z_j(\omega) - \max_{k+1 \le j \le n} Z_j(\omega) \right) \to 0 \quad \text{as } n \to \infty .$$

Thus we have verified condition (i) of Proposition 5.1, and $a_n \to \infty$ is used only in case 2 of the argument. Therefore, if $a_n \to \infty$, then for any sequence $(Z_j)_{j \in \mathbb{N}}$ with trivial tail-σ-field and $a_n^{-1}(\max_{1 \leq j \leq n} Z_j - b_n) \xrightarrow{d} \nu$ for some $\nu \in \mathcal{M}^1(\mathbb{R})$, this convergence is automatically mixing by Proposition 5.1.

(b) The simple argument from (a) to establish condition (i) of Proposition 5.1 does not work if a_n does not converge to infinity. For an example, let $(Z_j)_{j \in \mathbb{N}}$ be a sequence of independent random variables with $P^{Z_1} = U[3, 4]$, $P^{Z_2} = U[1, 2]$ and $P^{Z_j} = U[0, 1]$ for all $j \geq 3$. Then $\max_{1 \leq j \leq n} Z_j = Z_1$ for all $n \in \mathbb{N}$ so that

$$\frac{1}{a_n}\left(\max_{1 \leq j \leq n} Z_j - b_n\right) \xrightarrow{d} U[3, 4] \quad \text{as } n \to \infty$$

with $a_n = 1$ and $b_n = 0$ for all $n \in \mathbb{N}$. On the other hand, $\max_{2 \leq j \leq n} Z_j = Z_2$ for all $n \geq 2$ so that

$$\frac{1}{a_n}\left(f_n(Z_1, \ldots, Z_n) - f_{n-1}(Z_2, \ldots, Z_n)\right) = \max_{1 \leq j \leq n} Z_j - \max_{2 \leq j \leq n} Z_j = Z_1 - Z_2 \geq 1,$$

showing that condition (i) of Proposition 5.1 is not satisfied. Because $(Z_j)_{j \in \mathbb{N}}$ has a trivial tail-σ-field, all the other assumptions in Proposition 5.1 hold. In fact, we have $\max_{1 \leq j \leq n} Z_j \to \delta_{Z_1}$ stably.

(c) If the sequence $(Z_j)_{j \in \mathbb{N}}$ is independent and identically distributed and $a_n^{-1}(\max_{1 \leq j \leq n} Z_j - b_n) \xrightarrow{d} \nu$ for some $\nu \in \mathcal{M}^1(\mathbb{R})$ which is not a Dirac-measure, then condition (i) of Proposition 5.1 is satisfied for all sequences $(a_n)_{n \in \mathbb{N}}$, that is, also without the assumption that $a_n \to \infty$ for $n \to \infty$. Therefore, the convergence $a_n^{-1}(\max_{1 \leq j \leq n} Z_j - b_n) \to \nu$ is mixing by Proposition 5.1. For a proof, let F denote the distribution function of Z_1 and introduce the right endpoint $x^+ = \inf\{x \in \mathbb{R} : F(x) = 1\}$ of the support of F, where $\inf \emptyset = \infty$. Note that for all $x < x^+$ we have $F(x) < 1$ so that

$$P\left(\max_{1 \leq j \leq n} Z_j \leq x\right) = P\left(\bigcap_{j=1}^{n} \{Z_j \leq x\}\right) = F(x)^n \to 0 \quad \text{as } n \to \infty.$$

This proves $\max_{1 \leq j \leq n} Z_j \to x^+$ in probability as $n \to \infty$. But for non-decreasing sequences of random-variables like $\max_{1 \leq j \leq n} Z_j$ convergence in probability and almost sure convergence are equivalent so that we also have $\max_{1 \leq j \leq n} Z_j \to x^+$ almost surely. If $x^+ = \infty$, then the argument of case 1 in part (a), which does not require $a_n \to \infty$, applies and establishes the desired result. Therefore, assume $x^+ < \infty$. Then $F(x^+ - 0) = \lim_{x \uparrow x^+} F(x) = 1$. To see this, assume $F(x^+ - 0) < 1$. Because $\max_{1 \leq j \leq n} Z_j \uparrow x^+$ almost surely a nondegenerate weak limit means that there exist norming constants $a_n > 0$ such that $a_n \to 0$ and $\left(x^+ - \max_{1\leq j\leq n} Z_j\right)/a_n$ has a nondegenerate weak limit. Clearly, for all $x > 0$

$$
P\left(\frac{x^+ - \max_{1\leq j\leq n} Z_j}{a_n} \geq x\right) = F\left(x^+ - a_n x\right)^n \leq F\left(x^+ - 0\right)^n .
$$

If $F\left(x^+ - 0\right) < 1$, then the right-hand side converges to zero as $n \to \infty$, a contradiction.

Now $F\left(x^+ - 0\right) = 1 = F\left(x^+\right)$ implies that, for all $k \in \mathbb{N}$, $\max_{1\leq j\leq k} Z_j = x^+$ can occur only with probability zero so that $\max_{1\leq j\leq k} Z_j < x^+$ almost surely. In view of $\max_{1\leq j\leq n} Z_j \to x^+$ almost surely we see, by the argument used in case 1 of part (a), that for almost all $\omega \in \Omega$ there exists an $n_0\left(\omega, k\right) \in \mathbb{N}$ with $n_0\left(\omega, k\right) \geq k+1$ and $\max_{1\leq j\leq n} Z_j\left(\omega\right) = \max_{k+1\leq j\leq n} Z_j\left(\omega\right)$ for all $n \geq n_0\left(\omega, k\right)$ which gives

$$
\frac{1}{a_n}\left(\max_{1\leq j\leq n} Z_j\left(\omega\right) - \max_{k+1\leq j\leq n} Z_j\left(\omega\right)\right) = 0
$$

for all $n \geq n_0\left(\omega, k\right)$ and almost all $\omega \in \Omega$. This completes the proof. $\qquad\square$

Example 5.7 Let $\left(Z_j\right)_{j\geq 1}$ be a stationary and ergodic real process with $Z_1 \in \mathcal{L}^p\left(P\right)$ with $1 \leq p < \infty$, $EZ_1 = 0$, $\mathcal{Z} = \mathbb{R}$ and $\mathcal{X} = C\left(I\right)$ with $I = [0, T]$ or $\mathbb{R}_+$. If

$$
f_n\left(\left(z_j\right)_{j\geq 1}\right) := \left(\sum_{j=1}^{[nt]} z_j + \left(nt - [nt]\right) z_{[nt]+1}\right)_{t\in I} ,
$$

$X_n := f_n\left(\left(Z_j\right)_{j\geq 1}\right)/a_n \xrightarrow{d} \nu$ for some $\nu \in \mathcal{M}^1\left(C\left(I\right)\right)$ and $n^{1/p} = O\left(a_n\right)$, then $X_n \xrightarrow{d} \nu$ mixing. (As for $X_n \xrightarrow{d} \nu$ see e.g. [9], Theorems 20.1 and 23.1, [41], Sect. 5.4). In fact, if $I = [0, T]$, $T \in (0, \infty)$, we have

$$
\frac{1}{a_n}\left\|f_n\left(\left(Z_j\right)_{j\geq 1}\right) - f_n\left(\left(Z_j\right)_{j\geq 2}\right)\right\|_{\sup} \leq \frac{4}{a_n} \max_{1\leq i\leq nT+2} |Z_i| \to 0
$$

in probability, because for $\varepsilon > 0$,

$$
P\left(\frac{1}{m^{1/p}} \max_{1\leq i\leq m} |Z_i| > \varepsilon\right) = P\left(\bigcup_{i=1}^{m}\left\{|Z_i| > \varepsilon m^{1/p}\right\}\right) \leq m P\left(|Z_1| > \varepsilon m^{1/p}\right)
$$

$$
= m P\left(|Z_1|^p > \varepsilon^p m\right) \leq \frac{1}{\varepsilon^p} \int_{\{|Z_1|^p > \varepsilon^p m\}} |Z_1|^p \, dP \to 0
$$

as $m \to \infty$. The assertion follows from Proposition 5.3 and Remark 5.4.

In case $I = \mathbb{R}_+$ we obtain

$$Ed\left(X_n, \frac{1}{a_n} f_n\left((Z_j)_{j\geq 2}\right)\right)$$

$$= \sum_{k=1}^{\infty} 2^{-k} E\left(\sup_{t\in[0,k]} \left|X_{n,t} - \frac{1}{a_n} f_n\left((Z_j)_{j\geq 2}\right)_t\right| \wedge 1\right) \to 0$$

so that again Proposition 5.3 and Remark 5.4 yield the assertion. (The assertion also follows from Corollary 3.23.) □

Exercise 5.2 Let $(Z_n)_{n\in\mathbb{N}}$ be a stationary ergodic process, where $\mathcal{Z}$ is a separable metric space and $\mathcal{C} = \mathcal{B}(\mathcal{Z})$. Prove that $Z_n \to P^{Z_1}$ mixing as $n \to \infty$.

Exercise 5.3 Let $(Z_n)_{n\in\mathbb{N}}$ be a stationary ergodic process, $\mathcal{X} = \mathbb{R}$, $b_n \in \mathbb{R}$, $a_n \to \infty$ and $g : (\mathcal{Z}, \mathcal{C}) \to (\mathbb{R}, \mathcal{B}(\mathbb{R}))$ a measurable function satisfying $X_n := \left(\sum_{j=1}^n g(Z_j) - b_n\right)/a_n \xrightarrow{d} \nu$. Show that $X_n \to \nu$ mixing as $n \to \infty$.

Exercise 5.4 (U-statistics) Let $(Z_n)_{n\in\mathbb{N}}$ be an independent and identically distributed sequence of $(\mathcal{Z}, \mathcal{C})$-valued random variables, $g : (\mathcal{Z}^2, \mathcal{C}^2) \to (\mathbb{R}, \mathcal{B}(\mathbb{R}))$ a measurable symmetric function such that $g(Z_1, Z_2) \in \mathcal{L}^2(P)$,

$$U_n := \frac{1}{\binom{n}{2}} \sum_{1\leq i<j\leq n} g(Z_i, Z_j)$$

for $n \geq 2$ and $\vartheta := EU_n$. Furthermore, let $g_1(z_1) := Eg(z_1, Z_2), \sigma_1^2 := \text{Var } g_1(Z_1)$ and $\sigma_2^2 := \text{Var } g(Z_1, Z_2)$.

Prove that $n^{1/2}(U_n - \vartheta) \to N(0, 4\sigma_1^2)$ mixing and in case $\sigma_1^2 = 0 < \sigma_2^2$, $n(U_n - \vartheta) \to \nu$ mixing as $n \to \infty$ with the distribution ν of $\sum_{j\geq 1} \lambda_j (N_j^2 - 1)$, where $(N_j)_{j\geq 1}$ is an independent and identically distributed sequence of $N(0, 1)$-distributed random variables and $(\lambda_j)_{j\geq 1}$ are the nonzero eigenvalues of the operator $T : L^2(P^{Z_1}) \to L^2(P^{Z_1})$, $Th := Eh(Z_1)(g(Z_1, \cdot) - \vartheta)$.
Hint: [64], Kapitel 10.

The last result demonstrates the role of a nesting condition on filtrations for the stability of limit theorems in the case of a function space like $\mathcal{X} = C(\mathbb{R}_+)$. We consider the case of special identically distributed processes.

Theorem 5.8 *Let $\mathcal{X} = C(\mathbb{R}_+)$. For $n \in \mathbb{N}$, let $\mathbb{F}^n = (\mathcal{F}_t^n)_{t\geq 0}$ be a filtration in $\mathcal{F}$, $W^n = (W_t^n)_{t\geq 0}$ a (path-continuous) $\mathbb{F}^n$-Brownian motion and $\tau_n : \Omega \to \mathbb{R}_+$ a (finite) $\mathbb{F}^n$-stopping time such that $\tau_n \to 0$ in probability as $n \to \infty$. Let*

$$\mathcal{G} := \sigma\left(\bigcup_{n=1}^{\infty} \bigcap_{m\geq n} \mathcal{F}_{\tau_m}^m\right).$$

Then $W^n \to \nu$ $\mathcal{G}$-*mixing as* $n \to \infty$, *where* $\nu = P^{W^1} \in \mathcal{M}^1\left(C\left(\mathbb{R}_+\right)\right)$.

Recall that $\mathcal{F}_{\tau_n}^n = \left\{F \in \mathcal{F}_\infty^n : F \cap \{\tau_n \leq t\} \in \mathcal{F}_t^n \text{ for all } t \geq 0\right\}$, where $\mathcal{F}_\infty^n :=$ $\sigma\left(\bigcup_{t \geq 0} \mathcal{F}_t^n\right)$.

Proof Let $V_t^n := W_{\tau_n + t}^n - W_{\tau_n}^n$, $\varphi : C\left(\mathbb{R}_+\right) \times \mathbb{R}_+ \to \mathbb{R}$, $\varphi\left(x, t\right) := x\left(t\right)$ and $\psi : C\left(\mathbb{R}_+\right) \times \mathbb{R}_+ \to C\left(\mathbb{R}_+\right)$, $\psi\left(x, t\right) := x\left(t + \cdot\right) - x\left(\cdot\right)$. We identify $\mathbb{R}$ with the constant functions in $C\left(\mathbb{R}_+\right)$. Then φ and ψ are continuous and $W^n - V^n = \varphi\left(W^n, \tau_n\right) - \psi\left(W^n, \tau_n\right)$. Using $\tau_n \to 0$ in probability, we have $\left(W^n, \tau_n\right) \xrightarrow{d} \nu \otimes \delta_0$ so that $\varphi\left(W^n, \tau_n\right) \xrightarrow{d} \left(\nu \otimes \delta_0\right)^\varphi = \delta_0$ and $\psi\left(W^n, \tau_n\right) \xrightarrow{d} \left(\nu \otimes \delta_0\right)^\psi = \delta_0$. Consequently, $d\left(W^n, V^n\right) \to 0$ in probability as $n \to \infty$. Hence by Theorem 3.7 (a), it is enough to show that $V^n \to \nu$ $\mathcal{G}$-mixing. Note that by the strong Markov property V^n is a Brownian motion independent of $\mathcal{F}_{\tau_n}^n$ (see e.g. [51], Theorem 13.11). For $n \in \mathbb{N}$, let $\mathcal{G}_n := \bigcap_{m \geq n} \mathcal{F}_{\tau_m}^m$. Then $\left(\mathcal{G}_n\right)_{n \geq 1}$ is a filtration in $\mathcal{F}$ with $\mathcal{G}_n \subset \mathcal{F}_{\tau_n}^n$ and $\mathcal{G}_\infty = \sigma\left(\bigcup_{n=1}^\infty \mathcal{G}_n\right) = \mathcal{G}$. For $F \in \mathcal{G}$ with $P\left(F\right) > 0$ we have by the martingale convergence theorem $P\left(F|\mathcal{G}_n\right) \to 1_F$ in $\mathcal{L}^1\left(P\right)$ which implies

$$\left|E\left(h\left(V^n\right) P\left(F|\mathcal{G}_n\right)\right) - E\left(h\left(V^n\right) 1_F\right)\right| \leq \|h\|_{\sup} E\left|P\left(F|\mathcal{G}_n\right) - 1_F\right| \to 0$$

as $n \to \infty$ for every $h \in C_b\left(C\left(\mathbb{R}_+\right)\right)$. Now, using the independence of $\sigma\left(V^n\right)$ and $\mathcal{G}_n$, we have

$$E\left(h\left(V^n\right) P\left(F|\mathcal{G}_n\right)\right) = Eh\left(V^n\right) P\left(F\right) = \int h \, d\nu P\left(F\right)$$

for every $n \in \mathbb{N}$. Thus we obtain $P_F^{V^n} \to \nu$ weakly. The assertion follows from Corollary 3.3.   □

Corollary 5.9 *In the situation of Theorem 5.8 assume a nesting condition of the filtrations: For every $n \in \mathbb{N}$ there exists a (finite) $\mathbb{F}^n$-stopping time $\tau_n : \Omega \to \mathbb{R}_+$ such that*

(i) $\tau_n \to 0$ *in probability as* $n \to \infty$,

(ii) $\mathcal{F}_{\tau_n}^n \subset \mathcal{F}_{\tau_{n+1}}^{n+1}$ *for every* $n \in \mathbb{N}$, *that is,* $\left(\mathcal{F}_{\tau_n}^n\right)_{n \geq 1}$ *is a filtration in* $\mathcal{F}$,

(iii) $\sigma\left(\bigcup_{n=1}^\infty \mathcal{F}_{\tau_n}^n\right) = \sigma\left(\bigcup_{n=1}^\infty \mathcal{F}_\infty^n\right)$, *where* $\mathcal{F}_\infty^n := \sigma\left(\bigcup_{t \geq 0} \mathcal{F}_t^n\right)$.

Then $W^n \to \nu$ *mixing as* $n \to \infty$.

Proof Theorem 5.8 and Proposition 3.5.   □

Theorem 5.8 and Corollary 5.9 are very basic results on the stable convergence of semimartingales. Corollary 5.9 has been established in [99] while the generalization in Theorem 5.8 is contained in [71].

The above nesting condition is undoubtedly very restrictive. It is, however, met in the important case of the type of Example 4.3 where $W_t^n = n^{-1/2} W_{nt}$. If $\mathcal{F}_t :=$ $\sigma (W_s, s \leq t)$ and $\mathcal{F}_t^n := \mathcal{F}_{nt}$, then the nesting condition is met, for example, with $\tau_n = n^{-1/2}$.

General results on the stable convergence of sequences of semimartingales with applications to stable convergence of discretized processes (without any nesting condition) can be found in [60], Chap. 7, [50], Sections VIII.5 and IX.7, [46, 47, 49].

An application of the preceding corollary can be found in Chap. 6. Automatic stability also occurs in classical central limit theorems for martingale arrays under a nesting condition as is demonstrated in the next chapter.

Chapter 6
Stable Martingale Central Limit Theorems

Martingale central limit theorems are a generalization of classical central limit theorems for sums of independent random variables which have found a wide range of applications. In this chapter we will discuss the basic results with stable convergence in view and will illustrate them with some examples. Further applications will follow in subsequent chapters.

We begin with a fundamental stable central limit theorem for martingale difference arrays.

6.1 Martingale Arrays and the Nesting Condition

For every $n \in \mathbb{N}$ let $(X_{nk})_{1 \leq k \leq k_n}$ be a sequence of real random variables defined on a probability space $(\Omega, \mathcal{F}, P)$, and let $(\mathcal{F}_{nk})_{0 \leq k \leq k_n}$ be a filtration in $\mathcal{F}$, i.e. $\mathcal{F}_{n0} \subset \mathcal{F}_{n1} \subset \cdots \subset \mathcal{F}_{nk_n} \subset \mathcal{F}$. The sequence $(X_{nk})_{1 \leq k \leq k_n}$ is called adapted to the filtration $(\mathcal{F}_{nk})_{0 \leq k \leq k_n}$ if X_{nk} is measurable w.r.t. $\mathcal{F}_{nk}$ for all $1 \leq k \leq k_n$. The triangular array $(X_{nk})_{1 \leq k \leq k_n, n \in \mathbb{N}}$ of random variables is called *adapted to the triangular array* $(\mathcal{F}_{nk})_{0 \leq k \leq k_n, n \in \mathbb{N}}$ of σ-fields if the row $(X_{nk})_{1 \leq k \leq k_n}$ is adapted to the filtration $(\mathcal{F}_{nk})_{0 \leq k \leq k_n}$ for every $n \in \mathbb{N}$. Not all of the following results of a more technical nature require the assumption of adaptedness. Therefore, we will always state explicitly where adapted arrays are considered.

An array $(X_{nk})_{1 \leq k \leq k_n, n \in \mathbb{N}}$ adapted to $(\mathcal{F}_{nk})_{0 \leq k \leq k_n, n \in \mathbb{N}}$ is called a *martingale difference array* if $X_{nk} \in \mathcal{L}^1(P)$ with $E(X_{nk}|\mathcal{F}_{n,k-1}) = 0$ for all $1 \leq k \leq k_n$ and $n \in \mathbb{N}$, which means that for every $n \in \mathbb{N}$ the sequence $(X_{nk})_{1 \leq k \leq k_n}$ is a martingale difference sequence w.r.t. the filtration $(\mathcal{F}_{nk})_{0 \leq k \leq k_n}$. A martingale difference array is square integrable if $X_{nk} \in \mathcal{L}^2(P)$ for all $1 \leq k \leq k_n$ and $n \in \mathbb{N}$. Note that a martingale difference sequence or array is always by definition adapted to the σ-fields under consideration.

© Springer International Publishing Switzerland 2015
E. Häusler and H. Luschgy, *Stable Convergence and Stable Limit Theorems*,
Probability Theory and Stochastic Modelling 74,
DOI 10.1007/978-3-319-18329-9_6

From now on, we assume that the sequence $(k_n)_{n \in \mathbb{N}}$ is nondecreasing with $k_n \geq n$ for all $n \in \mathbb{N}$. We always set $\mathcal{F}_\infty = \sigma \left(\bigcup_{n=1}^\infty \mathcal{F}_{nk_n} \right)$. The array $(\mathcal{F}_{nk})_{0 \leq k \leq k_n, n \in \mathbb{N}}$ is called *nested* if $\mathcal{F}_{nk} \subset \mathcal{F}_{n+1,k}$ holds for all $n \in \mathbb{N}$ and $0 \leq k \leq k_n$. The subtle role of this property of the σ-fields in stable martingale central limit theorems will become evident in the sequel.

Our basic stable martingale central limit theorem reads as follows.

Theorem 6.1 *Let* $(X_{nk})_{1 \leq k \leq k_n, n \in \mathbb{N}}$ *be a square integrable martingale difference array adapted to the array* $(\mathcal{F}_{nk})_{0 \leq k \leq k_n, n \in \mathbb{N}}$. *Let* $\mathcal{G}_{nk} = \bigcap_{m \geq n} \mathcal{F}_{mk}$ *for* $n \in \mathbb{N}$ *and* $0 \leq k \leq k_n$, *and* $\mathcal{G} = \sigma \left(\bigcup_{n=1}^\infty \mathcal{G}_{nk_n} \right)$. *Assume that*

(N) $$\sum_{k=1}^{k_n} E \left(X_{nk}^2 | \mathcal{F}_{n,k-1} \right) \to \eta^2 \quad \text{in probability as } n \to \infty$$

for some $\mathcal{G}$-*measurable real random variable* $\eta \geq 0$

and

(CLB) $$\sum_{k=1}^{k_n} E \left(X_{nk}^2 \mathbf{1}_{\{|X_{nk}| \geq \varepsilon\}} | \mathcal{F}_{n,k-1} \right) \to 0 \quad \text{in probability as } n \to \infty$$

for every $\varepsilon > 0$

(conditional form of Lindeberg's condition). Then

$$\sum_{k=1}^{k_n} X_{nk} \to \eta N \quad \mathcal{G}\text{-stably as } n \to \infty,$$

where $P^N = N(0,1)$ *and* N *is independent of* $\mathcal{G}$.

The assertion may be read as

$$\sum_{k=1}^{k_n} X_{nk} \to N \left(0, \eta^2 \right) \quad \mathcal{G}\text{-stably as } n \to \infty.$$

Remark 6.2 (a) By construction $(\mathcal{G}_{nk})_{0 \leq k \leq k_n, n \in \mathbb{N}}$ is a nested array of σ-fields with $\mathcal{G}_{nk} \subset \mathcal{F}_{nk}$ for all $n \in \mathbb{N}$ and $0 \leq k \leq k_n$.

(b) If η^2 is constant, then $\mathcal{G}$-measurability of η^2 is immediate, and $\sum_{k=1}^{k_n} X_{nk} \to \eta N$ $\mathcal{G}$-stably implies $\sum_{k=1}^{k_n} X_{nk} \xrightarrow{d} \eta N$ as $n \to \infty$. Therefore, Theorem 6.1 contains the classical central limit theorem for martingale difference arrays in which η^2 is a constant as a special case.

(c) If η^2 is $\mathcal{F}_{n_0}$-measurable for all $n \geq n_0$ and some $n_0 \in \mathbb{N}$, then it is $\mathcal{G}_{n_0 0}$-measurable and hence $\mathcal{G}$-measurable. Measurability of η^2 w.r.t. $\bigcap_{n \geq n_0} \mathcal{F}_{n0}$ has sometimes been used as an assumption in stable martingale central limit theorems.

(d) The nesting condition which is satisfied in most applications yields full stable convergence. In fact, if $(\mathcal{F}_{nk})_{0 \leq k \leq k_n, n \in \mathbb{N}}$ is nested, then $\mathcal{G}_{nk} = \mathcal{F}_{nk}$ for all $n \in \mathbb{N}$ and $0 \leq k \leq k_n$, and measurability of η^2 w.r.t. $\mathcal{G} = \mathcal{F}_\infty$ can be assumed w.l.o.g.

Corollary 6.3 (Random norming) *In the situation of Theorem 6.1 assume* $P\left(\eta^2 > 0\right) > 0$. *Then*

$$\left(\sum_{k=1}^{k_n} E\left(X_{nk}^2 | \mathcal{F}_{n,k-1}\right)\right)^{-1/2} \sum_{k=1}^{k_n} X_{nk} \to N \quad \mathcal{G}\text{-mixing under } P_{\{\eta^2 > 0\}}$$

as $n \to \infty$.

Proof Applying Theorem 3.18 (b) to the assertion in Theorem 6.1 and condition (N), we obtain

$$\left(\sum_{k=1}^{k_n} X_{nk}, \sum_{k=1}^{k_n} E\left(X_{nk}^2 | \mathcal{F}_{n,k-1}\right)\right) \to \left(\eta N, \eta^2\right) \quad \mathcal{G}\text{-stably as } n \to \infty.$$

Because $\{\eta^2 > 0\} \in \mathcal{G}$ this implies

$$\left(\sum_{k=1}^{k_n} X_{nk}, \sum_{k=1}^{k_n} E\left(X_{nk}^2 | \mathcal{F}_{n,k-1}\right)\right) \to \left(\eta N, \eta^2\right) \quad \mathcal{G}\text{-stably under } P_{\{\eta^2 > 0\}}$$

as $n \to \infty$. The function $g : \mathbb{R} \times \mathbb{R} \to \mathbb{R}$ with

$$g(x, y) := \begin{cases} x/\sqrt{y}, & y > 0 \\ 0, & y \leq 0 \end{cases}$$

is Borel-measurable and $P_{\{\eta^2 > 0\}}^{(\eta N, \eta^2)}$-almost surely continuous so that by Theorem 3.18 (c)

$$g\left(\sum_{k=1}^{k_n} X_{nk}, \sum_{k=1}^{k_n} E\left(X_{nk}^2 | \mathcal{F}_{n,k-1}\right)\right) \to g\left(\eta N, \eta^2\right) = N$$

$$\mathcal{G}\text{-stably as } n \to \infty \text{ under } P_{\{\eta^2 > 0\}}.$$

Since N and $\mathcal{G}$ are independent, the convergence is $\mathcal{G}$-mixing. □

Corollary 6.4 (Random time change) *For every* $n \in \mathbb{N}$, *let* $(X_{nk})_{k \in \mathbb{N}}$ *be a square integrable martingale difference sequence w.r.t. the filtration* $(\mathcal{F}_{nk})_{k \geq 0}$, *and let* $\tau_n : \Omega \to \mathbb{N}_0$ *be a (finite) stopping time w.r.t.* $(\mathcal{F}_{nk})_{k \geq 0}$. *For* $n \in \mathbb{N}$ *and* $k \geq 0$ *set* $\mathcal{G}_{nk} = \bigcap_{m \geq n} \mathcal{F}_{mk}$ *and* $\mathcal{G} = \sigma\left(\bigcup_{n \in \mathbb{N}} \mathcal{G}_{n\infty}\right)$, *where* $\bar{\mathcal{G}}_{n\infty} = \sigma\left(\bigcup_{k=0}^{\infty} \mathcal{G}_{nk}\right)$. *If*

(N_{τ_n}) $\displaystyle\sum_{k=1}^{\tau_n} E\left(X_{nk}^2 | \mathcal{F}_{n,k-1}\right) \to \eta^2$ in probability as $n \to \infty$

for some $\mathcal{G}$-measurable real random variable $\eta \geq 0$

and

(CLB_{τ_n}) $\displaystyle\sum_{k=1}^{\tau_n} E\left(X_{nk}^2 1_{\{|X_{nk}| \geq \varepsilon\}} | \mathcal{F}_{n,k-1}\right) \to 0$ in probability as $n \to \infty$

for every $\varepsilon > 0$,

then

$$\sum_{k=1}^{\tau_n} X_{nk} \to \eta N \quad \mathcal{G}\text{-stably as } n \to \infty,$$

where $P^N = N(0,1)$ and N is independent of $\mathcal{G}$.

Proof Since τ_n is a finite random variable for every $n \in \mathbb{N}$, there exists some $k_n \in \mathbb{N}$ with $P(\tau_n > k_n) \leq 1/n$. Inductively, we can construct the k_n nondecreasing with $k_n \geq n$ for all $n \in \mathbb{N}$. Then $\sigma\left(\bigcup_{n \in \mathbb{N}} \mathcal{G}_{nk_n}\right) = \mathcal{G}$. For $n \in \mathbb{N}$ and $1 \leq k \leq k_n$ set $Y_{nk} := X_{nk} 1_{\{k \leq \tau_n\}}$. Since τ_n is a stopping time w.r.t. $(\mathcal{F}_{nk})_{k \geq 0}$, the array $(Y_{nk})_{1 \leq k \leq k_n, n \in \mathbb{N}}$ is a square integrable martingale difference array w.r.t. $(\mathcal{F}_{nk})_{0 \leq k \leq k_n, n \in \mathbb{N}}$. On the event $\{\tau_n \leq k_n\}$ we have

$$\sum_{k=1}^{k_n} E\left(Y_{nk}^2 | \mathcal{F}_{n,k-1}\right) = \sum_{k=1}^{k_n \wedge \tau_n} E\left(X_{nk}^2 | \mathcal{F}_{n,k-1}\right) = \sum_{k=1}^{\tau_n} E\left(X_{nk}^2 | \mathcal{F}_{n,k-1}\right)$$

so that, for every $\varepsilon > 0$,

$$P\left(\left|\sum_{k=1}^{k_n} E\left(Y_{nk}^2 | \mathcal{F}_{n,k-1}\right) - \sum_{k=1}^{\tau_n} E\left(X_{nk}^2 | \mathcal{F}_{n,k-1}\right)\right| \geq \varepsilon\right) \leq P(\tau_n > k_n) \leq \frac{1}{n}$$

which proves

$$\sum_{k=1}^{k_n} E\left(Y_{nk}^2 | \mathcal{F}_{n,k-1}\right) - \sum_{k=1}^{\tau_n} E\left(X_{nk}^2 | \mathcal{F}_{n,k-1}\right) \to 0$$ in probability as $n \to \infty$

and thus

$$\sum_{k=1}^{k_n} E\left(Y_{nk}^2 | \mathcal{F}_{n,k-1}\right) \to \eta^2$$ in probability as $n \to \infty$.

On the event $\{\tau_n \leq k_n\}$ we also have

$$\sum_{k=1}^{k_n} Y_{nk} = \sum_{k=1}^{k_n \wedge \tau_n} X_{nk} = \sum_{k=1}^{\tau_n} X_{nk}$$

so that, by the same type of argument as above,

$$\sum_{k=1}^{k_n} Y_{nk} - \sum_{k=1}^{\tau_n} X_{nk} \to 0 \quad \text{in probability as } n \to \infty .$$

Finally, for all $\varepsilon > 0$, using $|Y_{nk}| \leq |X_{nk}|$,

$$\sum_{k=1}^{k_n} E\left(Y_{nk}^2 1_{\{|Y_{nk}| \geq \varepsilon\}} | \mathcal{F}_{n,k-1}\right)$$

$$\leq \sum_{k=1}^{k_n \wedge \tau_n} E\left(X_{nk}^2 1_{\{|X_{nk}| \geq \varepsilon\}} | \mathcal{F}_{n,k-1}\right) \leq \sum_{k=1}^{\tau_n} E\left(X_{nk}^2 1_{\{|X_{nk}| \geq \varepsilon\}} | \mathcal{F}_{n,k-1}\right)$$

which implies

$$\sum_{k=1}^{k_n} E\left(Y_{nk}^2 1_{\{|Y_{nk}| \geq \varepsilon\}} | \mathcal{F}_{n,k-1}\right) \to 0 \quad \text{in probability as } n \to \infty$$

for all $\varepsilon > 0$. Therefore, Theorem 6.1 yields

$$\sum_{k=1}^{k_n} Y_{nk} \to \eta N \quad \mathcal{G}\text{-stably as } n \to \infty ,$$

and, using Theorem 3.18 (a), we conclude

$$\sum_{k=1}^{\tau_n} X_{nk} \to \eta N \quad \mathcal{G}\text{-stably as } n \to \infty . \qquad \square$$

The preceding corollary implies, for instance, the non-functional part of Example 4.8 for stopping times.

As for the proof of Theorem 6.1 we demonstrate that the Lindeberg method works in a basic general setting (see Step 1 of the proof). We require some technical results which will also be useful later. Note that in these statements adaptedness is not required.

Lemma 6.5 *Let $(X_{nk})_{1 \leq k \leq k_n, n \in \mathbb{N}}$ be an array of nonnegative integrable random variables, and let $(\mathcal{F}_{nk})_{0 \leq k \leq k_n, n \in \mathbb{N}}$ be an array of σ-fields. Then*

$$\sum_{k=1}^{k_n} E\left(X_{nk} | \mathcal{F}_{n,k-1}\right) \to 0 \quad \text{in probability as } n \to \infty$$

implies

$$\sum_{k=1}^{k_n} X_{nk} \to 0 \quad \text{in probability as } n \to \infty.$$

Proof For every $n \in \mathbb{N}$

$$\tau_n = \max\left\{ k \in \{0, 1, \ldots, k_n\} : \sum_{j=1}^{k} E\left(X_{nj} | \mathcal{F}_{n,j-1}\right) \leq 1 \right\}$$

is a stopping time w.r.t. the filtration $(\mathcal{F}_{nk})_{0 \leq k \leq k_n}$. From $\sum_{k=1}^{\tau_n} E\left(X_{nk} | \mathcal{F}_{n,k-1}\right) \leq \sum_{k=1}^{k_n} E\left(X_{nk} | \mathcal{F}_{n,k-1}\right)$ and $\sum_{k=1}^{\tau_n} E\left(X_{nk} | \mathcal{F}_{n,k-1}\right) \leq 1$ for all $n \in \mathbb{N}$ as well as the assumption of the lemma we obtain by dominated convergence that

$$E\left(\sum_{k=1}^{\tau_n} X_{nk}\right) = E\left(\sum_{k=1}^{\tau_n} E\left(X_{nk} | \mathcal{F}_{n,k-1}\right)\right) \to 0 \quad \text{as } n \to \infty$$

so that, in particular, $\sum_{k=1}^{\tau_n} X_{nk} \to 0$ in probability. For every $\varepsilon > 0$ we have

$$P\left(\left|\sum_{k=1}^{k_n} X_{nk} - \sum_{k=1}^{\tau_n} X_{nk}\right| \geq \varepsilon\right)$$

$$\leq P\left(\tau_n < k_n\right) = P\left(\sum_{k=1}^{k_n} E\left(X_{nk} | \mathcal{F}_{n,k-1}\right) > 1\right) \to 0$$

as $n \to \infty$, which completes the proof. $\square$

Exercise 6.1 Deduce Lemma 6.5 in the adapted case from the Lenglart inequality in Lemma A.8 (a).

Proposition 6.6 *Let $(X_{nk})_{1 \leq k \leq k_n, n \in \mathbb{N}}$ be an array of integrable random variables, and let $(\mathcal{F}_{nk})_{0 \leq k \leq k_n, n \in \mathbb{N}}$ be an array of σ-fields. Then*

(CLB$_1$) $\displaystyle\sum_{k=1}^{k_n} E\left(|X_{nk}|\,1_{\{|X_{nk}|\geq\varepsilon\}}|\mathcal{F}_{n,k-1}\right) \to 0$ *in probability as* $n \to \infty$

for every $\varepsilon > 0$

(an $\mathcal{L}^1$-version of (CLB)*) implies*

$$\max_{1\leq k\leq k_n} |X_{nk}| \to 0 \quad \text{in probability as } n \to \infty.$$

Proof From the assumption and Lemma 6.5 for all $\varepsilon > 0$ it follows that

$$\sum_{k=1}^{k_n} |X_{nk}|\,1_{\{|X_{nk}|\geq\varepsilon\}} \to 0 \quad \text{in probability as } n \to \infty.$$

But for all $\varepsilon > 0$ and $n \in \mathbb{N}$ we have

$$P\left(\max_{1\leq k\leq k_n}|X_{nk}| \geq \varepsilon\right) \leq P\left(\sum_{k=1}^{k_n}|X_{nk}|\,1_{\{|X_{nk}|\geq\varepsilon\}} \geq \varepsilon\right),$$

which completes the proof. □

Proposition 6.7 *Let* $(X_{nk})_{1\leq k\leq k_n, n\in\mathbb{N}}$ *be an array of square integrable random variables, and let* $(\mathcal{F}_{nk})_{0\leq k\leq k_n, n\in\mathbb{N}}$ *be an array of σ-fields. Then the conditional Lindeberg condition* (CLB) *implies*

$$\max_{1\leq k\leq k_n} E\left(X_{nk}^2|\mathcal{F}_{n,k-1}\right) \to 0 \quad \text{in probability as } n \to \infty.$$

Proof For every $\varepsilon > 0$ and $n \in \mathbb{N}$ we have

$$\max_{1\leq k\leq k_n} E\left(X_{nk}^2|\mathcal{F}_{n,k-1}\right)$$

$$= \max_{1\leq k\leq k_n} E\left(X_{nk}^2 1_{\{|X_{nk}|<\varepsilon\}} + X_{nk}^2 1_{\{|X_{nk}|\geq\varepsilon\}}|\mathcal{F}_{n,k-1}\right)$$

$$\leq \varepsilon^2 + \sum_{k=1}^{k_n} E\left(X_{nk}^2 1_{\{|X_{nk}|\geq\varepsilon\}}|\mathcal{F}_{n,k-1}\right),$$

which clearly implies the desired result. □

Now we are prepared to give the

Proof of Theorem 6.1. For brevity we write $\sigma_{nk}^2 = E\left(X_{nk}^2|\mathcal{F}_{n,k-1}\right)$ for all $n \in \mathbb{N}$ and $1 \leq k \leq k_n$.

The proof proceeds in several steps.

Step 1. In addition to conditions (N) and (CLB) we assume that

(i) $\displaystyle\sum_{k=1}^{k_n} \sigma_{nk}^2 = \eta^2$ a.s. for all $n \in \mathbb{N}$,

(ii) η^2 is $\mathcal{F}_{n0}$-measurable for all $n \in \mathbb{N}$, that is, η^2 is $\mathcal{G}_{10}$-measurable, and

(iii) $\eta^2 \le C < \infty$ a.s. for some constant C

and will show that $\sum_{k=1}^{k_n} X_{nk} \overset{d}{\to} \eta N$ as $n \to \infty$ holds.

W.l.o.g. we can assume that an array $(N_{nk})_{1 \le k \le k_n, n \in \mathbb{N}}$ of independent standard normal random variables is defined on $(\Omega, \mathcal{F}, P)$ such that $(N_{nk})_{1 \le k \le k_n, n \in \mathbb{N}}$ and $\mathcal{F}_\infty$ are independent. Then for every $n \in \mathbb{N}$ the conditional distribution of $\sum_{k=1}^{k_n} \sigma_{nk} N_{nk}$ given $\mathcal{F}_\infty$ is the normal distribution with mean zero and variance $\sum_{k=1}^{k_n} \sigma_{nk}^2 = \eta^2$, by assumption (i). Therefore, with $\overset{d}{=}$ denoting equality in distribution, $\sum_{k=1}^{k_n} \sigma_{nk} N_{nk} \overset{d}{=} \eta N$.

Let $f : \mathbb{R} \to \mathbb{R}$ be bounded and three times continuously differentiable with bounded derivatives. Taylor's formula implies

$$f(x+h) = f(x) + f'(x)h + \frac{1}{2}f''(x)h^2 + R_f(x, h) \quad \text{for all } x, h \in \mathbb{R}$$

with

$$\left| R_f(x, h) \right| \le C(f) \min\left\{ h^2, |h|^3 \right\} \quad \text{for all } x, h \in \mathbb{R},$$

where f', f'' and f''' are the derivatives of f and $C(f) = \max\{\frac{1}{6}\|f'''\|_\infty, \|f''\|_\infty\} < \infty$ with $\|g\|_\infty$ denoting the sup-norm of the bounded function $g : \mathbb{R} \to \mathbb{R}$.

Introducing

$$Y_{nk} := \sum_{j=1}^{k-1} X_{nj} + \sum_{j=k+1}^{k_n} \sigma_{nj} N_{nj}$$

for all $n \in \mathbb{N}$ and $1 \le k \le k_n$ we obtain

$$\left| E\left(f\left(\sum_{k=1}^{k_n} X_{nk} \right) \right) - E(f(\eta N)) \right|$$

$$= \left| E\left(f\left(\sum_{k=1}^{k_n} X_{nk} \right) \right) - E\left(f\left(\sum_{k=1}^{k_n} \sigma_{nk} N_{nk} \right) \right) \right|$$

$$= \left| \sum_{k=1}^{k_n} [E\left(f\left(Y_{nk} + X_{nk}\right)\right) - E\left(f\left(Y_{nk} + \sigma_{nk} N_{nk}\right)\right)] \right|$$

$$= \left| \sum_{k=1}^{k_n} \left[E\left(f'\left(Y_{nk}\right) X_{nk}\right) + \frac{1}{2} E\left(f''\left(Y_{nk}\right) X_{nk}^2\right) + E\left(R_f\left(Y_{nk}, X_{nk}\right)\right) \right. \right.$$

$$\left. \left. - E\left(f'\left(Y_{nk}\right) \sigma_{nk} N_{nk}\right) - \frac{1}{2} E\left(f''\left(Y_{nk}\right) \sigma_{nk}^2 N_{nk}^2\right) - E\left(R_f\left(Y_{nk}, \sigma_{nk} N_{nk}\right)\right) \right] \right|.$$

In the next crucial step of the proof we will show that the two expectations involving f' on the right-hand side of this chain of equations vanish individually whereas the two expectations involving f'' are equal and hence cancel out. Clearly, by independence of (Y_{nk}, σ_{nk}) and N_{nk} we have $E\left(f'\left(Y_{nk}\right) \sigma_{nk} N_{nk}\right) = E\left(f'\left(Y_{nk}\right) \sigma_{nk}\right) E\left(N_{nk}\right) = 0$. We note that by independence of $\mathcal{F}_\infty$ and $\left(N_{n1}, \ldots, N_{nk_n}\right)$ for all $n \in \mathbb{N}$ and $1 \leq k \leq k_n$ the conditional distribution of Y_{nk} given $\mathcal{F}_\infty$ is the normal distribution with mean $\sum_{j=1}^{k-1} X_{nj}$ and variance $\sum_{j=k+1}^{k_n} \sigma_{nj}^2 = \eta^2 - \sum_{j=1}^{k} \sigma_{nj}^2$, where the last equality follows from assumption (i). As a consequence of assumption (ii), this conditional distribution is measurable w.r.t. $\mathcal{F}_{n,k-1}$ up to $\mathcal{F}_\infty$-null sets, and this implies that the conditional expectations $E\left(f'\left(Y_{nk}\right) | \mathcal{F}_\infty\right)$ and $E\left(f''\left(Y_{nk}\right) | \mathcal{F}_\infty\right)$ are measurable w.r.t. $\mathcal{F}_{n,k-1}$ up to $\mathcal{F}_\infty$-null sets as well. Hence

$$E\left(f'\left(Y_{nk}\right) X_{nk}\right) = E\left(E\left(f'\left(Y_{nk}\right) X_{nk} | \mathcal{F}_\infty\right)\right)$$
$$= E\left(X_{nk} E\left(f'\left(Y_{nk}\right) | \mathcal{F}_\infty\right)\right) = E\left(E\left(X_{nk} | \mathcal{F}_{n,k-1}\right) E\left(f'\left(Y_{nk}\right) | \mathcal{F}_\infty\right)\right) = 0$$

because $E\left(X_{nk} | \mathcal{F}_{n,k-1}\right) = 0$ and

$$E\left(f''\left(Y_{nk}\right) X_{nk}^2\right) = E\left(E\left(X_{nk}^2 | \mathcal{F}_{n,k-1}\right) E\left(f''\left(Y_{nk}\right) | \mathcal{F}_\infty\right)\right)$$
$$= E\left(\sigma_{nk}^2 E\left(f''\left(Y_{nk}\right) | \mathcal{F}_\infty\right)\right) = E\left(f''\left(Y_{nk}\right) \sigma_{nk}^2\right) = E\left(f''\left(Y_{nk}\right) \sigma_{nk}^2 N_{nk}^2\right),$$

where the last equality holds by independence of (Y_{nk}, σ_{nk}) and N_{nk} combined with $E\left(N_{nk}^2\right) = 1$. Consequently, we obtain

$$\left| E\left(f\left(\sum_{k=1}^{k_n} X_{nk}\right)\right) - E\left(f\left(\eta N\right)\right) \right|$$

$$\leq \sum_{k=1}^{k_n} \left[E\left(\left|R_f\left(Y_{nk}, X_{nk}\right)\right|\right) + E\left(\left|R_f\left(Y_{nk}, \sigma_{nk} N_{nk}\right)\right|\right) \right]$$

$$\leq C\left(f\right) \sum_{k=1}^{k_n} \left[E\left(\min\left(X_{nk}^2, |X_{nk}|^3\right)\right) + E\left(\min\left(\sigma_{nk}^2 N_{nk}^2, \sigma_{nk}^3 |N_{nk}|^3\right)\right) \right],$$

where the last inequality follows from the bound on $|R_f(x, h)|$. For all $n \in \mathbb{N}$ and $\varepsilon > 0$ we have

$$\sum_{k=1}^{k_n} E\left(\min\left(X_{nk}^2, |X_{nk}|^3\right)\right)$$

$$\leq \sum_{k=1}^{k_n} E\left(X_{nk}^2 1_{\{|X_{nk}| \geq \varepsilon\}}\right) + \sum_{k=1}^{k_n} E\left(|X_{nk}|^3 1_{\{|X_{nk}| < \varepsilon\}}\right)$$

with

$$\sum_{k=1}^{k_n} E\left(|X_{nk}|^3 1_{\{|X_{nk}| < \varepsilon\}}\right) \leq \varepsilon \sum_{k=1}^{k_n} E\left(X_{nk}^2\right) = \varepsilon E\left(\sum_{k=1}^{k_n} \sigma_{nk}^2\right) \leq \varepsilon C$$

by assumptions (i) and (iii). Moreover,

$$\sum_{k=1}^{k_n} E\left(\min\left(\sigma_{nk}^2 N_{nk}^2, \sigma_{nk}^3 |N_{nk}|^3\right)\right) \leq \sum_{k=1}^{k_n} E\left(\sigma_{nk}^3 |N_{nk}|^3\right)$$

$$= \sum_{k=1}^{k_n} E\left(\sigma_{nk}^3\right) E\left(|N_{nk}|^3\right) = \sqrt{\frac{8}{\pi}} E\left(\sum_{k=1}^{k_n} \sigma_{nk}^3\right)$$

$$\leq \sqrt{\frac{8}{\pi}} E\left(\max_{1 \leq k \leq k_n} \sigma_{nk} \sum_{k=1}^{k_n} \sigma_{nk}^2\right) \leq \sqrt{\frac{8}{\pi}} C E\left(\max_{1 \leq k \leq k_n} \sigma_{nk}^2\right)^{1/2}$$

$$= \sqrt{\frac{8}{\pi}} C E\left(\max_{1 \leq k \leq k_n} E\left(X_{nk}^2 1_{\{|X_{nk}| < \varepsilon\}} + X_{nk}^2 1_{\{|X_{nk}| \geq \varepsilon\}} | \mathcal{F}_{n,k-1}\right)\right)^{1/2}$$

$$\leq \sqrt{\frac{8}{\pi}} C \left[\varepsilon + \left(\sum_{k=1}^{k_n} E\left(X_{nk}^2 1_{\{|X_{nk}| \geq \varepsilon\}}\right)\right)^{1/2}\right].$$

Combining these results, for all $n \in \mathbb{N}$ and $\varepsilon > 0$ we arrive at

$$\left| E\left(f\left(\sum_{k=1}^{k_n} X_{nk}\right)\right) - E\left(f\left(\eta N\right)\right) \right|$$

$$\leq C(f) \left[\sum_{k=1}^{k_n} E\left(X_{nk}^2 1_{\{|X_{nk}| \geq \varepsilon\}}\right) + \left(1 + \sqrt{\frac{8}{\pi}}\right) \varepsilon C\right.$$

$$\left. + \sqrt{\frac{8}{\pi}} C \left(\sum_{k=1}^{k_n} E\left(X_{nk}^2 1_{\{|X_{nk}| \geq \varepsilon\}}\right)\right)^{1/2}\right].$$

From (CLB) and assumptions (i) and (iii) we infer by dominated convergence that $\sum_{k=1}^{k_n} E\left(X_{nk}^2 1_{\{|X_{nk}|\geq\varepsilon\}}\right) \to 0$ as $n \to \infty$ for every $\varepsilon > 0$. This implies $\lim_{n\to\infty} E\left(f\left(\sum_{k=1}^{k_n} X_{nk}\right)\right) = E\left(f\left(\eta Z\right)\right)$ which proves $\sum_{k=1}^{k_n} X_{nk} \overset{d}{\to} \eta N$ as $n \to \infty$ and completes Step 1 of the proof.

Step 2. In the second step of the proof we assume (N), (CLB) and in addition

(iv) $\qquad \sum_{k=1}^{k_n} \sigma_{nk}^2 \leq C < \infty \quad$ for some constant C and all $n \in \mathbb{N}$

and will show that $\sum_{k=1}^{k_n} X_{nk} \overset{d}{\to} \eta N$ as $n \to \infty$ holds. For this, let $m \in \mathbb{N}$ be fixed. Note that for all $n \geq m + 1$ we have $k_n \geq n \geq m + 1$ and that $(X_{nk})_{m+1\leq k\leq k_n, n\geq m+1}$ is a square integrable martingale difference array adapted to the array $(\mathcal{F}_{nk})_{m\leq k\leq k_n, n\geq m+1}$. Clearly, for every $n \geq m + 1$

$$\tau_n(m) = \max\left\{k \in \{m, m+1, \ldots, k_n\} : \sum_{j=m+1}^{k} \sigma_{nj}^2 \leq E\left(\eta^2|\mathcal{G}_{mm}\right)\right\}$$

is a stopping time w.r.t. the filtration $(\mathcal{F}_{nk})_{m\leq k\leq k_n}$ (observe that $(\mathcal{G}_{nk})_{0\leq k\leq k_n, n\in\mathbb{N}}$ is a nested array with $\mathcal{G}_{nk} \subset \mathcal{F}_{nk}$ by Remark 6.2 (a)). For all $n \geq m + 1$ we introduce

$$\xi_n(m) := \left[E\left(\eta^2|\mathcal{G}_{mm}\right) - \sum_{k=m+1}^{\tau_n(m)} \sigma_{nk}^2\right]^{1/2}$$

and let $(Y_{nk})_{k_n+1\leq k\leq k_n+n}$ be independent random variables with $P(Y_{nk} = 1) = 1/2 = P(Y_{nk} = -1)$ for all $k_n + 1 \leq k \leq k_n + n$ which are independent of $\mathcal{F}_{nk_n}$. Define

$$Z_{nk}(m) := \begin{cases} X_{nk} 1_{\{k\leq\tau_n(m)\}} & , m + 1 \leq k \leq k_n \\ \dfrac{1}{\sqrt{n}}\xi_n(m) Y_{nk} & , k_n + 1 \leq k \leq k_n + n \end{cases}$$

and

$$\mathcal{H}_{nk}(m) := \begin{cases} \mathcal{F}_{nk} & , m \leq k \leq k_n \\ \sigma\left(\mathcal{F}_{nk_n}, Y_{n,k_n+1}, \ldots, Y_{nk}\right) & , k_n + 1 \leq k \leq k_n + n. \end{cases}$$

The sequence $(\mathcal{H}_{nk}(m))_{m\leq k\leq k_n+n}$ is nondecreasing and $(Z_{nk}(m))_{m+1\leq k\leq k_n+n}$ is adapted to $(\mathcal{H}_{nk}(m))_{m\leq k\leq k_n+n}$ for every $n \geq m + 1$. From (N) and assumption (iv) we infer that $\eta^2 \leq C$ almost surely so that $|\xi_n(m) Y_{nk}| \leq C^{1/2}$ almost surely and, consequently, all $Z_{nk}(m)$ are square integrable. For $n \geq m+1$ and $m+1 \leq k \leq k_n$ we have

$$E\left(Z_{nk}(m)\,|\mathcal{H}_{n,k-1}(m)\right) = E\left(X_{nk}1_{\{k\leq\tau_n(m)\}}|\mathcal{F}_{n,k-1}\right)$$
$$= 1_{\{k\leq\tau_n(m)\}}E\left(X_{nk}|\mathcal{F}_{n,k-1}\right) = 0$$

because $\tau_n(m)$ is a stopping time w.r.t. $(\mathcal{F}_{nk})_{m\leq k\leq k_n}$ and $E\left(X_{nk}|\mathcal{F}_{n,k-1}\right) = 0$, and for $k_n+1 \leq k \leq k_n+n$ we have

$$E\left(Z_{nk}(m)\,|\mathcal{H}_{n,k-1}(m)\right) = E\left(\frac{1}{\sqrt{n}}\xi_n(m)\,Y_{nk}\,\middle|\,\sigma\left(\mathcal{F}_{nk_n}, Y_{n,k_n+1}, \ldots, Y_{n,k-1}\right)\right)$$
$$= \frac{1}{\sqrt{n}}\xi_n(m)\,E\left(Y_{nk}|\sigma\left(\mathcal{F}_{nk_n}, Y_{n,k_n+1}, \ldots, Y_{n,k-1}\right)\right) = 0$$

because $\xi_n(m)$ is measurable w.r.t. $\mathcal{F}_{nk_n}$ and, by independence of Y_{nk} and $\sigma\left(\mathcal{F}_{nk_n}, Y_{n,k_n+1}, \ldots, Y_{n,k-1}\right)$, we also get $E\left(Y_{nk}|\sigma\left(\mathcal{F}_{nk_n}, Y_{n,k_n+1}, \ldots, Y_{n,k-1}\right)\right) = E(Y_{nk}) = 0$. Thus, $(Z_{nk}(m))_{m+1\leq k\leq k_n, n\geq m+1}$ is a square integrable martingale difference array with

$$\sum_{k=m+1}^{k_n+n} E\left(Z_{nk}^2(m)\,|\mathcal{H}_{n,k-1}(m)\right)$$

$$= \sum_{k=m+1}^{k_n} E\left(X_{nk}^2 1_{\{k\leq\tau_n(m)\}}|\mathcal{F}_{n,k-1}\right)$$

$$+ \sum_{k=k_n+1}^{k_n+n} E\left(\frac{1}{n}\xi_n^2(m)\,Y_{nk}^2\,\middle|\,\sigma\left(\mathcal{F}_{nk_n}, Y_{n,k_n+1}, \ldots, Y_{n,k-1}\right)\right)$$

$$= \sum_{k=m+1}^{\tau_n(m)} E\left(X_{nk}^2|\mathcal{F}_{n,k-1}\right) + \left[E\left(\eta^2|\mathcal{G}_{mm}\right) - \sum_{k=m+1}^{\tau_n(m)}\sigma_{nk}^2\right]\frac{1}{n}\sum_{k=k_n+1}^{k_n+n}E(Y_{nk}^2)$$

$$= E\left(\eta^2|\mathcal{G}_{mm}\right)$$

for $n \geq m+1$. Thus, the martingale difference array $(Z_{nk}(m))_{m+1\leq k\leq k_n+n, n\geq m+1}$ satisfies assumption (i) from Step 1 with $E\left(\eta^2|\mathcal{G}_{mm}\right)$ instead of η^2 if m is identified with 0. Trivially, $E\left(\eta^2|\mathcal{G}_{mm}\right) \leq C$ almost surely from $\eta^2 \leq C$ almost surely. If m is identified with 0 and since $\mathcal{G}_{mm} \subset \mathcal{G}_{nm}$ holds for all $n \geq m+1$, assumptions (ii) and (iii) are satisfied as well. Moreover, for all $\varepsilon > 0$ and $n \geq m+1$

$$\sum_{k=m+1}^{k_n+n} E\left(Z_{nk}^2\left(m\right) 1_{\{|Z_{nk}(m)|\geq\varepsilon\}}\right)$$

$$\leq \sum_{k=m+1}^{k_n} E\left(X_{nk}^2 1_{\{|X_{nk}|\geq\varepsilon\}}\right)$$

$$+\frac{1}{n}\sum_{k=k_n+1}^{k_n+n} E\left(\xi_n^2\left(m\right) Y_{nk}^2 1_{\{|\xi_n(m)Y_{nk}|\geq\varepsilon\sqrt{n}\}}\right).$$

The first summand on the right-hand side of this inequality converges to zero by (CLB), assumption (iv) and the dominated convergence theorem. The second summand is equal to zero for all sufficiently large n because $|\xi_n\left(m\right)Y_{nk}|\leq C^{1/2}$ almost surely for all $n\geq m+1$.

Thus we have shown that for every $m\in\mathbb{N}$ the square integrable martingale difference array $\left(Z_{nk}\left(m\right)\right)_{m+1\leq k\leq k_n+n,n\geq m+1}$ w.r.t. $\left(\mathcal{H}_{nk}\left(m\right)\right)_{m\leq k\leq k_n+n,n\geq m+1}$ fulfills all assumptions of Step 1 so that $\sum_{k=m+1}^{k_n+n} Z_{nk}\left(m\right) \xrightarrow{d} E\left(\eta^2|\mathcal{G}_{mm}\right)^{1/2} N$ as $n\to\infty$.

Because $\mathcal{G}_{nk}$ is nondecreasing in k and n the sequence $\left(\mathcal{G}_{mm}\right)_{m\in\mathbb{N}}$ is a nondecreasing sequence of σ-fields with $\sigma\left(\mathcal{G}_{mm}:m\in\mathbb{N}\right)=\sigma\left(\mathcal{G}_{nk_n}:n\in\mathbb{N}\right)=\mathcal{G}$. Consequently, thanks to the martingale convergence theorem, $E\left(\eta^2|\mathcal{G}_{mm}\right)\to E\left(\eta^2|\mathcal{G}\right)=\eta^2$ almost surely as $m\to\infty$. Hence $E\left(\eta^2|\mathcal{G}_{mm}\right)^{1/2} N \xrightarrow{d} \eta N$ as $m\to\infty$. In order to obtain $\sum_{k=1}^{k_n} X_{nk} \xrightarrow{d} \eta N$ as $n\to\infty$ it remains to verify in view of Theorem 3.21 that for every $\varepsilon>0$

$$\lim_{m\to\infty}\limsup_{n\to\infty} P\left(\left|\sum_{k=1}^{k_n} X_{nk} - \sum_{k=m+1}^{k_n+n} Z_{nk}\left(m\right)\right|\geq\varepsilon\right)=0.$$

For all $n\geq m+1$ we have

$$\sum_{k=1}^{k_n} X_{nk} - \sum_{k=m+1}^{k_n+n} Z_{nk}\left(m\right)$$

$$=\sum_{k=1}^{k_n} X_{nk} - \sum_{k=m+1}^{\tau_n(m)} X_{nk} - \sum_{k=k_n+1}^{k_n+n}\frac{1}{\sqrt{n}}\xi_n\left(m\right) Y_{nk}$$

$$=\sum_{k=1}^{m} X_{nk} + \sum_{k=\tau_n(m)+1}^{k_n} X_{nk} - \xi_n\left(m\right)\frac{1}{\sqrt{n}}\sum_{k=k_n+1}^{k_n+n} Y_{nk}$$

$$=V_1\left(m,n\right)+V_2\left(m,n\right)-V_3\left(m,n\right),$$

say. Clearly, it suffices to show for all $\varepsilon > 0$ and $j = 1, 2, 3$ that

$$\lim_{m\to\infty} \limsup_{n\to\infty} P\left(|V_j(m,n)| \geq \varepsilon\right) = 0.$$

Because for all $\varepsilon > 0$ and $n \in \mathbb{N}$

$$\sum_{k=1}^{k_n} E\left(|X_{nk}|\,1_{\{|X_{nk}|\geq\varepsilon\}}|\mathcal{F}_{n,k-1}\right) \leq \frac{1}{\varepsilon} \sum_{k=1}^{k_n} E\left(X_{nk}^2 1_{\{|X_{nk}|\geq\varepsilon\}}|\mathcal{F}_{n,k-1}\right),$$

condition (CLB) implies $\sum_{k=1}^{m} X_{nk} \to 0$ in probability as $n \to \infty$ for every $m \in \mathbb{N}$ via Proposition 6.6, and $\lim_{m\to\infty} \limsup_{n\to\infty} P\left(|V_1(m,n)| \geq \varepsilon\right) = 0$ is immediate. To handle $V_2(m,n)$ we write

$$E\left(\left(\sum_{k=\tau_n(m)+1}^{k_n} X_{nk}\right)^2\right) = E\left(\sum_{k=\tau_n(m)+1}^{k_n} \sigma_{nk}^2 1_{\{\tau_n(m)<k_n\}}\right)$$

$$= E\left(\left(\sum_{k=1}^{k_n} \sigma_{nk}^2 - E\left(\eta^2|\mathcal{G}_{mm}\right) + \xi_n^2(m) - \sum_{k=1}^{m} \sigma_{nk}^2\right) 1_{\{\tau_n(m)<k_n\}}\right)$$

$$\leq E\left(\left|\sum_{k=1}^{k_n} \sigma_{nk}^2 - \eta^2\right|\right) + E\left(\left|\eta^2 - E\left(\eta^2|\mathcal{G}_{mm}\right)\right|\right) + E\left(\sum_{k=1}^{m} \sigma_{nk}^2\right)$$

$$+ E\left(\xi_n^2(m) 1_{\{\tau_n(m)<k_n\}}\right) = I_n + II_m + III_{m,n} + IV_{m,n},$$

say. Clearly, $\lim_{n\to\infty} I_n = 0$ by (N), assumption (iv) and dominated convergence, and $\lim_{m\to\infty} II_m = 0$ by $E\left(\eta^2|\mathcal{G}_{mm}\right) \to \eta^2$ almost surely as $m \to \infty$, $\eta^2 \leq C$ and dominated convergence. Obviously, $III_{m,n} \leq mE\left(\max_{1\leq k\leq k_n} \sigma_{nk}^2\right)$ for all $n \geq m + 1$. But (CLB) implies $\max_{1\leq k\leq k_n} \sigma_{nk}^2 \to 0$ in probability as $n \to \infty$ via Proposition 6.7, whence $E\left(\max_{1\leq k\leq k_n} \sigma_{nk}^2\right) \to 0$ as $n \to \infty$ by assumption (iv) and dominated convergence. Therefore, $III_{m,n} \to 0$ as $n \to \infty$ for all $m \in \mathbb{N}$. Finally, by definition of $\tau_n(m)$, $IV_{m,n} \leq E\left(\max_{1\leq k\leq k_n} \sigma_{nk}^2\right) \to 0$ as $n \to \infty$ for every $m \in \mathbb{N}$. Thus we have shown that $\lim_{m\to\infty} \limsup_{n\to\infty} E\left(\left(\sum_{k=\tau_n(m)+1}^{k_n} X_{nk}\right)^2\right) = 0$, and $\lim_{m\to\infty} \limsup_{n\to\infty} P\left(|V_2(m,n)| \geq \varepsilon\right) = 0$ follows by Markov's inequality.

It remains to consider $V_3(m,n)$. Writing $\zeta_n = \frac{1}{\sqrt{n}} \sum_{k=k_n+1}^{k_n+n} Y_{nk}$ for all $n \in \mathbb{N}$ we note that (ζ_n) is bounded in probability (ζ_n in fact converges in distribution to the standard normal distribution by the classical central limit theorem). Then we obtain for all $m \in \mathbb{N}$ and $n \geq m + 1$

$$V_3\left(m,n\right) = \xi_n\left(m\right)\zeta_n 1_{\{\tau_n(m)<k_n\}} + \xi_n\left(m\right)\zeta_n 1_{\{\tau_n(m)=k_n\}}$$

$$\leq \left(\max_{1\leq k\leq k_n}\sigma_{nk}^2\right)^{1/2}|\zeta_n| + \left|E\left(\eta^2|\mathcal{G}_{mm}\right) - \sum_{k=1}^{k_n}\sigma_{nk}^2\right|^{1/2}|\zeta_n|$$

$$+ \left(\sum_{k=1}^{m}\sigma_{nk}^2\right)^{1/2}|\zeta_n| .$$

The first and the third summand on the right-hand side of this inequality converge to zero in probability as $n \to \infty$ because $\max_{1\leq k\leq k_n}\sigma_{nk}^2 \to 0$ in probability and (ζ_n) is bounded in probability. Hence for all $\varepsilon > 0$

$$\lim_{m\to\infty}\limsup_{n\to\infty} P\left(|V_3\left(m,n\right)| \geq \varepsilon\right)$$

$$\leq \lim_{m\to\infty}\limsup_{n\to\infty} P\left(\left|E\left(\eta^2|\mathcal{G}_{mm}\right) - \sum_{k=1}^{k_n}\sigma_{nk}^2\right|^{1/2}|\zeta_n| \geq \frac{\varepsilon}{2}\right).$$

Because (ζ_n) is bounded in probability the limit on the right-hand side of this inequality is zero provided that

$$\lim_{m\to\infty}\limsup_{n\to\infty} P\left(\left|E\left(\eta^2|\mathcal{G}_{mm}\right) - \sum_{k=1}^{k_n}\sigma_{nk}^2\right|^{1/2} \geq \varepsilon\right) = 0$$

for every $\varepsilon > 0$. But this follows from the inequality

$$\left|E\left(\eta^2|\mathcal{G}_{mm}\right) - \sum_{k=1}^{k_n}\sigma_{nk}^2\right|^{1/2} \leq \left|E\left(\eta^2|\mathcal{G}_{mm}\right) - \eta^2\right|^{1/2} + \left|\eta^2 - \sum_{k=1}^{k_n}\sigma_{nk}^2\right|^{1/2},$$

condition (N) and $E\left(\eta^2|\mathcal{G}_{mm}\right) \to \eta^2$ as $m \to \infty$ almost surely.

Summarizing our results we have shown that $\sum_{k=1}^{k_n} X_{nk} \overset{d}{\to} \eta N$ as $n \to \infty$. This completes Step 2 of the proof.

Step 3. To remove assumption (iv) from Step 2, let $0 < c < \infty$ be fixed. Then for every $n \in \mathbb{N}$

$$\tau_n\left(c\right) = \max\left\{k \in \{0, 1, \ldots, k_n\} : \sum_{j=1}^{k}\sigma_{nj}^2 < c\right\}$$

is a stopping time w.r.t. $(\mathcal{F}_{nk})_{0\leq k\leq k_n}$ and $W_{nk}\left(c\right) := X_{nk}1_{\{k\leq\tau_n(c)\}}$ for $1 \leq k \leq k_n$ defines a square integrable martingale difference sequence w.r.t. $(\mathcal{F}_{nk})_{0\leq k\leq k_n}$. For all $n \in \mathbb{N}$ we have

$$\sum_{k=1}^{k_n} E\left(W_{nk}^2(c)\,|\mathcal{F}_{n,k-1}\right) = \sum_{k=1}^{k_n} 1_{\{k\le \tau_n(c)\}} E\left(X_{nk}^2|\mathcal{F}_{n,k-1}\right) = \sum_{k=1}^{\tau_n(c)} \sigma_{nk}^2 < c$$

by definition of $\tau_n(c)$ so that the square integrable martingale difference array $(W_{nk}(c))_{1\le k\le k_n, n\in \mathbb{N}}$ w.r.t. $(\mathcal{F}_{nk})_{0\le k\le k_n, n\in \mathbb{N}}$ satisfies assumption (iv). From $|W_{nk}(c)| \le |X_{nk}|$ for all $n \in \mathbb{N}$, $1 \le k \le k_n$ and $0 < c < \infty$ and (CLB) it immediately follows that

$$\sum_{k=1}^{k_n} E\left(W_{nk}^2(c)\,1_{\{|W_{nk}(c)|\ge \varepsilon\}}|\mathcal{F}_{n,k-1}\right) \to 0 \quad \text{in probability as } n \to \infty$$

for all $\varepsilon > 0$. Now we set $\eta(c) := \eta 1_{\{\eta^2 < c\}} + \sqrt{c}\,1_{\{\eta^2 \ge c\}}$ for all $0 < c < \infty$ and will show that

$$\sum_{k=1}^{k_n} E\left(W_{nk}^2(c)\,|\mathcal{F}_{n,k-1}\right) \to \eta^2(c) = \eta^2 1_{\{\eta^2 < c\}} + c 1_{\{\eta^2 \ge c\}}$$

in probability as $n \to \infty$. To see this, for every $\varepsilon > 0$ and $n \in \mathbb{N}$ we write

$$P\left(\left|\sum_{k=1}^{k_n} E\left(W_{nk}^2(c)\,|\mathcal{F}_{n,k-1}\right) - \eta^2(c)\right| \ge \varepsilon\right)$$

$$= P\left(\left|\sum_{k=1}^{\tau_n(c)} \sigma_{nk}^2 - \eta^2(c)\right| \ge \varepsilon\right)$$

$$= P\left(\left\{\left|\sum_{k=1}^{\tau_n(c)} \sigma_{nk}^2 - \eta^2(c)\right| \ge \varepsilon\right\} \cap \left\{\sum_{k=1}^{k_n}\sigma_{nk}^2 < c\right\} \cap \left\{\eta^2 < c\right\}\right)$$

$$+ P\left(\left\{\left|\sum_{k=1}^{\tau_n(c)} \sigma_{nk}^2 - \eta^2(c)\right| \ge \varepsilon\right\} \cap \left\{\sum_{k=1}^{k_n}\sigma_{nk}^2 < c\right\} \cap \left\{\eta^2 \ge c\right\}\right)$$

$$+ P\left(\left\{\left|\sum_{k=1}^{\tau_n(c)} \sigma_{nk}^2 - \eta^2(c)\right| \ge \varepsilon\right\} \cap \left\{\sum_{k=1}^{k_n}\sigma_{nk}^2 \ge c\right\} \cap \left\{\eta^2 < c\right\}\right)$$

$$+ P\left(\left\{\left|\sum_{k=1}^{\tau_n(c)} \sigma_{nk}^2 - \eta^2(c)\right| \ge \varepsilon\right\} \cap \left\{\sum_{k=1}^{k_n}\sigma_{nk}^2 \ge c\right\} \cap \left\{\eta^2 \ge c\right\}\right)$$

$$= P_{1,n} + P_{2,n} + P_{3,n} + P_{4,n}\,,$$

say, and we will prove that $P_{j,n}$ converges to zero as $n \to \infty$ for $1 \le j \le 4$.

On the event $\left\{\sum_{k=1}^{k_n} \sigma_{nk}^2 < c\right\} \cap \{\eta^2 < c\}$ we have $\tau_n(c) = k_n$ and $\eta^2(c) = \eta^2$ so that

$$P_{1,n} \le P\left(\left|\sum_{k=1}^{k_n} \sigma_{nk}^2 - \eta^2\right| \ge \varepsilon\right) \to 0 \quad \text{as } \to \infty$$

by condition (N). On the event $\left\{\sum_{k=1}^{k_n} \sigma_{nk}^2 < c\right\} \cap \{\eta^2 \ge c\}$ we have $\tau_n(c) = k_n$ and $\eta^2(c) = c$ so that

$$\left|\sum_{k=1}^{\tau_n(c)} \sigma_{nk}^2 - \eta^2(c)\right| = c - \sum_{k=1}^{k_n} \sigma_{nk}^2 \le \eta^2 - \sum_{k=1}^{k_n} \sigma_{nk}^2 = \left|\eta^2 - \sum_{k=1}^{k_n} \sigma_{nk}^2\right|$$

whence

$$P_{2,n} \le P\left(\left|\sum_{k=1}^{k_n} \sigma_{nk}^2 - \eta^2\right| \ge \varepsilon\right) \to 0 \quad \text{as } n \to \infty.$$

On the event $\left\{\sum_{k=1}^{k_n} \sigma_{nk}^2 \ge c\right\} \cap \{\eta^2 < c\}$ we have $\eta^2(c) = \eta^2 < c \le \sum_{k=1}^{k_n} \sigma_{nk}^2$, and $\sum_{k=1}^{\tau_n(c)} \sigma_{nk}^2 < c$ holds by definition of $\tau_n(c)$. Therefore, $\tau_n(c) < k_n$ and $c \le \sum_{k=1}^{\tau_n(c)+1} \sigma_{nk}^2$, again by definition of $\tau_n(c)$. Now we consider two cases:

Case 1. $\sum_{k=1}^{\tau_n(c)} \sigma_{nk}^2 \le \eta^2$. Then

$$\left|\sum_{k=1}^{\tau_n(c)} \sigma_{nk}^2 - \eta^2(c)\right| = \eta^2 - \sum_{k=1}^{\tau_n(c)} \sigma_{nk}^2 < c - \sum_{k=1}^{\tau_n(c)} \sigma_{nk}^2$$

$$\le \sum_{k=1}^{\tau_n(c)+1} \sigma_{nk}^2 - \sum_{k=1}^{\tau_n(c)} \sigma_{nk}^2 \le \max_{1 \le k \le k_n} \sigma_{nk}^2.$$

Case 2. $\sum_{k=1}^{\tau_n(c)} \sigma_{nk}^2 > \eta^2$. Then

$$\left|\sum_{k=1}^{\tau_n(c)} \sigma_{nk}^2 - \eta^2(c)\right| = \sum_{k=1}^{\tau_n(c)} \sigma_{nk}^2 - \eta^2 \le \sum_{k=1}^{k_n} \sigma_{nk}^2 - \eta^2 = \left|\sum_{k=1}^{k_n} \sigma_{nk}^2 - \eta^2\right|.$$

Combining the two cases we see that

$$P_{3,n} \le P\left(\max_{1 \le k \le k_n} \sigma_{nk}^2 \ge \varepsilon\right) + P\left(\left|\sum_{k=1}^{k_n} \sigma_{nk}^2 - \eta^2\right| \ge \varepsilon\right) \to 0 \quad \text{as } n \to \infty$$

from $\max_{1 \le k \le k_n} \sigma_{nk}^2 \to 0$ in probability and condition (N).

Finally, on the event $\left\{ \sum_{k=1}^{k_n} \sigma_{nk}^2 \geq c \right\} \cap \{ \eta^2 \geq c \}$ we have $\eta^2(c) = c$ and $\tau_n(c) <$ k_n and therefore $\sum_{k=1}^{\tau_n(c)} \sigma_{nk}^2 < c \leq \sum_{k=1}^{\tau_n(c)+1} \sigma_{nk}^2$. Thus

$$\left| \sum_{k=1}^{\tau_n(c)} \sigma_{nk}^2 - \eta^2(c) \right| = c - \sum_{k=1}^{\tau_n(c)} \sigma_{nk}^2 \leq \sum_{k=1}^{\tau_n(c)+1} \sigma_{nk}^2 - \sum_{k=1}^{\tau_n(c)} \sigma_{nk}^2 \leq \max_{1 \leq k \leq k_n} \sigma_{nk}^2$$

and, consequently,

$$P_{4,n} \leq P\left(\max_{1 \leq k \leq k_n} \sigma_{nk}^2 \geq \varepsilon \right) \to 0 \quad \text{as } n \to \infty.$$

Now we can apply the result established in Step 2 to obtain, for all $0 < c < \infty$, $\sum_{k=1}^{k_n} W_{nk}(c) \xrightarrow{d} \eta(c) N$ as $n \to \infty$.

For every $\varepsilon > 0$ and $0 < c < \infty$ we have $P\left(\left| \eta^2(c) - \eta^2 \right| \geq \varepsilon \right) \leq P\left(\eta^2 \geq c \right) \to$ 0 as $c \to \infty$ so that $\eta(c) \to \eta$ in probability and hence $\eta(c) N \xrightarrow{d} \eta N$. In order to complete the proof of $\sum_{k=1}^{k_n} X_{nk} \xrightarrow{d} \eta N$ as $n \to \infty$ we have to show for all $\varepsilon > 0$ that

$$\lim_{c \to \infty} \limsup_{n \to \infty} P\left(\left| \sum_{k=1}^{k_n} X_{nk} - \sum_{k=1}^{k_n} W_{nk}(c) \right| \geq \varepsilon \right) = 0$$

(see Theorem 3.21). To see this, observe that

$$P\left(\left| \sum_{k=1}^{k_n} X_{nk} - \sum_{k=1}^{k_n} W_{nk}(c) \right| \geq \varepsilon \right)$$

$$= P\left(\left| \sum_{k=1}^{k_n} X_{nk} - \sum_{k=1}^{\tau_n(c)} X_{nk} \right| \geq \varepsilon \right) \leq P\left(\tau_n(c) < k_n \right) \leq P\left(\sum_{k=1}^{k_n} \sigma_{nk}^2 \geq c \right)$$

so that

$$\limsup_{n \to \infty} P\left(\left| \sum_{k=1}^{k_n} X_{nk} - \sum_{k=1}^{k_n} W_{nk}(c) \right| \geq \varepsilon \right) \leq \limsup_{n \to \infty} P\left(\sum_{k=1}^{k_n} \sigma_{nk}^2 \geq c \right) \to 0$$

as $c \to \infty$ because $\left(\sum_{k=1}^{k_n} \sigma_{nk}^2 \right)$ is bounded in probability by condition (N). This completes the proof of $\sum_{k=1}^{k_n} X_{nk} \xrightarrow{d} \eta N$ as $n \to \infty$ and of Step 3.

Step 4. Now we will show that the convergence in distribution established so far is $\mathcal{G}$-stable. By monotonicity of $\mathcal{G}_{nk}$ in k and n we have $\mathcal{G} = \sigma(\mathcal{E})$ for $\mathcal{E} = \bigcup_{m=1}^{\infty} \mathcal{G}_{mm}$, and $\mathcal{E}$ is a sub-field of $\mathcal{G}$ because $\mathcal{G}_{mm}$ is increasing in m. By Theorem 3.17 it is enough to show that $\sum_{k=1}^{k_n} X_{nk} \xrightarrow{d} \eta N$ as $n \to \infty$ under $P_F = P(\cdot|F)$ for all $F \in \mathcal{E}$ with $P(F) > 0$. For this, let $F \in \mathcal{E}$ be fixed. Then $F \in \mathcal{G}_{mm}$ for some $m \in \mathbb{N}$. For a sub-σ-field $\mathcal{H} \subset \mathcal{F}$ and an integrable random variable X on $(\Omega, \mathcal{F}, P)$, let $E_P(X|\mathcal{H})$ denote the conditional expectation of X w.r.t. $\mathcal{H}$ under P, whereas $E_{P_F}(X|\mathcal{H})$ is the conditional expectation of X w.r.t. $\mathcal{H}$ under P_F. Observe that for $F \in \mathcal{H}$ any version of $E_P(X|\mathcal{H})$ is also a version of $E_{P_F}(X|\mathcal{H})$, that is, $E_P(X|\mathcal{H}) = E_{P_F}(X|\mathcal{H})$ P_F-almost surely. Therefore, the array $(X_{nk})_{m+1 \le k \le k_n, n \ge m+1}$ is a square integrable martingale difference array adapted to $(\mathcal{F}_{nk})_{m \le k \le k_n, n \ge m+1}$ under P_F. Note that by (N), (CLB) and Proposition 6.7 we have

$$\sum_{k=m+1}^{k_n} E_P\left(X_{nk}^2|\mathcal{F}_{n,k-1}\right) \to \eta^2 \quad \text{in } P\text{-probability as } n \to \infty$$

from which by $E_P\left(X_{nk}^2|\mathcal{F}_{n,k-1}\right) = E_{P_F}\left(X_{nk}^2|\mathcal{F}_{n,k-1}\right)$ P_F-almost surely for all $m + 1 \le k \le k_n$ and $n \ge m + 1$ we obtain

$$\sum_{k=m+1}^{k_n} E_{P_F}\left(X_{nk}^2|\mathcal{F}_{n,k-1}\right) \to \eta^2 \quad \text{in } P_F\text{-probability as } n \to \infty.$$

Moreover,

$$\sum_{k=m+1}^{k_n} E_{P_F}\left(X_{nk}^2 1_{\{|X_{nk}| \ge \varepsilon\}}|\mathcal{F}_{n,k-1}\right) \to 0 \quad \text{in } P_F\text{-probability as } n \to \infty$$

for all $\varepsilon > 0$ is an immediate consequence of (CLB). Therefore $\sum_{k=m+1}^{k_n} X_{nk} \xrightarrow{d} \eta N$ under P_F as $n \to \infty$ by Step 3. Because $\max_{1 \le k \le k_n} |X_{nk}|$ converges to zero in P-probability by (CLB) and Proposition 6.6 and hence also in P_F-probability we arrive at $\sum_{k=1}^{k_n} X_{nk} \xrightarrow{d} \eta N$ under P_F as $n \to \infty$ so that the proof of Theorem 6.1 is complete. $\qquad\square$

Remark 6.8 (a) In applications of Theorem 6.1 stronger conditions than (CLB) may be used. Clearly, (CLB) is implied by the *classical Lindeberg condition*

(LB) $$\sum_{k=1}^{k_n} E\left(X_{nk}^2 1_{\{|X_{nk}| \ge \varepsilon\}}\right) \to 0 \quad \text{as } n \to \infty \text{ for every } \varepsilon > 0$$

as well as by the *conditional Lyapunov condition of order* $p \in (2, \infty)$, which requires for some $p \in (2, \infty)$ that

$$(\text{CLY}_p) \qquad \sum_{k=1}^{k_n} E\left(|X_{nk}|^p \mid \mathcal{F}_{n,k-1}\right) \to 0 \quad \text{in probability as } n \to \infty$$

and entails (CLB) through the inequality, valid for all $\varepsilon > 0$,

$$\sum_{k=1}^{k_n} E\left(X_{nk}^2 1_{\{|X_{nk}| \geq \varepsilon\}} \mid \mathcal{F}_{n,k-1}\right) \leq \frac{1}{\varepsilon^{p-2}} \sum_{k=1}^{k_n} E\left(|X_{nk}|^p \mid \mathcal{F}_{n,k-1}\right)$$

provided that $X_{nk} \in \mathcal{L}^p(P)$ for all $n \in \mathbb{N}$ and $1 \leq k \leq k_n$. In the latter case, (CLY_p) is obviously implied by the *classical Lyapunov condition of order* $p \in (2, \infty)$, that is,

$$(\text{LY}_p) \qquad \sum_{k=1}^{k_n} E\left(|X_{nk}|^p\right) \to 0 \quad \text{as } n \to \infty.$$

(b) For independent random variables N with $P^N = N(0, 1)$ and $\eta \geq 0$, the characteristic function $\phi_{\eta N}$ of ηN is given by, for all $t \in \mathbb{R}$,

$$\phi_{\eta N}(t) = E\left(\exp(it\eta N)\right) = \int_{[0,\infty)} E\left(\exp(itu N)\right) dP^\eta(u)$$

$$= \int_{[0,\infty)} \exp\left(-\frac{1}{2}t^2 u^2\right) dP^\eta(u) = E e^{-t^2 \eta^2/2}.$$

Thus $\phi_{\eta N}$ is real-valued and $P^{\eta N}$ is symmetric around zero. Therefore, all limit random variables in Theorem 6.1 are symmetric around zero. Furthermore, the distribution $P^{\eta N} = PN(0, \eta^2)$ satisfies $P^{\eta N} \ll \lambda$ if and only if $P(\eta^2 > 0) = 1$ and then

$$\frac{dP^{\eta N}}{d\lambda}(x) = E\left(\frac{1}{\sqrt{2\pi\eta^2}} e^{-x^2/\eta^2}\right), \quad x \in \mathbb{R}.$$

Exercise 6.2 (The case $k_n = \infty$) In the situation of Theorem 6.1 let $k_n = \infty$ for every $n \in \mathbb{N}$ and assume that for every $n \in \mathbb{N}$,

$$\sum_{k=1}^{\infty} X_{nk} \quad \text{converges a.s. in } \mathbb{R}$$

and

$$\sum_{k=1}^{\infty} E\left(X_{nk}^2 \mid \mathcal{F}_{n,k-1}\right) < \infty \quad \text{a.s.}$$

(By the martingale convergence theorem, both conditions are satisfied if

$$\sup_{k\in\mathbb{N}} E\left(\sum_{j=1}^{k} X_{nj}\right)^2 = \sum_{j=1}^{\infty} EX_{nj}^2 < \infty$$

for every $n \in \mathbb{N}$.)
    Show that

$$\sum_{j=1}^{\infty} X_{nj} \to N\left(0, \eta^2\right) \quad \mathcal{G}\text{-stably as } n \to \infty,$$

where $\mathcal{G}_{nk_n} = \mathcal{G}_{n\infty} = \sigma\left(\bigcup_{j=0}^{\infty} \mathcal{G}_{nj}\right)$.

Exercise 6.3 In the situation of Theorem 6.1 assume that $(\mathcal{F}_{nk})_{0\le k\le k_n, n\in\mathbb{N}}$ is a nested array and $P\left(\eta^2 > 0\right) > 0$. Show that the limit points satisfy

$$L\left(\left(\sum_{k=1}^{k_n} X_{nk}\right)_{n\in\mathbb{N}}\right) = \mathbb{R} \quad P_{\{\eta^2>0\}}\text{-a.s.}$$

6.2 Counterexamples

This section will shed some light on the role of the conditions in Theorem 6.1. The first result shows that the row sums of a square integrable martingale difference array have weak limit points if the row sums of the conditional variances are bounded in probability.

Proposition 6.9 *Let* $(X_{nk})_{1\le k\le k_n, n\in\mathbb{N}}$ *be a square integrable martingale differ-ence array adapted to an array* $(\mathcal{F}_{nk})_{0\le k\le k_n, n\in\mathbb{N}}$ *of* σ*-fields. If the sequence* $\left(\sum_{k=1}^{k_n} E\left(X_{nk}^2|\mathcal{F}_{n,k-1}\right)\right)_{n\in\mathbb{N}}$ *is bounded in probability, then the sequence* $\left(\sum_{k=1}^{k_n} X_{nk}\right)_{n\in\mathbb{N}}$ *is also bounded in probability.*

Note that for sequences of real (or $\mathbb{R}^d$-valued) random variables boundedness in probability is the same as tightness.

Proof For any fixed $n \in \mathbb{N}$, the process $\left(\sum_{k=1}^{j} E\left(X_{nk}^2|\mathcal{F}_{n,k-1}\right)\right)_{0\le j\le k_n}$ is the compensator (quadratic characteristic) of the positive submartingale $\left(\left(\sum_{k=1}^{j} X_{nk}\right)^2\right)_{0\le j\le k_n}$ so that, for all $0 < C, M < \infty$ by Lenglart's inequality of Theorem A.8 (a)

$$P\left(\left|\sum_{k=1}^{k_n} X_{nk}\right| \geq C\right) = P\left(\left(\sum_{k=1}^{k_n} X_{nk}\right)^2 \geq C^2\right)$$

$$\leq \frac{M}{C^2} + P\left(\sum_{k=1}^{k_n} E\left(X_{nk}^2 | \mathcal{F}_{n,k-1}\right) > M\right).$$

Therefore, for all $0 < C, M < \infty$,

$$\sup_{n \in \mathbb{N}} P\left(\left|\sum_{k=1}^{k_n} X_{nk}\right| \geq C\right) \leq \frac{M}{C^2} + \sup_{n \in \mathbb{N}} P\left(\sum_{k=1}^{k_n} E\left(X_{nk}^2 | \mathcal{F}_{n,k-1}\right) > M\right).$$

This inequality clearly implies the assertion by first letting $C \to \infty$ and then $M \to \infty$.   □

Boundedness in probability of $\left(\sum_{k=1}^{k_n} E\left(X_{nk}^2 | \mathcal{F}_{n,k-1}\right)\right)_{n \in \mathbb{N}}$ for a martingale difference array already entails the existence of weak limit points for the row sums by Proposition 6.9, the role of the much stronger condition (N) in conjunction with (CLB) is to ensure uniqueness of the weak limit points and their form as variance mixtures of centered normals. In the sequel we will show by examples that condition (N) is essential for obtaining stable convergence to a Gauss-kernel.

First, we will consider the special case of a non-random limit η^2 in condition (N). According to Remark 6.2 (b) conditions (CLB) and (N) with $\eta^2 = 1$ imply $\sum_{k=1}^{k_n} X_{nk} \overset{d}{\to} N$ as $n \to \infty$ with $P^N = N(0, 1)$ for any square integrable martingale difference array. This convergence, however, is in general not $\mathcal{F}_\infty$-stable, as shown by the following example.

Example 6.10 Let $\left(W^{(i)}(t)\right)_{t \geq 0}$ for $i = 1, 2$ be two independent Brownian motions. (Here and in the subsequent example it is convenient to write $W(t)$ instead of W_t.) For all $n \in \mathbb{N}$ and $1 \leq k \leq k_n = n$ we set

$$X_{nk} := \begin{cases} W^{(1)}\left(\dfrac{k}{n}\right) - W^{(1)}\left(\dfrac{k-1}{n}\right), & \text{if } n \text{ is even} \\[2mm] W^{(2)}\left(\dfrac{k}{n}\right) - W^{(2)}\left(\dfrac{k-1}{n}\right), & \text{if } n \text{ is odd} \end{cases}$$

and $\mathcal{F}_{nk} := \sigma\left(X_{nj}, 1 \leq j \leq k\right)$ with $\mathcal{F}_{n0} := \{\emptyset, \Omega\}$. Then, by independence of the increments of $W^{(i)}$, $(X_{nk})_{1 \leq k \leq n, n \in \mathbb{N}}$ is a square integrable martingale difference array w.r.t. $(\mathcal{F}_{nk})_{0 \leq k \leq n, n \in \mathbb{N}}$. For all $n \in \mathbb{N}$ we have $\sum_{k=1}^n E\left(X_{nk}^2 | \mathcal{F}_{n,k-1}\right) = 1$, again by independence of the increments of $W^{(i)}$. Moreover, for all $\varepsilon > 0$, all $n \in \mathbb{N}$ and N with $P^N = N(0, 1)$,

$$\sum_{k=1}^n E\left(X_{nk}^2 1_{\{|X_{nk}| \geq \varepsilon\}}\right) = E\left(N^2 1_{\{|N| \geq \varepsilon n^{1/2}\}}\right) \to 0 \quad \text{as } n \to \infty.$$

Finally, for all $n \in \mathbb{N}$

$$\sum_{k=1}^{n} X_{nk} = \begin{cases} W^{(1)}(1), & \text{if } n \text{ is even} \\ W^{(2)}(1), & \text{if } n \text{ is odd}, \end{cases}$$

so that $P^{\sum_{k=1}^{n} X_{nk}} = P^N$.

In this example, condition (N) is satisfied with $\eta^2 = 1$ (even with equality for every $n \in \mathbb{N}$ instead of convergence in probability as $n \to \infty$), the classical Lindeberg condition (LB) is satisfied which implies (CLB) by Remark 6.8 (a), and $\sum_{k=1}^{n} X_{nk} \overset{d}{\to} N$ as $n \to \infty$ for N with $P^N = N(0, 1)$ (again with equality (in distribution) for every n instead of convergence in distribution as $n \to \infty$). However, $\sum_{k=1}^{n} X_{nk}$ cannot converge $\sigma\left(W^{(1)}(1), W^{(2)}(1)\right)$-stably (and $\sigma\left(W^{(1)}(1), W^{(2)}(1)\right) \subset \mathcal{F}_\infty$). Otherwise, we have $\delta_{W^{(1)}(1)} = \delta_{W^{(2)}(1)}$ and thus $W^{(1)}(1) = W^{(2)}(1)$ almost surely, a contradiction.

One checks that $\mathcal{G}$ is trivial, that is, $P(\mathcal{G}) = \{0, 1\}$, hence Theorem 6.1 yields nothing else than distributional convergence in the present setting. For this, let $n \in \mathbb{N}$ and $0 \le k \le k_n = n$ be fixed. By definition $\mathcal{G}_{nk} \subset \mathcal{F}_{mk}$ for all $m \in \mathbb{N}$ with $m \ge n$ and

$$\mathcal{F}_{mk} = \sigma\left(W^{(i)}\left(\frac{j}{m}\right) - W^{(i)}\left(\frac{j-1}{m}\right); 1 \le j \le k\right) \subset \sigma\left(W^{(i)}(t); 0 \le t \le \frac{k}{m}\right)$$

with $i = 1$ if m is even and $i = 2$ if m is odd. For any $\varepsilon > 0$ we have $k/m \le \varepsilon$ for all large m so that $\mathcal{G}_{nk} \subset \sigma\left(W^{(1)}(t); 0 \le t \le \varepsilon\right) \cap \sigma\left(W^{(2)}(t); 0 \le t \le \varepsilon\right)$ which implies

$$\mathcal{G}_{nk} \subset \bigcap_{\varepsilon>0} \sigma\left(W^{(1)}(t); 0 \le t \le \varepsilon\right) \cap \bigcap_{\varepsilon>0} \sigma\left(W^{(2)}(t); 0 \le t \le \varepsilon\right).$$

Hence also

$$\mathcal{G} \subset \bigcap_{\varepsilon>0} \sigma\left(W^{(1)}(t); 0 \le t \le \varepsilon\right) \cap \bigcap_{\varepsilon>0} \sigma\left(W^{(2)}(t); 0 \le t \le \varepsilon\right).$$

But by Blumenthal's zero-one law for Brownian motion both σ-fields on the right-hand side of the last display are trivial, which proves the assertion. $\square$

Our next example shows what can happen for martingale difference arrays satisfying (CLB) and (N), except for the fact that the random variable η^2 is not measurable w.r.t. the σ-field $\mathcal{G}$.

Example 6.11 Let $(W(t))_{t \ge 0}$ be a Brownian motion. For every $n \in \mathbb{N}$ and $1 \le k \le k_n = 2n$ we define

$$
S_{nk} := \begin{cases} W\left(\dfrac{k}{n}\right) & , \ 1 \le k \le n \\[3ex] W\left(1 + \dfrac{k-n}{n} 1_{\{W(1)>0\}}\right) & , \ n+1 \le k \le 2n \text{ and } n \text{ even} \\[3ex] W\left(2 + \dfrac{k-n}{n} 1_{\{W(1)>0\}}\right) & , \ n+1 \le k \le 2n \text{ and } n \text{ odd} \end{cases}
$$

with $S_{n0} := 0$, $X_{nk} := S_{nk} - S_{n,k-1}$ and $\mathcal{F}_{nk} := \sigma(S_{n0}, \dots, S_{nk})$. Note that $S_{nk} = W\left(\frac{k}{n}\right) 1_{\{W(1)>0\}} + W(1) 1_{\{W(1)\le 0\}}$ for even $n \in \mathbb{N}$ and $n+1 \le k \le 2n$ and $S_{nk} = W\left(1 + \frac{k}{n}\right) 1_{\{W(1)>0\}} + W(2) 1_{\{W(1)\le 0\}}$ for odd $n \in \mathbb{N}$ and $n+1 \le k \le 2n$, which shows that the random variables S_{nk} are square integrable. Consequently, the random variables X_{nk} are also square integrable, and the array $(X_{nk})_{1 \le k \le 2n, n \in \mathbb{N}}$ is, by construction, adapted to the array $(\mathcal{F}_{nk})_{0 \le k \le 2n, n \in \mathbb{N}}$. For all $n \in \mathbb{N}$ and $1 \le k \le n$ we have, by independence of the increments of $(W(t))_{t \ge 0}$ and its moment properties,

$$
\begin{aligned}
E\left(X_{nk} | \mathcal{F}_{n,k-1}\right) &= E\left(W\left(\frac{k}{n}\right) - W\left(\frac{k-1}{n}\right) \,\middle|\, W\left(\frac{1}{n}\right), \dots, W\left(\frac{k-1}{n}\right)\right) \\
&= E\left(W\left(\frac{k}{n}\right) - W\left(\frac{k-1}{n}\right)\right) = 0
\end{aligned}
$$

and

$$
\begin{aligned}
&E\left(X_{nk}^2 | \mathcal{F}_{n,k-1}\right) \\
&= E\left(\left[W\left(\frac{k}{n}\right) - W\left(\frac{k-1}{n}\right)\right]^2 \,\middle|\, W\left(\frac{1}{n}\right), \dots, W\left(\frac{k-1}{n}\right)\right) \\
&= E\left(\left[W\left(\frac{k}{n}\right) - W\left(\frac{k-1}{n}\right)\right]^2\right) = \frac{1}{n}.
\end{aligned}
$$

For all even $n \in \mathbb{N}$ and $n+1 \le k \le 2n$ we have

$$
X_{nk} = S_{nk} - S_{n,k-1} = \left[W\left(\frac{k}{n}\right) - W\left(\frac{k-1}{n}\right)\right] 1_{\{W(1)>0\}}.
$$

Note that $W(1)$ is $\mathcal{F}_{n,k-1}$-measurable and that $\mathcal{F}_{n,k-1} \subset \sigma\left(W(t), 0 \le t \le \frac{k-1}{n}\right)$ so that $W\left(\frac{k}{n}\right) - W\left(\frac{k-1}{n}\right)$ is independent of $\mathcal{F}_{n,k-1}$, by independence of the increments of $(W(t))_{t \ge 0}$. This implies

$$
\begin{aligned}
E\left(X_{nk} | \mathcal{F}_{n,k-1}\right) &= 1_{\{W(1)>0\}} E\left(W\left(\frac{k}{n}\right) - W\left(\frac{k-1}{n}\right) \,\middle|\, \mathcal{F}_{n,k-1}\right) \\
&= 1_{\{W(1)>0\}} E\left(W\left(\frac{k}{n}\right) - W\left(\frac{k-1}{n}\right)\right) = 0
\end{aligned}
$$

and

$$E\left(X_{nk}^2|\mathcal{F}_{n,k-1}\right) = 1_{\{W(1)>0\}} E\left(\left[W\left(\frac{k}{n}\right) - W\left(\frac{k-1}{n}\right)\right]^2 \Big| \mathcal{F}_{n,k-1}\right)$$

$$= 1_{\{W(1)>0\}} E\left(\left[W\left(\frac{k}{n}\right) - W\left(\frac{k-1}{n}\right)\right]^2\right)$$

$$= \frac{1}{n} 1_{\{W(1)>0\}}.$$

For all odd $n \in \mathbb{N}$ and $n+1 \leq k \leq 2n$ we have

$$X_{nk} = S_{nk} - S_{n,k-1} = \left[W\left(1+\frac{k}{n}\right) - W\left(1+\frac{k-1}{n}\right)\right] 1_{\{W(1)>0\}}.$$

Note that again $W(1)$ is measurable w.r.t. $\mathcal{F}_{n,k-1}$, and that $\mathcal{F}_{n,k-1} \subset \sigma(W(t))$, $0 \leq t \leq 1 + \frac{k-1}{n})$ so that $W\left(1+\frac{k}{n}\right) - W\left(1+\frac{k-1}{n}\right)$ is independent of $\mathcal{F}_{n,k-1}$. This now implies

$$E\left(X_{nk}|\mathcal{F}_{n,k-1}\right) = 1_{\{W(1)>0\}} E\left(W\left(1+\frac{k}{n}\right) - W\left(1+\frac{k-1}{n}\right) \Big| \mathcal{F}_{n,k-1}\right)$$

$$= 1_{\{W(1)>0\}} E\left(W\left(1+\frac{k}{n}\right) - W\left(1+\frac{k-1}{n}\right)\right) = 0$$

and

$$E\left(X_{nk}^2|\mathcal{F}_{n,k-1}\right)$$

$$= 1_{\{W(1)>0\}} E\left(\left[W\left(1+\frac{k}{n}\right) - W\left(1+\frac{k-1}{n}\right)\right]^2 \Big| \mathcal{F}_{n,k-1}\right)$$

$$= 1_{\{W(1)>0\}} E\left(\left[W\left(1+\frac{k}{n}\right) - W\left(1+\frac{k-1}{n}\right)\right]^2\right)$$

$$= \frac{1}{n} 1_{\{W(1)>0\}}.$$

Thus we have shown that $(X_{nk})_{1\leq k\leq 2n, n\in\mathbb{N}}$ is a square integrable martingale difference array w.r.t. $(\mathcal{F}_{nk})_{0\leq k\leq 2n, n\in\mathbb{N}}$ with

$$\sum_{k=1}^{2n} E\left(X_{nk}^2|\mathcal{F}_{n,k-1}\right) = 1 + 1_{\{W(1)>0\}}$$

for all $n \in \mathbb{N}$. Moreover, for all $n \in \mathbb{N}$ and $1 \leq k \leq n$, $|X_{nk}|^3 = \left| W\left(\frac{k}{n}\right) - W\left(\frac{k-1}{n}\right) \right|^3$, whereas $|X_{nk}|^3 \leq \left| W\left(\frac{k}{n}\right) - W\left(\frac{k-1}{n}\right) \right|^3$ for all even $n \in \mathbb{N}$ and $n+1 \leq k \leq 2n$ and $|X_{nk}|^3 \leq \left| W\left(1+\frac{k}{n}\right) - W\left(1+\frac{k-1}{n}\right) \right|^3$ for all odd $n \in \mathbb{N}$ and $n+1 \leq k \leq 2n$. This yields $E\left(|X_{nk}|^3\right) \leq (8/\pi)^{1/2} n^{-3/2}$ for all $n \in \mathbb{N}$ and $1 \leq k \leq 2n$, because any increment of $(W(t))_{t \geq 0}$ of length $1/n$ has a centered normal distribution with variance $1/n$. Consequently, $\sum_{k=1}^{2n} E\left(|X_{nk}|^3\right) \leq 2(8/\pi)^{1/2} n^{-1/2}$ for all $n \in \mathbb{N}$ so that the array $(X_{nk})_{1 \leq k \leq 2n, n \in \mathbb{N}}$ satisfies the classical Ljapunov condition (LY_p) of order $p=3$ and hence (CLB) by Remark 6.8 (a). Thus, all conditions of Theorem 6.1 except $\mathcal{G}$-measurability of $\eta^2 = 1 + 1_{\{W(1)>0\}}$ are satisfied. For all $n \in \mathbb{N}$ we have

$$\sum_{k=1}^{2n} X_{nk} = S_{n,2n} = \begin{cases} W\left(1 + 1_{\{W(1)>0\}}\right), & n \text{ even} \\ W\left(2 + 1_{\{W(1)>0\}}\right), & n \text{ odd} \end{cases}$$

which shows that the sequences $\left(\sum_{k=1}^{2n} X_{nk}\right)_{n \in \mathbb{N}, n \text{ even}}$ and $\left(\sum_{k=1}^{2n} X_{nk}\right)_{n \in \mathbb{N}, n \text{ odd}}$ have two different limits in distribution. For a formal proof of $P^{W(1+1_{\{W(1)>0\}})} \neq P^{W(2+1_{\{W(1)>0\}})}$ note that

$$W\left(1 + 1_{\{W(1)>0\}}\right) = W(2)\, 1_{\{W(1)>0\}} + W(1)\, 1_{\{W(1)\leq 0\}}$$
$$= [W(2) - W(1)]\, 1_{\{W(1)>0\}} + W(1)$$

so that, by independence of the increments of $(W(t))_{t \geq 0}$,

$$E\left(W\left(1 + 1_{\{W(1)>0\}}\right)^2\right) = E\left([W(2) - W(1)]^2\right) P(W(1) > 0)$$
$$+ 2E(W(2) - W(1))\, E\left(W(1)\, 1_{\{W(1)>0\}}\right) + E\left(W(1)^2\right) = \frac{3}{2}$$

and

$$W\left(2 + 1_{\{W(1)>0\}}\right) = W(3)\, 1_{\{W(1)>0\}} + W(2)\, 1_{\{W(1)\leq 0\}}$$
$$= [W(3) - W(2)]\, 1_{\{W(1)>0\}} + W(2)$$

so that

$$E\left(W\left(2 + 1_{\{W(1)>0\}}\right)^2\right) = E\left([W(3) - W(2)]^2\right) P(W(1) > 0)$$
$$+ 2E(W(3) - W(2))\, E\left(W(2)\, 1_{\{W(1)>0\}}\right) + E\left(W(2)^2\right) = \frac{5}{2}.$$

Thus we have produced an example for which the sequence $\left(\sum_{k=1}^{2n} X_{nk}\right)_{n \in \mathbb{N}}$ does not converge in distribution. If we alter the construction by setting $S_{nk} = W\left(1 + \frac{k-n}{n} 1_{\{W(1)>0\}}\right)$ for all $n \in \mathbb{N}$ and $n+1 \leq k \leq 2n$, then we get

$\sum_{k=1}^{2n} X_{nk} = W \left(1 + 1_{\{W(1)>0\}}\right)$ for all $n \in \mathbb{N}$, that is, now $\sum_{k=1}^{2n} X_{nk}$ does converge (mixing) in distribution (and all the other assumptions of Theorem 6.1 remain satisfied, of course). The distribution of the limit random variable $W \left(1 + 1_{\{W(1)>0\}}\right)$ is not a variance mixture of centered normal distributions, however, because it is not symmetric around zero, see Remark 6.8 (b): In view of $W \left(1 + 1_{\{W(1)>0\}}\right) = W (2) \, 1_{\{W(1)>0\}} + W (1) \, 1_{\{W(1)\leq 0\}}$, it is clearly continuous, and

$$P \left(W \left(1 + 1_{\{W(1)>0\}}\right) > 0 \right)$$
$$= P \left(\{W (2) > 0\} \cap \{W (1) > 0\} \right) + P \left(\{W (1) > 0\} \cap \{W (1) \leq 0\} \right) < \frac{1}{2}.$$

Summarizing, we see that without $\mathcal{G}$-measurability of η^2 in Theorem 6.1 there may be several different distributional limit points for the whole sequence $\left(\sum_{k=1}^{2n} X_{nk}\right)_{n \in \mathbb{N}}$ of row sums so that this sequence does not converge in distribution, or there may be (mixing) convergence to a limit which is not a variance mixture of centered normal distributions.

For a direct proof of the fact that the random variable $\eta^2 = 1 + 1_{\{W(1)>0\}}$ is not $\mathcal{G}$-measurable, we show that as in Example 6.10 the σ-field $\mathcal{G}$ is trivial. For this, let $n \in \mathbb{N}$ and $0 \leq k \leq k_n = 2n$ be fixed. By definition, $\mathcal{G}_{nk} \subset \mathcal{F}_{mk}$ for all $m \in \mathbb{N}$ with $m \geq n$. If even $m \geq 2n$ holds, then $k \leq 2n \leq m$ so that for both definitions of the S_{nk}

$$\mathcal{F}_{mk} = \sigma \left(S_{m1}, \ldots, S_{mk}\right) = \sigma \left(W \left(\frac{j}{m}\right); 1 \leq j \leq k \right) \subset \sigma \left(W (t); 0 \leq t \leq \frac{k}{m} \right).$$

For any $\varepsilon > 0$ we have $k/m \leq \varepsilon$ for all large m so that $\mathcal{G}_{nk} \subset \sigma (W (t); 0 \leq t \leq \varepsilon)$ and hence $\mathcal{G}_{nk} \subset \bigcap_{\varepsilon>0} \sigma (W (t); 0 \leq t \leq \varepsilon) =: \mathcal{F}_W (0+)$ which finally implies $\mathcal{G} \subset \mathcal{F}_W (0+)$. By Blumenthal's zero-one law for Brownian motion, the σ-field $\mathcal{F}_W (0+)$ is trivial, which implies the assertion. $\qquad \square$

Our final example shows that convergence in probability in condition (N) in Theorem 6.1 cannot be replaced by $\mathcal{G}$-mixing convergence. Note that

$$\sum_{k=1}^{k_n} E \left(X_{nk}^2 | \mathcal{F}_{n,k-1} \right) \to \eta^2 \quad \mathcal{G}\text{-mixing as } n \to \infty$$

by definition requires independence of $\sigma (\eta^2)$ and $\mathcal{G}$ so that the assumption of $\mathcal{G}$-measurability of η^2 makes no sense now for a nonconstant η^2.

Example 6.12 Let $(N_k)_{k \in \mathbb{N}}$ be an independent sequence of standard normal random variables, and let $g, h : \mathbb{R} \to \mathbb{R}$ be two continuous functions with $|g| = |h|$ and $E \left(|g (N_1)|^3 \right) < \infty$. For all $n \in \mathbb{N}$ and $1 \leq k \leq k_n = 2n$ we set

$$
X_{nk} := \begin{cases}
n^{-1/2} N_k & , 1 \leq k \leq n \\[2ex]
n^{-1/2} g\left(n^{-1/2} \sum_{j=1}^{n} N_j \right) N_k & , n+1 \leq k \leq 2n \text{ and } n \text{ even} \\[3ex]
n^{-1/2} h\left(n^{-1/2} \sum_{j=1}^{n} N_j \right) N_k & , n+1 \leq k \leq 2n \text{ and } n \text{ odd}
\end{cases}
$$

and $\mathcal{F}_{nk} := \sigma(N_1, \ldots, N_k)$ with $\mathcal{F}_{n0} := \{\emptyset, \Omega\}$. Then $(X_{nk})_{1 \leq k \leq 2n, n \in \mathbb{N}}$ is an array of square integrable random variables adapted to $(\mathcal{F}_{nk})_{0 \leq k \leq 2n, n \in \mathbb{N}}$, and this array is nested because $\mathcal{F}_{n+1,k} = \sigma(N_1, \ldots, N_k) = \mathcal{F}_{nk}$ for all $n \in \mathbb{N}$ and $1 \leq k \leq 2n$. Since the N_k are independent standard normal random variables, for all $n \in \mathbb{N}$ and $1 \leq k \leq n$ we obtain

$$
E\left(X_{nk} | \mathcal{F}_{n,k-1} \right) = n^{-1/2} E\left(N_k | N_1, \ldots, N_{k-1} \right) = n^{-1/2} E\left(N_k \right) = 0,
$$

$$
E\left(X_{nk}^2 | \mathcal{F}_{n,k-1} \right) = n^{-1} E\left(N_k^2 | N_1, \ldots, N_{k-1} \right) = n^{-1} E\left(N_k^2 \right) = n^{-1}
$$

and

$$
E\left(|X_{nk}|^3 \right) = n^{-3/2} E\left(|N_k|^3 \right) = \left(\frac{8}{\pi} \right)^{1/2} n^{-3/2},
$$

whereas for all even $n \in \mathbb{N}$ and $n+1 \leq k \leq 2n$

$$
E\left(X_{nk} | \mathcal{F}_{n,k-1} \right) = n^{-1/2} E\left(N_k g\left(n^{-1/2} \sum_{j=1}^{n} N_j \right) \Bigg| N_1, \ldots, N_n, \ldots, N_{k-1} \right)
$$

$$
= n^{-1/2} g\left(n^{-1/2} \sum_{j=1}^{n} N_j \right) E\left(N_k | N_1, \ldots, N_{k-1} \right) = 0,
$$

$$
E\left(X_{nk}^2 | \mathcal{F}_{n,k-1} \right) = n^{-1} E\left(N_k^2 g^2\left(n^{-1/2} \sum_{j=1}^{n} N_j \right) \Bigg| N_1, \ldots, N_n, \ldots, N_{k-1} \right)
$$

$$
= n^{-1} g^2\left(n^{-1/2} \sum_{j=1}^{n} N_j \right) E\left(N_k^2 | N_1, \ldots, N_{k-1} \right)
$$

$$
= n^{-1} g^2\left(n^{-1/2} \sum_{j=1}^{n} N_j \right)
$$

and

$$E\left(|X_{nk}|^3\right) = n^{-3/2} E\left(\left|g\left(n^{-1/2}\sum_{j=1}^{n} N_j\right)\right|^3\right) E\left(|N_k|^3\right)$$

$$= \left(\frac{8}{\pi}\right)^{1/2} E\left(|g\left(N_1\right)|^3\right) n^{-3/2},$$

while for all odd n and $n+1 \leq k \leq 2n$, replacing g by h, clearly $E\left(X_{nk}|\mathcal{F}_{n,k-1}\right) = 0$ and

$$E\left(X_{nk}^2|\mathcal{F}_{n,k-1}\right) = n^{-1}h^2\left(n^{-1/2}\sum_{j=1}^{n} N_j\right) = n^{-1}g^2\left(n^{-1/2}\sum_{j=1}^{n} N_j\right)$$

as well as

$$E\left(|X_{nk}|^3 |\mathcal{F}_{n,k-1}\right) = \left(\frac{8}{\pi}\right)^{1/2} E\left(|h\left(N_1\right)|^3\right) n^{-3/2} = \left(\frac{8}{\pi}\right)^{1/2} E\left(|g\left(N_1\right)|^3\right) n^{-3/2},$$

using $|h| = |g|$.

From the above results we see that $(X_{nk})_{1\leq k\leq 2n,n\in\mathbb{N}}$ is a square integrable martingale difference array w.r.t. $(\mathcal{F}_{nk})_{0\leq k\leq 2n,n\in\mathbb{N}}$ with $\sum_{k=1}^{2n} E\left(|X_{nk}|^3\right) = \left(\frac{8}{\pi}\right)^{1/2}$ $\left(1 + E\left(|g\left(N_1\right)|^3\right)\right) n^{-1/2}$ for all $n \in \mathbb{N}$ so that the classical Ljapunov condition (LY_p) of order $p = 3$ holds. Moreover,

$$\sum_{k=1}^{2n} E\left(X_{nk}^2|\mathcal{F}_{n,k-1}\right) = 1 + g^2\left(n^{-1/2}\sum_{j=1}^{n} N_j\right) \overset{d}{=} 1 + g^2\left(N\right)$$

for all $n \in \mathbb{N}$ and a random variable N with $P^N = N\left(0,1\right)$ which is independent of $\mathcal{F}_\infty$. It follows from the classical stable central limit theorem (see Example 3.13 (b)) and the continuity of g, using Theorem 3.18 (c), that

$$1 + g^2\left(n^{-1/2}\sum_{j=1}^{n} N_j\right) \to 1 + g^2(N) \quad \mathcal{F}_\infty\text{-mixing as } n \to \infty,$$

which shows that condition (N) holds with mixing convergence instead of convergence in probability. Note that for all even $n \in \mathbb{N}$

$$\sum_{k=1}^{2n} X_{nk} = n^{-1/2}\sum_{k=1}^{n} N_k + g\left(n^{-1/2}\sum_{j=1}^{n} N_j\right) n^{-1/2}\sum_{k=n+1}^{2n} N_k \overset{d}{=} N + g\left(N\right) N',$$

where N and N' are independent random variables with $P^N = P^{N'} = N(0,1)$ which are independent of $\mathcal{F}_\infty$, whereas for all odd $n \in \mathbb{N}$, by the same argument,

$$\sum_{k=1}^{2n} X_{nk} \overset{d}{=} N + h(N)N'.$$

For the functions $g(x) = x$ and $h(x) = |x|$ for all $x \in \mathbb{R}$ the above assumptions on g and h are satisfied, and $P^{N+NN'} \neq P^{N+|N|N'}$. To verify the latter, observe that, by independence of N and N', $E\left((N + NN')^4\right) = E(N^4) E\left((1 + N')^4\right) = 30$ because $E(N^4) = 3$ and $E\left((1 + N')^4\right) = 10$, whereas $E\left((N + |N|N')^4\right) = 24$. Therefore, our construction yields a square integrable martingale difference array with different distributional limits for $\left(\sum_{k=1}^{2n} X_{nk}\right)_{n \in \mathbb{N}, n \text{ even}}$ and $\left(\sum_{k=1}^{2n} X_{nk}\right)_{n \in \mathbb{N}, n \text{ odd}}$.

If we take $g = h$, then we have $\sum_{k=1}^{2n} X_{nk} \overset{d}{=} N + g(N)N'$ for all $n \in \mathbb{N}$ and, by the same reasoning as above for $\sum_{k=1}^{2n} E\left(X_{nk}^2 | \mathcal{F}_{n,k-1}\right)$, we see that

$$\sum_{k=1}^{2n} X_{nk} \to N + g(N)N' \quad \mathcal{F}_\infty\text{-mixing as } n \to \infty.$$

For the function $g(x) := x 1_{[0,\infty)}(x), x \in \mathbb{R}$, we obtain $E\left((N + g(N)N')^3\right) = 3E\left(N^3 1_{[0,\infty)}(N)\right) > 0$, showing that $P^{N+g(N)N'}$ is not symmetric around zero and hence no mixture of centered normal distributions by Remark 6.8 (b).

Consequently, if in condition (N) of Theorem 6.1 convergence in probability is replaced by $\mathcal{F}_\infty$-mixing convergence, there may occur several subsequential weak limits for the row sums, or the row sums may converge $\mathcal{F}_\infty$-mixing, but to a limit which is not a variance mixture of centered normals. □

6.3 Further Sufficient Conditions

The conditions (N) and (CLB) in Theorem 6.1 may be replaced by several other sets of sufficient conditions. Some of these will be introduced and discussed in this section, which is partly based on [34]. We always consider an array $(X_{nk})_{1 \leq k \leq k_n, n \in \mathbb{N}}$ of random variables and an array $(\mathcal{F}_{nk})_{0 \leq k \leq k_n, n \in \mathbb{N}}$ of sub-σ-fields of $\mathcal{F}$ for some basic probability space $(\Omega, \mathcal{F}, P)$. The σ-fields $\mathcal{G}_{nk}$ and $\mathcal{G}$ are defined as in Theorem 6.1.

For a square integrable array $(X_{nk})_{1 \leq k \leq k_n, n \in \mathbb{N}}$ we introduce the condition

(M$_2$) $$E\left(\max_{1\le k\le k_n} X_{nk}^2\right) \to 0 \quad \text{as } n \to \infty$$

whereas the conditions

(M$_1$) $$E\left(\max_{1\le k\le k_n} |X_{nk}|\right) \to 0 \quad \text{as } n \to \infty$$

and

(CLB$_1$) $$\sum_{k=1}^{k_n} E\left(|X_{nk}|\,1_{\{|X_{nk}|\ge\varepsilon\}}|\mathcal{F}_{n,k-1}\right) \to 0 \quad \text{in probability as } n \to \infty$$

for every $\varepsilon > 0$

can be imposed on any array $(X_{nk})_{1\le k\le k_n,\,n\in\mathbb{N}}$ of integrable random variables. *Raikov's condition*

(R) $$\sum_{k=1}^{k_n} X_{nk}^2 \to \eta^2 \quad \text{in probability as } n \to \infty \text{ for some}$$

$\mathcal{G}$-measurable real random variable $\eta \ge 0$,

which may replace condition (N), and, for any $a > 0$, the conditions

(T$_a$) $$\sum_{k=1}^{k_n} X_{nk}1_{\{|X_{nk}|>a\}} + E\left(X_{nk}1_{\{|X_{nk}|\le a\}}|\mathcal{F}_{n,k-1}\right) \to 0$$

in probability as $n \to \infty$,

(TM$_a$) $$\max_{1\le k\le k_n} \left|X_{nk}1_{\{|X_{nk}|\le a\}} - E\left(X_{nk}1_{\{|X_{nk}|\le a\}}|\mathcal{F}_{n,k-1}\right)\right| \to 0$$

in probability as $n \to \infty$,

and

(TR$_a$) $$\sum_{k=1}^{k_n} \left[X_{nk}1_{\{|X_{nk}|\le a\}} - E\left(X_{nk}1_{\{|X_{nk}|\le a\}}|\mathcal{F}_{n,k-1}\right)\right]^2 \to \eta^2$$

in probability as $n \to \infty$ for some $\mathcal{G}$-measurable real random variable $\eta \ge 0$

are meaningful without any integrability assumption on $(X_{nk})_{1\le k\le k_n,\,n\in\mathbb{N}}$.

We will first disclose the relationship between these conditions without reference to the central limit theorem. As a technical tool, we need the following partial converse of Lemma 6.5.

Lemma 6.13 *Let* $(X_{nk})_{1 \leq k \leq k_n, n \in \mathbb{N}}$ *be an array of nonnegative integrable random variables adapted to the array* $(\mathcal{F}_{nk})_{0 \leq k \leq k_n, n \in \mathbb{N}}$ *of* σ-*fields. Assume that* $\{\max_{1 \leq k \leq k_n} X_{nk} : n \in \mathbb{N}\}$ *is uniformly integrable. Then*

$$\sum_{k=1}^{k_n} X_{nk} \to 0 \quad \text{in probability as } n \to \infty$$

implies

$$\sum_{k=1}^{k_n} E\left(X_{nk} | \mathcal{F}_{n,k-1}\right) \to 0 \quad \text{in probability as } n \to \infty.$$

Proof For every $n \in \mathbb{N}$ the process $\left(\sum_{k=1}^{j} E\left(X_{nk} | \mathcal{F}_{n,k-1}\right)\right)_{0 \leq j \leq k_n}$ is the compensator of the nonnegative submartingale $\left(\sum_{k=1}^{j} X_{nk}\right)_{0 \leq j \leq k_n}$ so that, for every $\varepsilon, \delta > 0$ by Lenglart's inequality in Lemma A.8 (b)

$$P\left(\sum_{k=1}^{k_n} E\left(X_{nk} | \mathcal{F}_{n,k-1}\right) \geq \varepsilon\right) \leq \frac{1}{\varepsilon}\left(\delta + E\left(\max_{1 \leq k \leq k_n} X_{nk}\right)\right) + P\left(\sum_{k=1}^{k_n} X_{nk} > \delta\right).$$

Consequently,

$$\limsup_{n \to \infty} P\left(\sum_{k=1}^{k_n} E\left(X_{nk} | \mathcal{F}_{n,k-1}\right) > \varepsilon\right) \leq \frac{\delta}{\varepsilon} + \frac{1}{\varepsilon} \limsup_{n \to \infty} E\left(\max_{1 \leq k \leq k_n} X_{nk}\right).$$

Letting δ tend to zero and since $0 \leq \max_{1 \leq k \leq k_n} X_{nk} \leq \sum_{k=1}^{k_n} X_{nk} \to 0$ in probability, and hence $E\left(\max_{1 \leq k \leq k_n} X_{nk}\right) \to 0$ using uniform integrability, the assertion follows. □

As a second technical tool, we need the following lemma.

Lemma 6.14 *Let* $(X_{nk})_{1 \leq k \leq k_n, n \in \mathbb{N}}$ *be an array of random variables with* $\sum_{k=1}^{k_n} E\left(X_{nk}^2\right) \leq C < \infty$ *for some constant* C *and all* $n \in \mathbb{N}$ *and with*

(LB) $$\sum_{k=1}^{k_n} E\left(X_{nk}^2 \mathbf{1}_{\{|X_{nk}| \geq \varepsilon\}}\right) \to 0 \quad \text{as } n \to \infty \text{ for every } \varepsilon > 0.$$

Then for every array $(\mathcal{F}_{nk})_{0 \le k \le k_n, n \in \mathbb{N}}$ *of* σ*-fields*

$$\sum_{k=1}^{k_n} E\left(X_{nk}^2 | \mathcal{F}_{n,k-1}\right) - \sum_{k=1}^{k_n} X_{nk}^2 \to 0 \quad in \ \mathcal{L}^1(P) \ as \ n \to \infty.$$

Proof For every $\varepsilon > 0$ and $n \in \mathbb{N}$ we have

$$\left| \sum_{k=1}^{k_n} E\left(X_{nk}^2 | \mathcal{F}_{n,k-1}\right) - \sum_{k=1}^{k_n} X_{nk}^2 \right|$$

$$\le \sum_{k=1}^{k_n} E\left(X_{nk}^2 1_{\{|X_{nk}|>\varepsilon\}} | \mathcal{F}_{n,k-1}\right) + \sum_{k=1}^{k_n} X_{nk}^2 1_{\{|X_{nk}|>\varepsilon\}}$$

$$+ \left| \sum_{k=1}^{k_n} \left[X_{nk}^2 1_{\{|X_{nk}|\le\varepsilon\}} - E\left(X_{nk}^2 1_{\{|X_{nk}|\le\varepsilon\}} | \mathcal{F}_{n,k-1}\right) \right] \right|$$

$$= I_n + II_n + III_n,$$

say. Assumption (LB) implies $I_n \to 0$ and $II_n \to 0$ in $\mathcal{L}^1(P)$ as $n \to \infty$, whereas for III_n we have

$$E\left(III_n^2\right) = \sum_{k=1}^{k_n} E\left(\left[X_{nk}^2 1_{\{|X_{nk}|\le\varepsilon\}} - E\left(X_{nk}^2 1_{\{|X_{nk}|\le\varepsilon\}} | \mathcal{F}_{n,k-1}\right) \right]^2\right)$$

$$\le \sum_{k=1}^{k_n} E\left(X_{nk}^4 1_{\{|X_{nk}|\le\varepsilon\}}\right) \le \varepsilon^2 \sum_{k=1}^{k_n} E\left(X_{nk}^2\right) \le \varepsilon^2 C$$

by assumption so that $E(III_n) \le \varepsilon C^{1/2}$. Because $\varepsilon > 0$ is arbitrary, this clearly implies the assertion of the lemma. $\qquad\qquad\square$

Now we are prepared to analyze the relationship between the conditions (N) and (CLB) and the additional conditions formulated above.

Proposition 6.15 *Let* $(X_{nk})_{1 \le k \le k_n, n \in \mathbb{N}}$ *be an array of square integrable random variables adapted to the array* $(\mathcal{F}_{nk})_{0 \le k \le k_n, n \in \mathbb{N}}$ *of* σ*-fields. Then* (M$_2$) *implies* (M$_1$) *and* (CLB).

Proof Clearly, (M$_2$) implies (M$_1$). For the proof of (CLB) we note that for all $\varepsilon, \delta > 0$ and $n \in \mathbb{N}$ we have

$$P\left(\sum_{k=1}^{k_n} X_{nk}^2 1_{\{|X_{nk}|\ge\varepsilon\}} \ge \delta\right) \le P\left(\max_{1\le k\le k_n} |X_{nk}| \ge \varepsilon\right).$$

Consequently, (M_2) implies

$$\sum_{k=1}^{k_n} X_{nk}^2 1_{\{|X_{nk}|\geq\varepsilon\}} \to 0 \quad \text{in probability as } n \to \infty$$

for every $\varepsilon > 0$ as well as uniform integrability of $\{\max_{1\leq k\leq k_n} X_{nk}^2 : n \in \mathbb{N}\}$, and (CLB) follows by an application of Lemma 6.13. $\square$

Proposition 6.16 *Let $(X_{nk})_{1\leq k\leq k_n, n\in\mathbb{N}}$ be an array of square integrable random variables, and let $(\mathcal{F}_{nk})_{0\leq k\leq k_n, n\in\mathbb{N}}$ be an array of σ-fields. If (CLB) is satisfied and $\left(\sum_{k=1}^{k_n} E\left(X_{nk}^2|\mathcal{F}_{n,k-1}\right)\right)_{n\in\mathbb{N}}$ is bounded in probability, then*

$$\sum_{k=1}^{k_n} E\left(X_{nk}^2|\mathcal{F}_{n,k-1}\right) - \sum_{k=1}^{k_n} X_{nk}^2 \to 0 \quad \text{in probability as } n \to \infty.$$

In particular, (CLB) and (N) imply (R).

Proof For $0 < c < \infty$ and $n \in \mathbb{N}$ we define the stopping time

$$\tau_n(c) = \max\left\{k \in \{0, 1, \ldots, k_n\} : \sum_{j=1}^{k} E\left(X_{nj}^2|\mathcal{F}_{n,j-1}\right) \leq c\right\}$$

w.r.t. the filtration $(\mathcal{F}_{nk})_{0\leq k\leq k_n}$ and introduce the random variables

$$X_{nk}(c) := X_{nk} 1_{\{k\leq\tau_n(c)\}}, \ 1 \leq k \leq k_n, \ n \in \mathbb{N}.$$

Then for all $0 < c < \infty$ and $n \in \mathbb{N}$

$$\sum_{k=1}^{k_n} E\left(X_{nk}^2|\mathcal{F}_{n,k-1}\right) - \sum_{k=1}^{k_n} X_{nk}^2 = \sum_{k=1}^{k_n} E\left(X_{nk}^2|\mathcal{F}_{n,k-1}\right) - \sum_{k=1}^{k_n} E\left(X_{nk}^2(c)|\mathcal{F}_{n,k-1}\right)$$

$$+ \sum_{k=1}^{k_n} E\left(X_{nk}^2(c)|\mathcal{F}_{n,k-1}\right) - \sum_{k=1}^{k_n} X_{nk}^2(c) + \sum_{k=1}^{k_n} X_{nk}^2(c) - \sum_{k=1}^{k_n} X_{nk}^2$$

$$= I_n(c) + II_n(c) + III_n(c) ,$$

say. Because

$$I_n(c) = \sum_{k=1}^{k_n} E\left(X_{nk}^2|\mathcal{F}_{n,k-1}\right) - \sum_{k=1}^{\tau_n(c)} E\left(X_{nk}^2|\mathcal{F}_{n,k-1}\right)$$

we have

$$P\left(|I_n(c)| \geq \varepsilon\right) \leq P\left(\tau_n(c) < k_n\right) \leq P\left(\sum_{k=1}^{k_n} E\left(X_{nk}^2 | \mathcal{F}_{n,k-1}\right) > c\right)$$

for every $\varepsilon > 0$ and, similarly,

$$P\left(|III_n(c)| \geq \varepsilon\right) \leq P\left(\tau_n(c) < k_n\right) \leq P\left(\sum_{k=1}^{k_n} E\left(X_{nk}^2 | \mathcal{F}_{n,k-1}\right) > c\right).$$

By definition of $\tau_n(c)$ we have

$$\sum_{k=1}^{k_n} E\left(X_{nk}^2(c) | \mathcal{F}_{n,k-1}\right) = \sum_{k=1}^{\tau_n(c)} E\left(X_{nk}^2 | \mathcal{F}_{n,k-1}\right) \leq c$$

for all $n \in \mathbb{N}$, so that $\sum_{k=1}^{k_n} E\left(X_{nk}^2(c)\right) \leq c$. Moreover, from $|X_{nk}(c)| \leq |X_{nk}|$ we see that (CLB) implies

$$\sum_{k=1}^{k_n} E\left(X_{nk}^2(c) 1_{\{|X_{nk}(c)| \geq \varepsilon\}} | \mathcal{F}_{n,k-1}\right) \to 0 \quad \text{in probability as } n \to \infty$$

for all $\varepsilon > 0$, so that, by dominated convergence,

$$\sum_{k=1}^{k_n} E\left(X_{nk}^2(c) 1_{\{|X_{nk}(c)| \geq \varepsilon\}}\right) \to 0 \quad \text{as } n \to \infty.$$

Therefore, Lemma 6.14 yields

$$II_n(c) = \sum_{k=1}^{k_n} E\left(X_{nk}^2(c) | \mathcal{F}_{n,k-1}\right) - \sum_{k=1}^{k_n} X_{nk}^2(c) \to 0 \quad \text{in } \mathcal{L}^1(P) \text{ as } n \to \infty.$$

Now, for every $\varepsilon > 0$ and $n \in \mathbb{N}$ we have

$$P\left(\left|\sum_{k=1}^{k_n} E\left(X_{nk}^2 | \mathcal{F}_{n,k-1}\right) - \sum_{k=1}^{k_n} X_{nk}^2\right| \geq 3\varepsilon\right)$$
$$\leq P\left(|I_n(c)| \geq \varepsilon\right) + P\left(|II_n(c)| \geq \varepsilon\right) + P\left(|III_n(c)| \geq \varepsilon\right)$$
$$\leq 2P\left(\sum_{k=1}^{k_n} E\left(X_{nk}^2 | \mathcal{F}_{n,k-1}\right) > c\right) + P\left(|II_n(c)| \geq \varepsilon\right).$$

Here, $P\left(|II_n\left(c\right)| \geq \varepsilon\right) \to 0$ as $n \to \infty$ because $II_n\left(c\right) \to 0$ in $\mathcal{L}^1\left(P\right)$ as $n \to \infty$, and the sequence $\sum_{k=1}^{k_n} E\left(X_{nk}^2 | \mathcal{F}_{n,k-1}\right)$, $n \in \mathbb{N}$, is bounded in probability by assumption. This proves

$$\sum_{k=1}^{k_n} E\left(X_{nk}^2 | \mathcal{F}_{n,k-1}\right) - \sum_{k=1}^{k_n} X_{nk}^2 \to 0 \quad \text{in probability as } n \to \infty. \qquad \square$$

Proposition 6.17 *Let* $(X_{nk})_{1 \leq k \leq k_n, n \in \mathbb{N}}$ *be an array of square integrable random variables, and let* $(\mathcal{F}_{nk})_{0 \leq k \leq k_n, n \in \mathbb{N}}$ *be an array of σ-fields. Then* (CLB) *implies* (CLB$_1$).

Proof For every $\varepsilon > 0$ and $n \in \mathbb{N}$ we have

$$\sum_{k=1}^{k_n} E\left(|X_{nk}| \, 1_{\{|X_{nk}| \geq \varepsilon\}} | \mathcal{F}_{n,k-1}\right) \leq \frac{1}{\varepsilon} \sum_{k=1}^{k_n} E\left(X_{nk}^2 1_{\{|X_{nk}| \geq \varepsilon\}} | \mathcal{F}_{n,k-1}\right),$$

which proves the proposition. $\qquad\qquad\square$

Proposition 6.18 *Let* $(X_{nk})_{1 \leq k \leq k_n, n \in \mathbb{N}}$ *be an array of integrable random variables adapted to the array* $(\mathcal{F}_{nk})_{0 \leq k \leq k_n, n \in \mathbb{N}}$ *of σ-fields. Then* (M$_1$) *implies* (CLB$_1$).

Proof For all $\varepsilon, \delta > 0$ and $n \in \mathbb{N}$ we have

$$P\left(\sum_{k=1}^{k_n} |X_{nk}| \, 1_{\{|X_{nk}| \geq \varepsilon\}} \geq \delta\right) \leq P\left(\max_{1 \leq k \leq k_n} |X_{nk}| \geq \varepsilon\right).$$

Consequently, (M$_1$) implies

$$\sum_{k=1}^{k_n} |X_{nk}| \, 1_{\{|X_{nk}| \geq \varepsilon\}} \to 0 \quad \text{in probability as } n \to \infty$$

for every $\varepsilon > 0$ as well as uniform integrability of $\{\max_{1 \leq k \leq k_n} |X_{nk}| : n \in \mathbb{N}\}$, and (CLB$_1$) follows by an application of Lemma 6.13. $\qquad\square$

Proposition 6.19 *Let* $(X_{nk})_{1 \leq k \leq k_n, n \in \mathbb{N}}$ *be a martingale difference array w.r.t. an array* $(\mathcal{F}_{nk})_{0 \leq k \leq k_n, n \in \mathbb{N}}$ *of σ-fields. Then* (CLB$_1$) *and* (R) *imply* (T$_a$), (TM$_a$) *and* (TR$_a$) *for every* $a > 0$.

Proof Fix $a > 0$. Because $(X_{nk})_{1 \leq k \leq k_n}$ is a martingale difference sequence w.r.t. the filtration $(\mathcal{F}_{nk})_{0 \leq k \leq k_n}$, we have

$$E\left(X_{nk} 1_{\{|X_{nk}| \leq a\}} | \mathcal{F}_{n,k-1}\right) = -E\left(X_{nk} 1_{\{|X_{nk}| > a\}} | \mathcal{F}_{n,k-1}\right)$$

for all $n \in \mathbb{N}$ and $1 \le k \le k_n$. This fact will be crucial several times in the sequel. For the proof of (T_a) we use it to obtain for all $n \in \mathbb{N}$

$$\sum_{k=1}^{k_n} \left| E\left(X_{nk}1_{\{|X_{nk}|\le a\}}|\mathcal{F}_{n,k-1}\right)\right| = \sum_{k=1}^{k_n} \left| E\left(X_{nk}1_{\{|X_{nk}|>a\}}|\mathcal{F}_{n,k-1}\right)\right|$$

$$\le \sum_{k=1}^{k_n} E\left(|X_{nk}|\,1_{\{|X_{nk}|>a\}}|\mathcal{F}_{n,k-1}\right)$$

so that by condition (CLB_1)

$$\sum_{k=1}^{k_n} \left| E\left(X_{nk}1_{\{|X_{nk}|\le a\}}|\mathcal{F}_{n,k-1}\right)\right| \to 0 \quad \text{in probability as } n \to \infty.$$

Moreover, according to Lemma 6.5 condition (CLB_1) implies

$$\sum_{k=1}^{k_n} |X_{nk}|\,1_{\{|X_{nk}|>a\}} \to 0 \quad \text{in probability as } n \to \infty,$$

which completes the proof of (T_a).

To verify (TM_a) we use Proposition 6.6 to obtain $\max_{1\le k\le k_n} |X_{nk}| \to 0$ in probability as $n \to \infty$ from (CLB_1), and the inequality

$$\max_{1\le k\le k_n} \left|X_{nk}1_{\{|X_{nk}|\le a\}} - E\left(X_{nk}1_{\{|X_{nk}|\le a\}}|\mathcal{F}_{n,k-1}\right)\right|$$

$$\le \max_{1\le k\le k_n} |X_{nk}| + \sum_{k=1}^{k_n} E\left(|X_{nk}|\,1_{\{|X_{nk}|>a\}}|\mathcal{F}_{n,k-1}\right)$$

completes the proof by another application of condition (CLB_1).

It remains to verify (TR_a). Note that for all $n \in \mathbb{N}$

$$\left|\sum_{k=1}^{k_n} \left[X_{nk}1_{\{|X_{nk}|\le a\}} - E\left(X_{nk}1_{\{|X_{nk}|\le a\}}|\mathcal{F}_{n,k-1}\right)\right]^2 - \sum_{k=1}^{k_n} X_{nk}^2\right|$$

$$\le \sum_{k=1}^{k_n} X_{nk}^2 1_{\{|X_{nk}|>a\}}$$

$$+ 2\sum_{k=1}^{k_n} |X_{nk}|\,1_{\{|X_{nk}|\le a\}} \left| E\left(X_{nk}1_{\{|X_{nk}|\le a\}}|\mathcal{F}_{n,k-1}\right)\right|$$

$$+ \sum_{k=1}^{k_n} E\left(X_{nk} 1_{\{|X_{nk}| \le a\}} | \mathcal{F}_{n,k-1}\right)^2$$

$$\le \sum_{k=1}^{k_n} X_{nk}^2 1_{\{|X_{nk}| > a\}} + 3a \sum_{k=1}^{k_n} E\left(|X_{nk}| 1_{\{|X_{nk}| > a\}} | \mathcal{F}_{n,k-1}\right).$$

Now

$$\sum_{k=1}^{k_n} X_{nk}^2 1_{\{|X_{nk}| > a\}} \to 0 \quad \text{in probability as } n \to \infty$$

follows from $\max_{1 \le k \le k_n} |X_{nk}| \to 0$ in probability, which when combined with
(CLB$_1$) gives

$$\sum_{k=1}^{k_n} \left[X_{nk} 1_{\{|X_{nk}| \le a\}} - E\left(X_{nk} 1_{\{|X_{nk}| \le a\}} | \mathcal{F}_{n,k-1}\right)\right]^2 - \sum_{k=1}^{k_n} X_{nk}^2 \to 0$$

in probability as $n \to \infty$. Now (TR$_a$) follows from (R).   □

As a consequence of Propositions 6.15–6.19 we see that for an array
$(X_{nk})_{1 \le k \le k_n, n \in \mathbb{N}}$ of random variables adapted to an array $(\mathcal{F}_{nk})_{0 \le k \le k_n, n \in \mathbb{N}}$ of σ-fields
the implications in the following display are true under appropriate moment assump-
tions and if $(X_{nk})_{1 \le k \le k_n, n \in \mathbb{N}}$ is a martingale difference array w.r.t. $(\mathcal{F}_{nk})_{0 \le k \le k_n, n \in \mathbb{N}}$
for the implication $\overset{(*)}{\Rightarrow}$:

6.20 *Conditions in the martingale central limit theorem*:

$$\begin{array}{ccc}
\text{(M}_2\text{) and (N)} & \Rightarrow & \text{(M}_1\text{) and (R)} \\
\Downarrow & & \Downarrow \\
\text{(CLB) and (N)} \Rightarrow \text{(CLB}_1\text{) and (R)} & \overset{(*)}{\Rightarrow} & \text{(T}_a\text{), (TM}_a\text{) and (TR}_a\text{)}
\end{array}$$

The conditions in the left column require square integrable random variables,
in the middle integrability is sufficient, and on the right-hand side no moments are
needed at all. The role of these conditions as sufficient conditions in a stable central
limit theorem is disclosed by the following proposition which shows that for any
array $(X_{nk})_{1 \le k \le k_n, n \in \mathbb{N}}$ of random variables adapted to the array $(\mathcal{F}_{nk})_{0 \le k \le k_n, n \in \mathbb{N}}$ of
σ-fields which satisfies (T$_a$), (TM$_a$) and (TR$_a$) for *some* $a > 0$ there exists a bounded
martingale difference array which satisfies the strongest set of conditions (M$_2$) and
(N) and has asymptotically equivalent row sums.

Proposition 6.21 *Let* $(X_{nk})_{1 \le k \le k_n, n \in \mathbb{N}}$ *be an array of random variables adapted
to an array* $(\mathcal{F}_{nk})_{0 \le k \le k_n, n \in \mathbb{N}}$ *of σ-fields. Assume that there exists some $a > 0$ for
which the conditions* (T$_a$), (TM$_a$) *and* (TR$_a$) *are satisfied. Then for the (bounded)
martingale difference array*

$$X_{nk}(a) := X_{nk}1_{\{|X_{nk}|\leq a\}} - E\left(X_{nk}1_{\{|X_{nk}|\leq a\}}|\mathcal{F}_{n,k-1}\right), \quad 1 \leq k \leq k_n, \; n \in \mathbb{N},$$

w.r.t. $(\mathcal{F}_{nk})_{0\leq k\leq k_n, n\in\mathbb{N}}$ the conditions (M_2) and (N) are satisfied and

$$\sum_{k=1}^{k_n} X_{nk} - \sum_{k=1}^{k_n} X_{nk}(a) \to 0 \quad \text{in probability as } n \to \infty.$$

Proof Note that condition (TM_a) is tantamount to $\max_{1\leq k\leq k_n} |X_{nk}(a)| \to 0$ in probability as $n \to \infty$ and hence to $\max_{1\leq k\leq k_n} X_{nk}^2(a) \to 0$ in probability. Because $|X_{nk}(a)| \leq 2a$ for all $n \in \mathbb{N}$ and $1 \leq k \leq k_n$ we obtain $E\left(\max_{1\leq k\leq k_n} X_{nk}^2(a)\right) \to 0$ as $n \to \infty$ by dominated convergence, which is condition (M_2) for the array $(X_{nk}(a))_{1\leq k\leq k_n, n\in\mathbb{N}}$. By definition, we have

$$\sum_{k=1}^{k_n} X_{nk} - \sum_{k=1}^{k_n} X_{nk}(a) = \sum_{k=1}^{k_n} X_{nk}1_{\{|X_{nk}|>a\}} + E\left(X_{nk}1_{\{|X_{nk}|\leq a\}}|\mathcal{F}_{n,k-1}\right),$$

which converges to zero in probability as $n \to \infty$ by condition (T_a). Therefore, it remains to show that the array $(X_{nk}(a))_{1\leq k\leq k_n, n\in\mathbb{N}}$ satisfies condition (N). For this, we define the stopping time

$$\tau_n(c) = \min\left\{k \in \{1,\ldots,k_n\} : \sum_{i=1}^{k} X_{ni}^2(a) > c\right\} \wedge k_n$$

with $\min\emptyset := \infty$ for all $n \in \mathbb{N}$ and $0 < c < \infty$ and set

$$Y_{nk}(c) := X_{nk}(a)\,1_{\{k\leq\tau_n(c)\}}, \quad 1 \leq k \leq k_n, \; n \in \mathbb{N}.$$

Then for all $n \in \mathbb{N}$

$$\sum_{k=1}^{k_n} E\left(X_{nk}^2(a)|\mathcal{F}_{n,k-1}\right) - \sum_{k=1}^{k_n} X_{nk}^2(a)$$

$$= \sum_{k=1}^{k_n} E\left(X_{nk}^2(a)|\mathcal{F}_{n,k-1}\right) - \sum_{k=1}^{k_n} E\left(Y_{nk}^2(c)|\mathcal{F}_{n,k-1}\right)$$

$$+ \sum_{k=1}^{k_n} E\left(Y_{nk}^2(c)|\mathcal{F}_{n,k-1}\right) - \sum_{k=1}^{k_n} Y_{nk}^2(c)$$

$$+ \sum_{k=1}^{k_n} Y_{nk}^2(c) - \sum_{k=1}^{k_n} X_{nk}^2(a) = I_n(c) + II_n(c) + III_n(c),$$

say. Because

$$I_n(c) = \sum_{k=1}^{k_n} E\left(X_{nk}^2(a)\,|\mathcal{F}_{n,k-1}\right) - \sum_{k=1}^{\tau_n(c)} E\left(X_{nk}^2(a)\,|\mathcal{F}_{n,k-1}\right)$$

we have for all $n \in \mathbb{N}$ and $\varepsilon > 0$

$$P\left(|I_n(c)| \geq \varepsilon\right) \leq P\left(\tau_n(c) < k_n\right) \leq P\left(\sum_{k=1}^{k_n} X_{nk}^2(a) > c\right).$$

Similarly, for all $n \in \mathbb{N}$ and $\varepsilon > 0$

$$P\left(|III_n(c)| \geq \varepsilon\right) \leq P\left(\tau_n(c) < k_n\right) \leq P\left(\sum_{k=1}^{k_n} X_{nk}^2(a) > c\right).$$

To obtain a bound for $II_n(c)$, note that for all $\varepsilon, \delta > 0$ and $n \in \mathbb{N}$ we have

$$P\left(\sum_{k=1}^{k_n} Y_{nk}^2(c)\,1_{\{|Y_{nk}(c)| \geq \varepsilon\}} \geq \delta\right) \leq P\left(\max_{1 \leq k \leq k_n} |Y_{nk}(c)| \geq \varepsilon\right)$$

so that

$$\sum_{k=1}^{k_n} Y_{nk}^2(c)\,1_{\{|Y_{nk}(c)| \geq \varepsilon\}} \to 0 \quad \text{in probability as } n \to \infty$$

for every $\varepsilon > 0$ because

$$\max_{1 \leq k \leq k_n} |Y_{nk}(c)| \leq \max_{1 \leq k \leq k_n} |X_{nk}(a)| \to 0 \quad \text{in probability as } n \to \infty.$$

Moreover, by definition of $\tau_n(c)$, for all $n \in \mathbb{N}$

$$\sum_{k=1}^{k_n} Y_{nk}^2(c) = \sum_{k=1}^{\tau_n(c)} X_{nk}^2(a) \leq c + \max_{1 \leq k \leq k_n} X_{nk}^2(a) \leq c + 4a^2$$

so that $\sum_{k=1}^{k_n} E\left(Y_{nk}^2(c)\right) \leq c + 4a^2$ and, by dominated convergence,

$$\sum_{k=1}^{k_n} E\left(Y_{nk}^2(c)\,1_{\{|Y_{nk}(c)| \geq \varepsilon\}}\right) \to 0 \quad \text{as } n \to \infty$$

for all $\varepsilon > 0$. Hence by Lemma 6.14 we see that $II_n(c) \to 0$ in $\mathcal{L}^1(P)$ as $n \to \infty$. Now, for every $\varepsilon > 0$ and $n \in \mathbb{N}$ we get

$$P\left(\left|\sum_{k=1}^{k_n} E\left(X_{nk}^2(a) \,|\, \mathcal{F}_{n,k-1}\right) - \sum_{k=1}^{k_n} X_{nk}^2(a)\right| \geq 3\varepsilon\right)$$

$$\leq P\left(|I_n(c)| \geq \varepsilon\right) + P\left(|II_n(c)| \geq \varepsilon\right) + P\left(|III_n(c)| \geq \varepsilon\right)$$

$$\leq 2P\left(\sum_{k=1}^{k_n} X_{nk}^2(a) > c\right) + P\left(|II_n(c)| \geq \varepsilon\right).$$

Here, $P\left(|II_n(c)| \geq \varepsilon\right) \to 0$ as $n \to \infty$, and the sequence $\sum_{k=1}^{k_n} X_{nk}^2(a)$, $n \in \mathbb{N}$, is bounded in probability because condition (TR$_a$) is tantamount to

$$\sum_{k=1}^{k_n} X_{nk}^2(a) \to \eta^2 \quad \text{in probability as } n \to \infty.$$

This proves

$$\sum_{k=1}^{k_n} E\left(X_{nk}^2(a) \,|\, \mathcal{F}_{n,k-1}\right) - \sum_{k=1}^{k_n} X_{nk}^2(a) \to 0 \quad \text{in probability as } n \to \infty,$$

and another application of (TR$_a$) gives

$$\sum_{k=1}^{k_n} E\left(X_{nk}^2(a) \,|\, \mathcal{F}_{n,k-1}\right) \to \eta^2 \quad \text{in probability as } n \to \infty,$$

which is condition (N) for the array $(X_{nk}(a))_{1 \leq k \leq k_n, n \in \mathbb{N}}$ so that the proof is complete. $\qquad\square$

Corollary 6.22 *Let* $(X_{nk})_{1 \leq k \leq k_n, n \in \mathbb{N}}$ *be an array of random variables adapted to an array* $(\mathcal{F}_{nk})_{0 \leq k \leq k_n, n \in \mathbb{N}}$ *of σ-fields. Assume that there exists some $a > 0$ for which the conditions* (T$_a$), (TM$_a$) *and* (TR$_a$) *are satisfied. Then*

$$\sum_{k=1}^{k_n} X_{nk} \to \eta N \quad \mathcal{G}\text{-stably as } n \to \infty,$$

where $P^N = N(0, 1)$ *and N is independent of $\mathcal{G}$.*

Proof Let the random variables $X_{nk}(a)$, $1 \leq k \leq k_n$, $n \in \mathbb{N}$, be defined as in Proposition 6.21. Then according to Proposition 6.21, $(X_{nk}(a))_{1 \leq k \leq k_n, n \in \mathbb{N}}$ is a bounded martingale difference array w.r.t. $(\mathcal{F}_{nk})_{0 \leq k \leq k_n, n \in \mathbb{N}}$ which satisfies (M$_2$) and (N).

According to Proposition 6.15 condition (CLB) is satisfied as well. Therefore by Theorem 6.1

$$\sum_{k=1}^{k_n} X_{nk}(a) \to \eta N \quad \mathcal{G}\text{-stably as } n \to \infty.$$

Since by Proposition 6.21

$$\sum_{k=1}^{k_n} X_{nk} - \sum_{k=1}^{k_n} X_{nk}(a) \to 0 \quad \text{in probability as } n \to \infty,$$

the proof is completed by an application of part (a) of Theorem 3.18.   □

The self-evident consequence of 6.20 and of Corollary 6.22 is the fact that for a martingale difference array $(X_{nk})_{1 \le k \le k_n, n \in \mathbb{N}}$ w.r.t. an array $(\mathcal{F}_{nk})_{0 \le k \le k_n, n \in \mathbb{N}}$ of σ-fields any set of conditions occurring in 6.20 implies $\mathcal{G}$-stable convergence of the row sums to ηN. In the sense made precise by 6.20 and Proposition 6.21, as sufficient conditions in the $\mathcal{G}$-stable martingale central limit theorem, all these conditions are tantamount to each other, though not mathematically equivalent.

A version of Corollary 6.22 for martingale difference arrays under the conditions (M_1) and (R) is contained in [58].

Exercise 6.4 ([58]) Let $(X_{nk})_{1 \le k \le k_n, n \in \mathbb{N}}$ be a martingale difference array w.r.t. $(\mathcal{F}_{nk})_{0 \le k \le k_n, n \in \mathbb{N}}$. Under conditions (M_1) and (R) we have

$$\sum_{k=1}^{k_n} X_{nk} \to \eta N \quad \mathcal{G}\text{-stably}$$

where $P^N = N(0, 1)$ and N is independent of $\mathcal{G}$ (see Corollary 6.22 and 6.20). Show that, in general, this assertion is not true if (M_1) is replaced by the weaker condition $\max_{1 \le k \le k_n} |X_{nk}| \to 0$ in probability. To this end, consider an array $(X_{nk})_{1 \le k \le k_n, n \in \mathbb{N}}$ with $X_{n1}, \ldots, X_{nn}$ being independent and identically distributed, $P\left(X_{n1} = \frac{1}{n}\right) = \left(1 - \frac{1}{n}\right)^{1/n}$ and $P(X_{n1} = x_n) = 1 - \left(1 - \frac{1}{n}\right)^{1/n}$, where $x_n < 0$ is such that $EX_{n1} = 0$. Furthermore, let $\mathcal{F}_{nk} = \sigma\left(X_{nj}, 1 \le j \le k\right)$ with $\mathcal{F}_{n0} = \{\emptyset, \Omega\}$.

6.4 Martingales

Let $(\Omega, \mathcal{F}, P)$ be a probability space and $\mathbb{F} = (\mathcal{F}_k)_{k \ge 0}$ a filtration, that is, a non-decreasing sequence of sub-σ-fields of $\mathcal{F}$. Set $\mathcal{F}_\infty := \sigma\left(\bigcup_{k=0}^\infty \mathcal{F}_k\right)$. A sequence $(X_k)_{k \ge 1}$ of random variables on $(\Omega, \mathcal{F}, P)$ is called adapted to $\mathbb{F}$ if X_k is measurable w.r.t. $\mathcal{F}_k$ for every $k \in \mathbb{N}$, and a sequence $(X_k)_{k \ge 1}$ of integrable random variables adapted to $\mathbb{F}$ is called a martingale difference sequence w.r.t. $\mathbb{F}$, if $E\left(X_k|\mathcal{F}_{k-1}\right) = 0$ for all $k \in \mathbb{N}$.

Let $(X_k)_{k\geq 1}$ be a martingale difference sequence w.r.t. the filtration $\mathbb{F}$, and let $(a_n)_{n\geq 1}$ be a sequence of positive real numbers. Then

$$X_{nk} := \frac{1}{a_n}X_k \quad \text{for } 1 \leq k \leq n \qquad \text{and} \qquad \mathcal{F}_{nk} := \mathcal{F}_k \quad \text{for } 0 \leq k \leq n,\ n \in \mathbb{N}$$

defines a martingale difference array $(X_{nk})_{1\leq k\leq n, n\in\mathbb{N}}$ w.r.t. $(\mathcal{F}_{nk})_{0\leq k\leq n, n\in\mathbb{N}}$, and the σ-fields are nested because $\mathcal{F}_{n+1,k} = \mathcal{F}_k = \mathcal{F}_{nk}$ for all $n \in \mathbb{N}$ and $0 \leq k \leq n$. Therefore, Theorem 6.1 and the sufficient conditions of Sect. 6.3 can be applied with $\mathcal{G} = \mathcal{F}_\infty$ and yield stable central limit theorems for the normalized partial sums $a_n^{-1}\sum_{k=1}^n X_k$ of $(X_k)_{k\geq 1}$ under appropriate moment conditions. For ease of reference we explicitly formulate here the two sets of sufficient conditions for martingale difference sequences that will be applied later on.

Theorem 6.23 *Let $(X_k)_{k\geq 1}$ be a martingale difference sequence w.r.t. the filtration $\mathbb{F}$, and let $(a_n)_{n\in\mathbb{N}}$ be a sequence of positive real numbers with $a_n \to \infty$. If*

(R_{a_n})
$$\frac{1}{a_n^2}\sum_{k=1}^n X_k^2 \to \eta^2 \quad \text{in probability as } n \to \infty$$

for some real random variable $\eta \geq 0$

and

(M_{1,a_n})
$$\frac{1}{a_n}E\left(\max_{1\leq k\leq n}|X_k|\right) \to 0 \quad \text{as } n \to \infty,$$

or if $(X_k)_{k\geq 1}$ is square integrable with

(N_{a_n})
$$\frac{1}{a_n^2}\sum_{k=1}^n E\left(X_k^2|\mathcal{F}_{k-1}\right) \to \eta^2 \quad \text{in probability as } n \to \infty$$

for some real random variable $\eta \geq 0$

and

(CLB_{a_n})
$$\frac{1}{a_n^2}\sum_{k=1}^n E\left(X_k^2 1_{\{|X_k|\geq\varepsilon a_n\}}|\mathcal{F}_{k-1}\right) \to 0 \quad \text{in probability as } n \to \infty$$

for all $\varepsilon > 0$,

then

$$\frac{1}{a_n} \sum_{k=1}^{n} X_k \to \eta N \quad \mathcal{F}_{\infty}\text{-}\textit{stably as } n \to \infty,$$

where $P^N = N(0, 1)$ and N is independent of $\mathcal{F}_{\infty}$.

Proof Proposition 6.20 and Corollary 6.22. Note that η^2 in conditions (N_{a_n}) and (R_{a_n}) is w.l.o.g. $\mathcal{F}_{\infty}$-measurable.   □

Condition (R_{a_n}) and slightly stronger conditions than (M_{1,a_n}) appear in Theorem 2 of [15] and Theorem 2 of [4].

Corollary 6.24   (Random norming) *Under the assumptions of Theorem 6.23 in case $P\left(\eta^2 > 0\right) > 0$ conditions (R_{a_n}) and (M_{1,a_n}) imply*

$$\left(\sum_{k=1}^{n} X_k^2\right)^{-1/2} \sum_{k=1}^{n} X_k \to N \quad \mathcal{F}_{\infty}\text{-}\textit{mixing under } P_{\{\eta^2 > 0\}} \textit{ as } n \to \infty,$$

and conditions (N_{a_n}) and (CLB_{a_n}) imply

$$\left(\sum_{k=1}^{n} E\left(X_k^2 | \mathcal{F}_{k-1}\right)\right)^{-1/2} \sum_{k=1}^{n} X_k \to N \quad \mathcal{F}_{\infty}\text{-}\textit{mixing under } P_{\{\eta^2 > 0\}} \textit{ as } n \to \infty,$$

where $P^N = N(0, 1)$ and N is independent of $\mathcal{F}_{\infty}$.

Proof Replace Theorem 6.1 by Theorem 6.23 and condition (N) by conditions (R_{a_n}) or (N_{a_n}) in the proof of Corollary 6.3.   □

An immediate consequence of the preceding theorem is the classical stable central limit theorem of Examples 3.13 (b) or 3.16.

Remark 6.25   (a) In Theorem 6.23 we do require explicitly that $a_n \to \infty$ as $n \to \infty$. However, if $P\left(E\left(X_k^2 | \mathcal{F}_{k-1}\right) > 0\right) > 0$ for some $k \in \mathbb{N}$, which means that not all X_k vanish almost surely, then (N_{a_n}) and (CLB_{a_n}) as well as (M_{1,a_n}) already imply $a_n \to \infty$ as $n \to \infty$. For martingales $X_0 + \sum_{k=1}^{n} X_k$ with $X_0 \neq 0$ the condition $a_n \to \infty$ assures the validity of Theorem 6.23.
(b) Just as in Remark 6.8, condition (CLB_{a_n}) is implied by its classical form

$$(LB_{a_n}) \qquad \frac{1}{a_n^2} \sum_{k=1}^{n} E\left(X_k^2 1_{\{|X_k| \geq \varepsilon a_n\}}\right) \to 0 \quad \text{as } n \to \infty \text{ for all } \varepsilon > 0$$

and by the *conditional Lyapunov condition of order* $p \in (2, \infty)$, which requires for some $p \in (2, \infty)$ that

$(\mathrm{CLY}_{a_n,p})$ $\dfrac{1}{a_n^p} \displaystyle\sum_{k=1}^{n} E\left(|X_k|^p \,|\mathcal{F}_{k-1}\right) \to 0$ in probability as $n \to \infty$.

Condition $(\mathrm{CLY}_{a_n,p})$ itself is implied by its classical form

$(\mathrm{LY}_{a_n,p})$ $\dfrac{1}{a_n^p} \displaystyle\sum_{k=1}^{n} E\left(|X_k|^p\right) \to 0$ as $n \to \infty$.

Corollary 6.26 (Stationary martingale differences) *Let $X = (X_n)_{n \in \mathbb{N}}$ be a stationary sequence of real random variables with σ-field $\mathcal{I}_X = X^{-1}\left(\mathcal{B}(\mathbb{R})^{\mathbb{N}}(S)\right)$ induced by invariant sets $(S : \mathbb{R}^{\mathbb{N}} \to \mathbb{R}^{\mathbb{N}}$ being the shift operator; see Chap. 5). If $X_1 \in \mathcal{L}^2(P)$ and if X is a martingale difference sequence w.r.t. $\mathbb{F} = (\mathcal{F}_k)_{k \geq 0}$, then*

$$\frac{1}{\sqrt{n}} \sum_{k=1}^{n} X_k \to E\left(X_1^2 | \mathcal{I}_X\right)^{1/2} N \quad \mathcal{F}_\infty\text{-stably as } n \to \infty,$$

where N is independent of $\mathcal{F}_\infty$ with $P^N = N(0,1)$. If X is also ergodic, that is, $P(\mathcal{I}_X) = \{0, 1\}$, then

$$\frac{1}{\sqrt{n}} \sum_{k=1}^{n} X_k \to E\left(X_1^2\right)^{1/2} N \quad \mathcal{F}_\infty\text{-mixing as } n \to \infty.$$

The distributional convergence in this result goes back to [8, 45].

Proof The ergodic theorem implies

$$\frac{1}{n} \sum_{k=1}^{n} X_k^2 \to E\left(X_1^2 | \mathcal{I}_X\right) \quad \text{a.s. and in } \mathcal{L}^1(P) \text{ as } n \to \infty$$

so that condition (R_{a_n}) is satisfied with $a_n = \sqrt{n}$ and $\eta = E\left(X_1^2 | \mathcal{I}_X\right)^{1/2}$. Since the X_k are identically distributed, the classical Lindeberg condition (LB_{a_n}) is also satisfied with $a_n = \sqrt{n}$ because for all $\varepsilon > 0$

$$\frac{1}{n} \sum_{k=1}^{n} E\left(X_k^2 \mathbf{1}_{\{|X_k| \geq \varepsilon n^{1/2}\}}\right) = E\left(X_1^2 \mathbf{1}_{\{|X_1| \geq \varepsilon n^{1/2}\}}\right) \to 0 \quad \text{as } n \to \infty,$$

which through the inequality, valid for all $\varepsilon > 0$ and $n \in \mathbb{N}$,

$$\left(\frac{1}{\sqrt{n}} \max_{1 \leq k \leq n} E\left(|X_k|\right)\right)^2 \leq \frac{1}{n} E\left(\max_{1 \leq k \leq n} X_k^2\right) \leq \varepsilon + \frac{1}{n} \sum_{k=1}^{n} E\left(X_1^2 \mathbf{1}_{\{|X_1| \geq \varepsilon n^{1/2}\}}\right)$$

implies (M_{1,a_n}). Therefore, Theorem 6.23 implies the first assertion. If $\mathcal{I}_X$ is trivial, then $E\left(X_1^2|\mathcal{I}_X\right) = E\left(X_1^2\right)$ almost surely, whence the second assertion. □

Let $X = (X_n)_{n\in\mathbb{N}}$ be an exchangeable sequence of real random variables on $(\Omega, \mathcal{F}, P)$, that is, $P^{(X_{\pi_1},\dots,X_{\pi_n})} = P^{(X_1,\dots,X_n)}$ for all permutations $(\pi_1, \dots, \pi_n)$ of $(1, \dots, n)$ and all $n \in \mathbb{N}$. Then $P^{(X_1,X_2,\dots,X_n,X_{n+1})} = P^{(X_2,X_3,\dots,X_{n+1},X_1)}$ so that $P^{(X_1,X_2,\dots,X_n)} = P^{(X_2,X_3,\dots,X_{n+1})}$ for all $n \in \mathbb{N}$, which shows the $(X_n)_{n\in\mathbb{N}}$ is stationary. Moreover, the σ-field induced in Ω by symmetric events is almost surely equal to the tail-σ-field $\mathcal{T}_X$ and almost surely equal to the invariant σ-field $\mathcal{I}_X$ of the stationary process X; see e.g. [52], Corollary 1.6.

Corollary 6.27 (Exchangeable processes) *If $X = (X_n)_{n\in\mathbb{N}}$ is exchangeable with $X_1 \in \mathcal{L}^2$, then*

$$\frac{1}{\sqrt{n}} \sum_{k=1}^{n} (X_k - E(X_1|\mathcal{T}_X)) \to \mathrm{Var}(X_1|\mathcal{T}_X)^{1/2} N \quad \mathcal{F}_\infty\text{-stably as } n \to \infty,$$

where N is independent of $\mathcal{F}_\infty = \sigma(X_k, k \in \mathbb{N})$ and $P^N = N(0,1)$.

In [16] this result was obtained with $\mathcal{T}_X$ instead of $\mathcal{F}_\infty$ under the assumptions $E(X_1|\mathcal{T}_X) = 0$ and $E\left(X_1^2|\mathcal{T}_X\right) \leq C$ for some finite constant C. The general result is stated in [3], p. 59.

Proof Exchangeability implies that the conditional distribution of X_n given $\mathcal{T}_X$ is independent of $n \in \mathbb{N}$. This yields $E(X_n|\mathcal{T}_X) = E(X_1|\mathcal{T}_X)$ almost surely for all $n \in \mathbb{N}$. The random variables $Y_n := X_n - E(X_1|\mathcal{T}_X), n \in \mathbb{N}$, form a martingale difference sequence w.r.t. the σ-fields $\mathcal{F}_n := \sigma(\mathcal{T}_X \cup \sigma(X_1, \dots, X_n)), n \geq 0$: Clearly, Y_n is $\mathcal{F}_n$-measurable for all $n \in \mathbb{N}$, and $E(Y_1|\mathcal{F}_0) = E(X_1 - E(X_1|\mathcal{T}_X)|\mathcal{T}_X) = 0$ almost surely. Moreover, for all $n \geq 1$, the σ-fields $\sigma(X_1, \dots, X_{n-1})$ and $\sigma(X_n)$ are conditionally independent given $\mathcal{T}_X$, and Theorem 7.3.1 in [17] implies for all $n \geq 2$ almost surely

$$E(Y_n|\mathcal{F}_{n-1}) = E(X_n|\mathcal{F}_{n-1}) - E(X_1|\mathcal{T}_X) = E(X_n|\mathcal{T}_X) - E(X_1|\mathcal{T}_X) = 0.$$

Furthermore, because $\mathcal{T}_X = \mathcal{I}_X$ almost surely, we have $E(X_1|\mathcal{T}_X) = E(X_1|\mathcal{I}_X)$ almost surely, from which it follows that $(Y_n)_{n\in\mathbb{N}}$ is a stationary process. Clearly, $X_1 \in \mathcal{L}^2(P)$ implies $Y_1 \in \mathcal{L}^2(P)$, and an application of Corollary 6.26 yields the assertion. □

For arbitrary stationary sequences $(X_n)_{n\in\mathbb{N}}$ it is often possible to approximate the partial sums $\left(\sum_{i=1}^{n} X_i\right)_{n\in\mathbb{N}}$ by a martingale with stationary differences so that under suitable conditions on the error term, Corollary 6.26 also yields a stable central limit theorem in this general setting. In the ergodic case this approach is due to Gordin [36] with generalization to the non-ergodic case in [27] (see also e.g. [22, 37, 41, 72]).

In order to check the assumptions of limit theorems, the following lemma is very useful.

Lemma 6.28 (Toeplitz) *Let* $(b_n)_{n \geq 1}$ *be a sequence in* $[0, \infty)$ *such that* $b_1 > 0$ *and* $\sum_{n=1}^{\infty} b_n = \infty$.
(a) *Let* $(x_n)_{n \geq 1}$ *be a sequence in* $\mathbb{R}$. *If* $\lim_{n \to \infty} x_n = x$ *with* $x \in \mathbb{R}$, *then*

$$\lim_{n \to \infty} \frac{\sum_{j=1}^{n} b_j x_j}{\sum_{j=1}^{n} b_j} = x.$$

(b) *Assume* $b_n > 0$ *for every* $n \geq 1$ *and let* $(a_n)_{n \geq 1}$ *be a sequence in* $\mathbb{R}$. *If* $\lim_{n \to \infty} a_n / b_n = c$ *with* $c \in \mathbb{R}$, *then*

$$\lim_{n \to \infty} \frac{\sum_{j=1}^{n} a_j}{\sum_{j=1}^{n} b_j} = c.$$

The assumption in (b) can be read as $\Delta \left(\sum_{j=1}^{n} a_j \right) / \Delta \left(\sum_{j=1}^{n} b_j \right) \to c$. Therefore, the variant (b) is called the *discrete rule of de l'Hospital*.

Proof (a) Let $\varepsilon > 0$ and $n_0 \in \mathbb{N}$ be such that $|x_n - x| \leq \varepsilon$ for every $n > n_0$. Then for $n > n_0$

$$\left| \frac{\sum_{j=1}^{n} b_j x_j}{\sum_{j=1}^{n} b_j} - x \right| \leq \frac{\sum_{j=1}^{n} b_j |x_j - x|}{\sum_{j=1}^{n} b_j} = \frac{\sum_{j=1}^{n_0} b_j |x_j - x|}{\sum_{j=1}^{n} b_j} + \frac{\sum_{j=n_0+1}^{n} b_j |x_j - x|}{\sum_{j=1}^{n} b_j}$$

$$\leq \frac{\sum_{j=1}^{n_0} b_j |x_j - x|}{\sum_{j=1}^{n} b_j} + \varepsilon.$$

This implies

$$\limsup_{n \to \infty} \left| \frac{\sum_{j=1}^{n} b_j x_j}{\sum_{j=1}^{n} b_j} - x \right| \leq \varepsilon.$$

(b) follows from (a) by setting $x_n := a_n / b_n$. $\quad\square$

Example 6.29 (Adaptive Monte Carlo estimators) For $X \in \mathcal{L}^1(P)$, one wishes to compute $\vartheta := EX$.
(a) ([6]) We assume that there are a measurable space $(\mathcal{Z}, \mathcal{C})$, a measurable map $F : (\mathbb{R}^d \times \mathcal{Z}, \mathcal{B}(\mathbb{R}^d) \otimes \mathcal{C}) \to (\mathbb{R}, \mathcal{B}(\mathbb{R}))$ and a $(\mathcal{Z}, \mathcal{C})$-valued random variable Z such that $F(\lambda, Z) \in \mathcal{L}^1(P)$ and $EX = EF(\lambda, Z)$ for every $\lambda \in \mathbb{R}^d$. Now let $(Z_n)_{n \geq 1}$ be an independent and identically distributed sequence of $(\mathcal{Z}, \mathcal{C})$-valued random variables with $Z_1 \overset{d}{=} Z$, $Z_0 := 0$, $\mathcal{F}_n := \sigma(Z_0, \ldots, Z_n)$, $\mathbb{F} := (\mathcal{F}_n)_{n \geq 0}$ and $(\lambda_n)_{n \geq 0}$ an $\mathbb{F}$-adapted sequence of $\mathbb{R}^d$-valued random variables with $\lambda_0 = 0$. In this abstract setting we investigate the adaptive Monte Carlo estimators

$$\widehat{\vartheta}_n := \frac{1}{n} \sum_{j=1}^{n} F\left(\lambda_{j-1}, Z_j\right), \quad n \geq 1,$$

of ϑ. For all $n \geq 0$ define $M_n := \sum_{j=1}^{n} \left(F\left(\lambda_{j-1}, Z_j\right) - \vartheta\right)$ with $M_0 = 0$. Then $\widehat{\vartheta}_n - \vartheta = M_n/n$ for all $n \geq 1$. For $p \in [1, \infty)$, let $f_p : \mathbb{R}^d \to [0, \infty]$, $f_p(\lambda) := E |F(\lambda, Z)|^p$. If $Ef_1(\lambda_n) < \infty$ for every $n \geq 0$, then $M = (M_n)_{n \geq 0}$ is an $\mathbb{F}$-martingale. In fact, since λ_{n-1} is $\mathcal{F}_{n-1}$-measurable and $\sigma(Z_n)$ and $\mathcal{F}_{n-1}$ are independent, for all $n \geq 1$,

$$E |F(\lambda_{n-1}, Z_n)| = \int \int |F(\lambda, z)| \, dP^Z(z) \, dP^{\lambda_{n-1}}(\lambda) = Ef_1(\lambda_{n-1}) < \infty,$$

so that M is an $\mathbb{F}$-adapted $\mathcal{L}^1$-process, and moreover, for all $n \geq 1$,

$$E(F(\lambda_{n-1}, Z_n)|\mathcal{F}_{n-1}) = \int F(\lambda_{n-1}, z) \, dP^Z(z) = \vartheta,$$

which implies

$$E(M_n|\mathcal{F}_{n-1}) = M_{n-1} + E(F(\lambda_{n-1}, Z_n)|\mathcal{F}_{n-1}) - \vartheta = M_{n-1}.$$

If additionally $\sup_{n \geq 0} f_p(\lambda_n) < \infty$ almost surely for some $p > 1$, then it follows from the strong law of large numbers for martingales in Theorem A.9 that $M_n/n \to 0$ almost surely and hence $\widehat{\vartheta}_n \to \vartheta$ almost surely as $n \to \infty$. Now we assume

(i) $\lambda_n \to \lambda_\infty$ a.s. for some $\mathbb{R}^d$-valued random variable λ_∞,
(ii) $f_2 < \infty$ and f_2 is continuous,
(iii) $Ef_2(\lambda_n) < \infty$ for every $n \geq 0$,
(iv) $F(\cdot, z) : \mathbb{R}^d \to \mathbb{R}$ is continuous for all $z \in \mathcal{Z}$ or
(iv') $\sup_{n \geq 0} f_p(\lambda_n) < \infty$ a.s. for some $p > 2$.

Then an application of Corollary 6.23 yields

$$\sqrt{n}\left(\widehat{\vartheta}_n - \vartheta\right) = n^{-1/2} M_n \to N\left(0, f_2(\lambda_\infty) - \vartheta^2\right) \quad \text{stably}.$$

In view of (iii), M is an $\mathcal{L}^2$-martingale with quadratic characteristic

$$\langle M \rangle_n = \sum_{j=1}^{n} E\left((\Delta M_j)^2 |\mathcal{F}_{j-1}\right) = \sum_{j=1}^{n} \left(f_2(\lambda_{j-1}) - \vartheta^2\right)$$

because, for all $n \geq 1$,

$$E\left((\Delta M_n)^2 |\mathcal{F}_{n-1}\right) = \int (F(\lambda_{n-1}, z) - \vartheta)^2 \, dP^Z(z) = f_2(\lambda_{n-1}) - \vartheta^2.$$

Since $f_2(\lambda_{n-1}) \to f_2(\lambda_\infty)$ almost surely by (i) and (ii), the Toeplitz Lemma 6.28 yields $\langle M \rangle_n/n \to f_2(\lambda_\infty) - \vartheta^2$ almost surely as $n \to \infty$, which is condition (N_{a_n}) with $a_n = n^{1/2}$. To verify the conditional Lindeberg condition (CLB_{a_n}) with $a_n = n^{1/2}$ note that for all $n \in \mathbb{N}$ and $\varepsilon > 0$

$$\frac{1}{n} \sum_{j=1}^{n} E\left((\Delta M_j)^2 1_{\{|\Delta M_j| \geq \varepsilon n^{1/2}\}} | \mathcal{F}_{j-1}\right)$$

$$\leq \frac{1}{n} \sum_{j=1}^{n} E\left((\Delta M_j)^2 1_{\{|\Delta M_j| \geq \varepsilon j^{1/2}\}} | \mathcal{F}_{j-1}\right)$$

so that (CLB_{a_n}) follows from

$$E\left((\Delta M_j)^2 1_{\{|\Delta M_j| \geq \varepsilon j^{1/2}\}} | \mathcal{F}_{j-1}\right) \to 0 \quad \text{a.s. as } j \to \infty$$

and the Toeplitz Lemma 6.28. Now

$$E\left((\Delta M_j)^2 1_{\{|\Delta M_j| \geq \varepsilon j^{1/2}\}} | \mathcal{F}_{j-1}\right)$$
$$= \int \left(F(\lambda_{j-1}, z) - \vartheta\right)^2 1_{\{|F(\lambda_{j-1}, z) - \vartheta| \geq \varepsilon j^{1/2}\}} dP^Z(z) ,$$

and from (i) and (iv) it follows almost surely as $j \to \infty$ for all $z \in \mathcal{Z}$ that $F(\lambda_{j-1}, z) \to F(\lambda_\infty, z)$ and hence

$$\left(F(\lambda_{j-1}, z) - \vartheta\right)^2 1_{\{|F(\lambda_{j-1}, z) - \vartheta| \geq \varepsilon j^{1/2}\}} \to 0$$

with an exceptional null set which is independent of $z \in \mathcal{Z}$. Moreover, almost surely for all $j \in \mathbb{N}$ and $z \in \mathcal{Z}$

$$\left(F(\lambda_{j-1}, z) - \vartheta\right)^2 1_{\{|F(\lambda_{j-1}, z) - \vartheta| \geq \varepsilon j^{1/2}\}}$$
$$\leq \left(F(\lambda_{j-1}, z) - \vartheta\right)^2 \to \left(F(\lambda_\infty, z) - \vartheta\right)^2 \quad \text{as } j \to \infty$$

and

$$\int \left(F(\lambda_{j-1}, z) - \vartheta\right)^2 dP^Z(z) = f_2(\lambda_{j-1}) - \vartheta^2$$
$$\to f_2(\lambda_\infty) - \vartheta^2 = \int \left(F(\lambda_\infty, z) - \vartheta\right)^2 dP^Z(z)$$

from which almost surely, by Pratt's dominated convergence theorem,

$$\int \left(F\left(\lambda_{j-1}, z\right) - \vartheta \right)^2 1_{\{|F(\lambda_{j-1},z)-\vartheta|\geq \varepsilon j^{1/2}\}} \, dP^Z(z) \to 0 \quad \text{as } j \to \infty.$$

Under condition (iv') we have

$$\begin{aligned}
E\left(|\Delta M_n|^p \,|\mathcal{F}_{n-1}\right) &= E\left(|F\left(\lambda_{n-1}, Z_n\right) - \vartheta|^p \,|\mathcal{F}_{n-1}\right) \\
&\leq 2^{p-1} E\left(|F\left(\lambda_{n-1}, Z_n\right)|^p \,|\mathcal{F}_{n-1}\right) + 2^{p-1} |\vartheta|^p \\
&\leq 2^{p-1} \sup_{j\geq 0} f_p\left(\lambda_j\right) + 2^{p-1} |\vartheta|^p < \infty
\end{aligned}$$

almost surely for all $n \geq 1$, hence the conditional Lyapunov condition

$$\frac{1}{n^{p/2}} \sum_{j=1}^{n} E\left(|\Delta M_j|^p \,|\mathcal{F}_{j-1}\right) \to 0 \quad \text{a.s.}$$

Of course, one is mainly interested in estimators λ_n of the parameter λ which provide minimal variance, that is $\lambda_n \to \lambda_{\min}$ almost surely with $\lambda_{\min} \in \mathbb{R}^d$ such that

$$f_2\left(\lambda_{\min}\right) - \vartheta^2 = \operatorname{Var} F\left(\lambda_{\min}, Z\right) = \min_{\lambda\in\mathbb{R}^d} \operatorname{Var} F\left(\lambda, Z\right).$$

(b) ([68]) Assume $X \in \mathcal{L}^2(P)$ and $\operatorname{Var} X > 0$. Let $Y \in \mathcal{L}^2(P)$ be another random variable with $EX = EY$, $\operatorname{Var} Y > 0$ and $\operatorname{Var}(X - Y) > 0$. For $\lambda \in \mathbb{R}$, let $U(\lambda) := X - \lambda(X - Y)$. Then $EU(\lambda) = \vartheta$, and for

$$g(\lambda) := \operatorname{Var} U(\lambda) = \operatorname{Var} X - 2\lambda \operatorname{Cov}(X, X - Y) + \lambda^2 \operatorname{Var}(X - Y)$$

we get

$$\min_{\lambda\in\mathbb{R}} g(\lambda) = g(\lambda_{\min}) \quad \text{with} \quad \lambda_{\min} := \frac{\operatorname{Cov}(X, X - Y)}{\operatorname{Var}(X - Y)}$$

and

$$\sigma_{\min}^2 := g(\lambda_{\min}) = \operatorname{Var} X - \frac{\operatorname{Cov}(X, X - Y)^2}{\operatorname{Var}(X - Y)} = \operatorname{Var} X \left(1 - \rho_{X,X-Y}^2\right),$$

where

$$\rho_{X,X-Y} := \frac{\operatorname{Cov}(X, X - Y)}{\left(\operatorname{Var} X \operatorname{Var}(X - Y)\right)^{1/2}}$$

denotes the correlation coefficient.

Now let $((X_n, Y_n))_{n \geq 1}$ be an independent and identically distributed sequence of $\mathbb{R}^2$-valued random variables with $(X_1, Y_1) \overset{d}{=} (X, Y)$ and set

$$\widehat{\lambda}_n := \frac{\sum_{j=1}^{n} X_j (X_j - Y_j)}{\sum_{j=1}^{n} (X_j - Y_j)^2}$$

for all $n \geq 1$ with $\widehat{\lambda}_0 := 0$ and $\widetilde{\lambda}_n := (-n) \vee (\widehat{\lambda}_n \wedge n)$ for all $n \geq 0$. We consider the adaptive Monte Carlo estimator

$$\widehat{\vartheta}_n := \frac{1}{n} \sum_{j=1}^{n} \left(X_j - \widetilde{\lambda}_{j-1} (X_j - Y_j) \right), \quad n \geq 1,$$

of ϑ.

This setting is a special case of (a) with $d = 1$, $Z = (X, Y)$, $\mathcal{Z} = \mathbb{R}^2$, $F(\lambda, z) = z_1 - \lambda (z_1 - z_2)$ and $\lambda_n = \widetilde{\lambda}_n$. The strong law of large numbers of Kolmogorov implies that $\widehat{\lambda}_n \to \lambda_{\min}$ almost surely and hence $\widetilde{\lambda}_n \to \lambda_{\min}$ almost surely as well. Furthermore, $f_2(\lambda) = \operatorname{Var} F(\lambda, Z) + \vartheta^2 = g(\lambda) + \vartheta^2 < \infty$ so that f_2 is continuous and thus $\sup_{n \geq 0} f_2(\widetilde{\lambda}_n) < \infty$ almost surely. Since $|\widetilde{\lambda}_n| \leq n$, we have $E f_2(\widetilde{\lambda}_n) < \infty$ for every $n \geq 0$. In particular, by (a), $\widehat{\vartheta}_n \to \vartheta$ almost surely. Clearly, $F(\lambda, z)$ is continuous in λ for all $z \in \mathbb{R}^2$. Thus (i)–(iv) are satisfied, and it follows from (a) that

$$\sqrt{n} \left(\widehat{\vartheta}_n - \vartheta \right) \to N \left(0, \sigma_{\min}^2 \right) \quad \text{mixing as } n \to \infty$$

and therefore, the estimator $\widehat{\vartheta}_n$ provides the optimal variance reduction.   □

Example 6.30 (The Pólya urn) Assume that an urn contains initially (at time 0) r red balls and s black balls, $r, s \in \mathbb{N}$. At every time n one draws at random a ball from the urn and then puts it back into the urn with another m balls of the same colour, $m \in \mathbb{N}$. Then, at time n, the urn contains (once the new balls have been put into the urn) $r + s + mn$ balls. Let Y_n and $X_n = Y_n / (r + s + mn)$ denote the number and the proportion of red balls inside the urn at time n, respectively. One models the drawings using an independent and identically distributed sequence $(U_n)_{n \geq 1}$ of $U(0, 1)$-distributed random variables as follows: If $U_{n+1} \leq X_n$, the ball drawn at time $n + 1$ is red, otherwise it is black. Then the dynamics of $Y = (Y_n)_{n \geq 0}$ and $X = (X_n)_{n \geq 0}$ are given by

$$Y_0 = r, \quad Y_{n+1} = Y_n + m 1_{\{U_{n+1} \leq X_n\}}$$

and

$$X_0 = \frac{r}{r + s}, \quad X_{n+1} = X_n + \frac{m}{r + s + m(n + 1)} \left(1_{\{U_{n+1} \leq X_n\}} - X_n \right).$$

The process X is a $[0, 1]$-valued martingale with respect to the filtration $\mathbb{F} = (\mathcal{F}_n)_{n\geq 0}$, $\mathcal{F}_n := \sigma(U_1, \ldots, U_n)$ with $\mathcal{F}_0 = \{\emptyset, \Omega\}$, so that $X_n \to X_\infty$ almost surely as $n \to \infty$ by the martingale convergence theorem, where the limit X_∞ is $\mathcal{T}_X$-measurable. Furthermore, for fixed $p \in \mathbb{N}$, the process $Z = (Z_n)_{n\geq 0}$ defined by

$$Z_n := \prod_{i=0}^{p-1} \frac{Y_n + mi}{r + s + m(n + i)}$$

satisfies $Z_n \to X_\infty^p$ almost surely and one checks that Z is also an $\mathbb{F}$-martingale. This implies

$$E X_\infty^p = E Z_0 = Z_0 = \prod_{i=0}^{p-1} \frac{r + mi}{r + s + mi}.$$

Hence, the distribution of X_∞ has the moments of a beta distribution with parameters r/m and s/m. Both distributions have compact support, hence, they are equal.

Now, for $n \geq 1$ introduce $V_n := 1_{\{U_n \leq X_{n-1}\}}$. It is well known that $(V_n)_{n\geq 1}$ is exchangeable and

$$\frac{1}{n} \sum_{i=1}^{n} V_i \to W := E(V_1 | \mathcal{T}_V) \quad \text{a.s.}$$

(see e.g. [64], Beispiel 10.15 and Satz 10.9). Since $\mathrm{Var}(V_1 | \mathcal{T}_V) = W - W^2 = W(1 - W)$, Corollary 6.27 yields

$$\sqrt{n}\left(\frac{1}{n} \sum_{i=1}^{n} V_i - W\right) = \frac{1}{\sqrt{n}} \sum_{i=1}^{n}(V_i - W) \to N(0, W(1 - W)) \quad \text{stably}.$$

We obtain

$$X_n = \frac{r}{r + s + mn} + \frac{m \sum_{i=1}^{n} V_i}{r + s + mn} \to W \quad \text{a.s.}$$

implying $X_\infty = W$ and

$$\sqrt{n}(X_n - X_\infty) \to N(0, X_\infty(1 - X_\infty)) \quad \text{stably}$$

using Theorem 3.7 (a) because

$$\sqrt{n}\left(X_n - \frac{1}{n} \sum_{i=1}^{n} V_i\right) = \frac{r\sqrt{n}}{r + s + mn} - \frac{r + s}{(r + s + mn)\sqrt{n}} \sum_{i=1}^{n} V_i \to 0 \quad \text{a.s.}$$

Distributional convergence of the randomly centered X_n has been investigated in [41], pp. 80–81 and stable convergence is contained in [20], Example 6. (See also [19], Corollary 4.2 and [39], Example 4.2 for an even stronger convergence result.) ▢

Exercise 6.5 Let $(Y_n)_{n\geq1}$ be an independent and identically distributed sequence with $Y_1 \in \mathcal{L}^2(P)$, $EY_1 = 0$ and let X_0 be a $\{-1, 0+1\}$-valued random variable independent of $(Y_n)_{n\geq1}$. Set $X_n := Y_n 1_{\{X_0\neq0\}}$, $M_n := \sum_{j=0}^n X_j$, $\mathcal{F}_n := \sigma(X_0, Y_1, \ldots, Y_n)$ and $\mathbb{F} = (\mathcal{F}_n)_{n\geq0}$. Prove that M is an $\mathbb{F}$-martingale,

$$n^{-1/2} M_n \to N\left(0, \sigma^2 X_0^2\right) \quad \text{stably}$$

and

$$n^{-1/2} M_n \overset{d}{\to} P(X_0 = 0)\,\delta_0 + P(X_0 \neq 0)\,N\left(0, \sigma^2\right),$$

where $\sigma^2 := \text{Var}\, Y_1$.

Exercise 6.6 Let $(Z_n)_{n\geq1}$ be an independent and identically distributed sequence with $Z_1 \in \mathcal{L}^p(P)$ for some $p > 2$ and $EZ_1 = 0$. Set $M_n := \sum_{j=1}^n \left(\sum_{i=1}^{j-1} Z_i/i\right) Z_j$ with $M_0 = M_1 = 0$, $\sigma^2 := \text{Var}\, Z_1$ and $V := \sum_{i=1}^\infty Z_i/i$. Show that

$$n^{-1/2} M_n \to N\left(0, \sigma^2 V^2\right) \quad \text{stably}.$$

Exercise 6.7 (Martingale tail sums) Let $M = (M_n)_{n\geq0}$ be an $\mathcal{L}^2$-bounded martingale with respect to the filtration $\mathbb{F} = (\mathcal{F}_n)_{n\geq0}$, $M_n = X_0 + \sum_{k=1}^n X_k$ and let $a_n > 0$. Assume

$$a_n^2 \sum_{j>n} E\left(X_j^2 | \mathcal{F}_{j-1}\right) \to \eta^2 \quad \text{in probability as } n \to \infty$$

for some random variable $\eta \geq 0$

and

$$a_n^2 \sum_{j>n} E\left(X_j^2 1_{\{|X_j|\geq\varepsilon/a_n\}} | \mathcal{F}_{j-1}\right) \to 0 \quad \text{in probability as } n \to \infty$$

for all $\varepsilon > 0$.

Show that

$$a_n \sum_{j=n+1}^\infty X_j \to N\left(0, \eta^2\right) \quad \text{stably as } n \to \infty.$$

Exercise 6.8 (Stabilizing time change) Let $\mathbb{F} = (\mathcal{F}_n)_{n\geq 0}$ be a filtration in $\mathcal{F}$ and let $M = (M_n)_{n\geq 0}$ be an $\mathbb{F}$-martingale satisfying $|\Delta M_n| \leq c < \infty$ almost surely for every $n \geq 1$ and $\langle M\rangle_\infty = \infty$ almost surely. Consider the $\mathbb{F}$-stopping times $\tau_n := \inf\{k \geq 1 : \langle M\rangle_k \geq n\}, n \in \mathbb{N}$. Show that

$$n^{-1/2}M_{\tau_n} \;\rightarrow\; N(0, 1) \quad \text{mixing as } n \rightarrow \infty.$$

Exercise 6.9 Show that the numbers Y_n of red balls in the Pólya urn scheme of Example 6.30 satisfy

$$n^{-1/2}(Y_n - (r + s + mn)X_\infty) \rightarrow N\left(0, m^2 X_\infty(1 - X_\infty)\right) \quad \text{stably}.$$

Exercise 6.10 Let $X = (X_n)_{n\geq 1}$ be an exchangeable $(\mathcal{X}, \mathcal{B}(\mathcal{X}))$-valued process, where $\mathcal{X}$ is polish. Show that $X_n \rightarrow P^{X_1|\mathcal{T}_X}$ stably.

6.5 A Continuous Time Version

We finally present a continuous-time version of Theorem 6.23 and Corollary 6.24 for path-continuous (local) martingales. Its proof is obtained by using the associated Dambis-Dubins-Schwarz Brownian motion.

Theorem 6.31 *Let $M = (M_t)_{t\geq 0}$ be a path-continuous local $\mathbb{F}$-martingale, where $\mathbb{F} = (\mathcal{F}_t)_{t\geq 0}$ denotes a right-continuous filtration in $\mathcal{F}$, and let $a : (0, \infty) \rightarrow (0, \infty)$ be a nondecreasing function with $a(t) \rightarrow \infty$ as $t \rightarrow \infty$. Assume for the (continuous) quadratic characteristic*

$$\frac{\langle M\rangle_t}{a(t)^2} \rightarrow \eta^2 \quad \text{in probability as } t \rightarrow \infty$$

for some $\mathbb{R}_+$-valued random variable η. Then

$$\frac{M_t}{a(t)} \rightarrow N\left(0, \eta^2\right) \quad \text{stably as } t \rightarrow \infty$$

and if $P(\eta^2 > 0) > 0$,

$$\frac{M_t}{\langle M\rangle_t^{1/2}} \rightarrow N(0, 1) \quad \text{mixing under } P_{\{\eta^2 > 0\}} \text{ as } t \rightarrow \infty.$$

$(M_t/0 := 0.)$

Proof Since $\langle M - M_0 \rangle = \langle M \rangle$, we may assume $M_0 = 0$. Let $(s_n)_{n \geq 1}$ be an arbitrary sequence in $(0, \infty)$ with $s_n \uparrow \infty$. The assertions reduce to

$$\frac{M_{s_n}}{a(s_n)} \to N\left(0, \eta^2\right) \quad \text{stably as } n \to \infty$$

and

$$\frac{M_{s_n}}{\langle M \rangle_{s_n}^{1/2}} \to N(0, 1) \quad \text{mixing under } P_{\{\eta^2 > 0\}} \text{ as } n \to \infty.$$

By the Dambis-Dubins-Schwarz time-change theorem there exists (possibly after a suitable extension of the underlying probability space) a (continuous) Brownian motion W such that $M = W_{\langle M \rangle}$ ([51], Theorem 18.4). For $n \in \mathbb{N}$, define

$$W_t^n := \frac{1}{a(s_n)} W_{a(s_n)^2 t}, \quad t \geq 0$$

and $\mathbb{G}^n := \left(\mathcal{G}_{a(s_n)^2 t}\right)_{t \geq 0}$, where $\mathcal{G}_t = \sigma(W_s, s \leq t)$. Then, by the scaling invariance of Brownian motion, W^n is a $\mathbb{G}^n$-Brownian motion and the filtrations $\mathbb{G}^n$ satisfy the nesting condition from Corollary 5.9 with $t_n := 1/a(s_n)$: We have $t_n \to 0$, $\left(\mathcal{G}_{a(s_n)^2 t_n}\right)_{n \geq 1}$ is a filtration and $\sigma\left(\bigcup_{n=1}^\infty \mathcal{G}_{a(s_n)^2 t_n}\right) = \mathcal{G}_\infty$. Consequently, it follows from Corollary 5.9 that $W^n \to \nu$ mixing, where $\nu = P^W \in \mathcal{M}^1(C(\mathbb{R}_+))$. Therefore, by Theorem 3.7 (b),

$$\left(W^n, \frac{\langle M \rangle_{s_n}}{a(s_n)^2}\right) \to \nu \otimes \delta_{\eta^2} \quad \text{stably}$$

and using the continuity of $\varphi : C(\mathbb{R}_+) \times \mathbb{R}_+ \to \mathbb{R}$, $\varphi(x, t) = x(t)$, Theorem 3.7 (c) yields

$$\frac{M_{s_n}}{a(s_n)} = \frac{1}{a(s_n)} W_{\langle M \rangle_{s_n}} = W_{\langle M \rangle_{s_n}/a(s_n)^2}^n = \varphi\left(W^n, \frac{\langle M \rangle_{s_n}}{a(s_n)^2}\right)$$
$$\to (\nu \otimes \delta_{\eta^2})^\varphi = N\left(0, \eta^2\right) \quad \text{stably}$$

as $n \to \infty$.

As for the second assertion, observe that by Theorem 3.7 (b)

$$\left(\frac{M_{s_n}}{a(s_n)}, \frac{\langle M \rangle_{s_n}}{a(s_n)^2}\right) \to K_{\eta^2} := N\left(0, \eta^2\right) \otimes \delta_{\eta^2} \quad \text{stably},$$

in particular we have stable convergence under $P_{\{\eta^2 > 0\}}$, the function $g : \mathbb{R}^2 \to \mathbb{R}$, $g(x, y) := x/\sqrt{y}$ if $y > 0$ and $g(x, y) := 0$ if $y \leq 0$ is Borel-measurable and $P_{\{\eta^2 > 0\}} K_{\eta^2}$-almost surely continuous because

$$P_{\{\eta^2 > 0\}} K_{\eta^2} (\mathbb{R} \times \{0\}) = \int N\left(0, \eta^2\right) (\mathbb{R}) \, \delta_{\eta^2} (\{0\}) \, dP_{\{\eta^2 > 0\}}$$

$$= P_{\{\eta^2 > 0\}} \left(\eta^2 = 0\right) = 0$$

and moreover, $K_{\eta^2} (\omega, \cdot)^g = N(0, 1)$ for $\omega \in \{\eta^2 > 0\}$. Thus, it follows from Theorem 3.7 (c) that

$$\frac{M_{s_n}}{\langle M \rangle_{s_n}^{1/2}} = g\left(\frac{M_{s_n}}{a(s_n)}, \frac{\langle M \rangle_{s_n}}{a(s_n)^2}\right) \to N(0, 1) \quad \text{mixing under } P_{\{\eta^2 > 0\}}. \qquad \square$$

Chapter 7
Stable Functional Martingale Central Limit Theorems

This chapter is devoted to stable functional central limit theorems for partial sum processes based on martingale differences which correspond to the results for partial sums presented in Sects. 6.1, 6.3 and 6.4. As in Chap. 6 it is convenient to consider arrays of martingale differences, but to keep technicalities as simple as possible, we consider a fixed filtration $\mathbb{F} = (\mathcal{F}_k)_{k \geq 0}$ on the basic probability space $(\Omega, \mathcal{F}, P)$. As usual, $\mathcal{F}_\infty = \sigma \left(\bigcup_{k=0}^{\infty} \mathcal{F}_k \right)$. For every $n \in \mathbb{N}$, let $(X_{nk})_{k \geq 1}$ be a martingale difference sequence w.r.t. $\mathbb{F}$, and for every $n \in \mathbb{N}$ and $t \in [0, \infty)$ set

$$S_{(n)}(t) := \sum_{k=1}^{[nt]} X_{nk} + (nt - [nt]) X_{n,[nt]+1} \, .$$

Then $\left(S_{(n)}(t) \right)_{t \in [0,\infty)}$ is a random process with sample paths in $C(\mathbb{R}_+)$. Note that the array $\left(\mathcal{F}_{n,k} \right)_{k \geq 0, n \in \mathbb{N}}$ with $\mathcal{F}_{n,k} := \mathcal{F}_k$ is obviously nested.

For a nonnegative stochastic process $(\eta(t))_{t \in [0,\infty)}$ with paths in $C(\mathbb{R}_+)$ and square integrable X_{nk} we introduce the conditions

(N$_t$) $\displaystyle \sum_{k=1}^{[nt]} E\left(X_{nk}^2 | \mathcal{F}_{k-1} \right) \to \eta^2(t)$ in probability as $n \to \infty$ for all $t \in [0, \infty)$

and

(CLB$_t$) $\displaystyle \sum_{k=1}^{[nt]} E\left(X_{nk}^2 1_{\{|X_{nk}| \geq \varepsilon\}} | \mathcal{F}_{k-1} \right) \to 0$ in probability as $n \to \infty$

for all $\varepsilon > 0$ and all $t \in [0, \infty)$.

Note that any process η^2 appearing in (N$_t$) is nonnegative with almost surely nondecreasing paths and $\eta^2(0) = 0$. The conditions (N$_t$) and (CLB$_t$) are our basic

© Springer International Publishing Switzerland 2015
E. Häusler and H. Luschgy, *Stable Convergence and Stable Limit Theorems*,
Probability Theory and Stochastic Modelling 74,
DOI 10.1007/978-3-319-18329-9_7

conditions which ensure stable convergence of $S_{(n)}$ to a Brownian motion with time change η^2.

Theorem 7.1 *Let $(X_{nk})_{k \in \mathbb{N}}$ be a square integrable martingale difference sequence w.r.t. $\left(\mathcal{F}_{n,k}\right)_{k \geq 0}$ for every $n \in \mathbb{N}$. Under (N_t) and (CLB_t),*

$$S_{(n)} \to \left(W\left(\eta^2\left(t\right)\right)\right)_{t \in [0,\infty)} \quad \mathcal{F}_\infty\text{-stably as } n \to \infty \text{ in } C\left(\mathbb{R}_+\right),$$

where $W = (W(t))_{t \geq 0}$ is a Brownian motion which is independent of $\mathcal{F}_\infty$.

According to Proposition 3.20 we have to show that the finite dimensional distributions of $S_{(n)}$ converge $\mathcal{F}_\infty$-stably to the finite dimensional distributions of $\left(W\left(\eta^2\left(t\right)\right)\right)_{t \in [0,\infty)}$ and that the sequence $\left(S_{(n)}\right)_{n \in \mathbb{N}}$ is tight in $C\left(\mathbb{R}_+\right)$.

Proof of stable convergence of the finite dimensional distributions. For all $0 < t_1 < t_2 < \cdots < t_r < \infty$ we have to show

$$\left(S_{(n)}\left(t_1\right), \ldots, S_{(n)}\left(t_r\right)\right) \to \left(W\left(\eta^2\left(t_1\right)\right), \ldots, W\left(\eta^2\left(t_r\right)\right)\right) \quad \mathcal{F}_\infty\text{-stably as } n \to \infty.$$

Clearly, this is equivalent to

$$\left(S_{(n)}\left(t_1\right), S_{(n)}\left(t_2\right) - S_{(n)}\left(t_1\right), \ldots, S_{(n)}\left(t_r\right) - S_{(n)}\left(t_{r-1}\right)\right) \to$$
$$\left(W\left(\eta^2\left(t_1\right)\right), W\left(\eta^2\left(t_2\right)\right) - W\left(\eta^2\left(t_1\right)\right), \ldots, W\left(\eta^2\left(t_r\right)\right) - W\left(\eta^2\left(t_{r-1}\right)\right)\right)$$

$\mathcal{F}_\infty$-stably as $n \to \infty$. Putting $t_0 = 0$ and observing that $S_{(n)}\left(t_0\right) = W\left(\eta^2\left(t_0\right)\right) = 0$, by the Cramér-Wold technique, Corollary 3.19, (i) $\Leftrightarrow$ (iii), the last convergence is equivalent to

$$\sum_{q=1}^{r} \lambda_q \left(S_{(n)}\left(t_q\right) - S_{(n)}\left(t_{q-1}\right)\right) \to \sum_{q=1}^{r} \lambda_q \left(W\left(\eta^2\left(t_q\right)\right) - W\left(\eta^2\left(t_{q-1}\right)\right)\right)$$

$\mathcal{F}_\infty$-stably as $n \to \infty$ for all $\lambda_1, \ldots, \lambda_r \in \mathbb{R}$.

First, note that for all $t \in [0, \infty)$

$$\left| S_{(n)}\left(t\right) - \sum_{k=1}^{[nt]} X_{nk} \right| \leq \left| X_{n,[nt]+1} \right| \leq \max_{1 \leq k \leq [n(t+1)]} |X_{nk}| \to 0$$

in probability as $n \to \infty$, where the convergence to zero follows from (CLB_t) and Proposition 6.6 (note that $[nt] + 1 \leq [n(t+1)]$). Therefore, by Theorem 3.18 (a) it is sufficient to show that

$$\sum_{q=1}^{r} \lambda_q \left(\sum_{k=1}^{[nt_q]} X_{nk} - \sum_{k=1}^{[nt_{q-1}]} X_{nk} \right) \to \sum_{q=1}^{r} \lambda_q \left(W\left(\eta^2\left(t_q\right)\right) - W\left(\eta^2\left(t_{q-1}\right)\right)\right)$$

$\mathcal{F}_\infty$-stably as $n \to \infty$. Setting $I\left(\left[nt_{q-1}\right] + 1 \le k \le \left[nt_q\right]\right) = 1$ if $\left[nt_{q-1}\right] + 1 \le k \le \left[nt_q\right]$ is true and $= 0$ otherwise, we have, for all $n \in \mathbb{N}$ and $1 \le k \le [nt_r]$,

$$
\sum_{q=1}^{r} \lambda_q \left(\sum_{k=1}^{[nt_q]} X_{nk} - \sum_{k=1}^{[nt_{q-1}]} X_{nk} \right)
$$

$$
= \sum_{q=1}^{r} \sum_{k=1}^{[nt_r]} \lambda_q I\left(\left[nt_{q-1}\right] + 1 \le k \le \left[nt_q\right]\right) X_{nk} = \sum_{k=1}^{[nt_r]} a_{nk} X_{nk} \,,
$$

with

$$
a_{nk} := \sum_{q=1}^{r} \lambda_q I\left(\left[nt_{q-1}\right] + 1 \le k \le \left[nt_q\right]\right) .
$$

We see that $(a_{nk} X_{nk})_{1 \le k \le [nt_r], n \in \mathbb{N}}$ is a square integrable martingale difference array w.r.t. the nested array $\left(\mathcal{F}_{n,k}\right)_{0 \le k \le [nt_r], n \in \mathbb{N}}$ (where $\mathcal{F}_{n,k} = \mathcal{F}_k$) and

$$
\sum_{k=1}^{[nt_r]} E\left(a_{nk}^2 X_{nk}^2 | \mathcal{F}_{n,k-1} \right) = \sum_{q=1}^{r} \sum_{k=[nt_{q-1}]+1}^{[nt_q]} a_{nk}^2 E\left(X_{nk}^2 | \mathcal{F}_{k-1} \right)
$$

$$
= \sum_{q=1}^{r} \sum_{k=[nt_{q-1}]+1}^{[nt_q]} \lambda_q^2 E\left(X_{nk}^2 | \mathcal{F}_{k-1} \right)
$$

$$
= \sum_{q=1}^{r} \lambda_q^2 \left(\sum_{k=1}^{[nt_q]} E\left(X_{nk}^2 | \mathcal{F}_{k-1} \right) - \sum_{k=1}^{[nt_{q-1}]} E\left(X_{nk}^2 | \mathcal{F}_{k-1} \right) \right)
$$

$$
\to \sum_{q=1}^{r} \lambda_q^2 \left(\eta^2\left(t_q\right) - \eta^2\left(t_{q-1}\right) \right)
$$

in probability as $n \to \infty$ by (N_t). Moreover, for all $\varepsilon > 0$,

$$
\sum_{k=1}^{[nt_r]} E\left(a_{nk}^2 X_{nk}^2 1_{\{|a_{nk}||X_{nk}| \ge \varepsilon\}} | \mathcal{F}_{n,k-1} \right)
$$

$$
\le \left(\sum_{q=1}^{r} |\lambda_q| \right)^2 \sum_{k=1}^{[nt_r]} E\left(X_{nk}^2 1_{\left\{ |X_{nk}| \ge \varepsilon / \sum_{q=1}^{r} |\lambda_q| \right\}} | \mathcal{F}_{k-1} \right) \to 0
$$

in probability as $n \to \infty$ by (CLB$_t$). Here, we assume w.l.o.g. that not all λ_q are equal to zero. Therefore, Theorem 6.1 and Remark 6.2 imply

$$\sum_{k=1}^{[nt_r]} a_{nk} X_{nk} \rightarrow \left(\sum_{q=1}^{r} \lambda_q^2 \left(\eta^2 (t_q) - \eta^2 (t_{q-1})\right)\right)^{1/2} N \quad \mathcal{F}_\infty\text{-stably as } n \rightarrow \infty,$$

where N is independent of $\mathcal{F}_\infty$ with $P^N = N(0, 1)$. But, by independence of the increments of W, independence of W and $\mathcal{F}_\infty$, and $\mathcal{F}_\infty$-measurability of η^2, using Lemmas A.4 (c) and A.5 (a), the conditional distributions of

$$\left(\sum_{q=1}^{r} \lambda_q^2 \left(\eta^2 (t_q) - \eta^2 (t_{q-1})\right)\right)^{1/2} N \quad \text{and} \quad \sum_{q=1}^{r} \lambda_q \left(W \left(\eta^2 (t_q)\right) - W \left(\eta^2 (t_{q-1})\right)\right)$$

given $\mathcal{F}_\infty$ both coincide with

$$N\left(0, \sum_{q=1}^{r} \lambda_q^2 \left(\eta^2 (t_q) - \eta^2 (t_{q-1})\right)\right)$$

which gives

$$\sum_{q=1}^{r} \lambda_q \left(\sum_{k=1}^{[nt_q]} X_{nk} - \sum_{k=1}^{[nt_{q-1}]} X_{nk}\right) \rightarrow \sum_{q=1}^{r} \lambda_q \left(W \left(\eta^2 (t_q)\right) - W \left(\eta^2 (t_{q-1})\right)\right)$$

$\mathcal{F}_\infty$-stably as $n \rightarrow \infty$ and completes the proof of the finite dimensional distributions.

Proof of tightness. We prove tightness of the sequence $\left(S_{(n)} (t)\right)_{t \in [0,T]}$, $n \in \mathbb{N}$, in $C([0, T])$ for every $T \in \mathbb{N}$, that is, for every $T \in \mathbb{N}$ and $\varepsilon > 0$ we show

$$\lim_{\delta \downarrow 0} \limsup_{n \rightarrow \infty} P \left(\sup_{\substack{0 \leq s,t \leq T \\ |s-t| \leq \delta}} \left|S_{(n)} (s) - S_{(n)} (t)\right| \geq \varepsilon\right) = 0$$

(cf. [51], Theorem 16.5). Then the assertion follows from Proposition 3.20 and Corollary 3.23. Let $T \in \mathbb{N}$ be fixed from now on.

Step 1. For $n \in \mathbb{N}$ and $0 \leq k \leq nT$, $(X_{nk})_{1 \leq k \leq nT, n \in \mathbb{N}}$ is a square integrable martingale difference array w.r.t. $(\mathcal{F}_{n,k})_{0 \leq k \leq nT, n \in \mathbb{N}}$. We augment this array by independent random variables X_{nk} for $n \in \mathbb{N}$ and $k \geq nT + 1$ which are independent of $\mathcal{F}_\infty$ and satisfy $P\left(X_{nk} = 1/\sqrt{n}\right) = 1/2 = P\left(X_{nk} = -1/\sqrt{n}\right)$. (These new random variables X_{nk} should not be confused with the original random variables X_{nk} for $k \geq nT + 1$, which play no role in the current proof for fixed T.) If we set $\mathcal{F}_{n,k}^T := \mathcal{F}_{n,k}$ for $n \in \mathbb{N}$ and $0 \leq k \leq nT$ and $\mathcal{F}_{n,k}^T := \sigma\left(\mathcal{F}_{nT} \cup \sigma\left(X_{n,nT+1}, \ldots, X_{nk}\right)\right)$ for $n \in \mathbb{N}$ and $k \geq nT + 1$, then $(X_{nk})_{k, n \in \mathbb{N}}$ is a square integrable martingale difference array w.r.t. $\left(\mathcal{F}_{n,k}^T\right)_{k \geq 0, n \in \mathbb{N}}$ with

$$\sum_{k=1}^{\infty} E\left(X_{nk}^2 | \mathcal{F}_{n,k-1}^T\right) = \infty \quad \text{a.s. for all } n \in \mathbb{N}.$$

For all $n \in \mathbb{N}$ and $t \in [0, \infty)$ we define the almost surely finite stopping times

$$\tau_n(t) := \max\left\{ j \geq 0 : \sum_{k=1}^{j} E\left(X_{nk}^2 | \mathcal{F}_{n,k-1}^T\right) \leq t \right\}$$

w.r.t. $\left(\mathcal{F}_{n,k}^T\right)_{k \geq 0}$ and

$$T_n(t) := \sum_{k=1}^{\tau_n(t)} X_{nk}.$$

Our first aim is to show that the process $(T_n(t))_{t \in [0,\infty)}$ satisfies for every $\widetilde{T} \in \mathbb{N}$ and $\varepsilon > 0$,

$$\lim_{\delta \downarrow 0} \limsup_{n \to \infty} P\left(\sup_{\substack{0 \leq s,t \leq \widetilde{T} \\ |s-t| \leq \delta}} |T_n(s) - T_n(t)| \geq \varepsilon \right) = 0.$$

By monotonicity it is sufficient to show

$$\lim_{M \to \infty} \limsup_{n \to \infty} P\left(\sup_{\substack{0 \leq s,t \leq \widetilde{T} \\ |s-t| \leq 1/M}} |T_n(s) - T_n(t)| \geq \varepsilon \right) = 0.$$

To prove this, we use a classical discretization technique. Clearly,

$$\sup_{\substack{0 \leq s,t \leq \widetilde{T} \\ |s-t| \leq 1/M}} |T_n(s) - T_n(t)|$$

$$\leq 3 \max_{0 \leq m \leq \widetilde{T}M-1} \sup_{m/M \leq t \leq (m+1)/M} \left| T_n(t) - T_n\left(\frac{m}{M}\right) \right|$$

$$\leq 3 \max_{0 \leq m \leq \widetilde{T}M-1} \max_{\tau_n(m/M)+1 \leq j \leq \tau_n((m+1)/M)} \left| \sum_{k=\tau_n(m/M)+1}^{j} X_{nk} \right|$$

so that

$$
P\left(\sup_{\substack{0\le s,t\le\tilde{T}\\|s-t|\le 1/M}}|T_n(s)-T_n(t)|\ge\varepsilon\right)
$$

$$
\le\sum_{m=0}^{\tilde{T}M-1}P\left(\max_{\tau_n(m/M)+1\le j\le\tau_n((m+1)/M)}\left|\sum_{k=\tau_n(m/M)+1}^{j}X_{nk}\right|\ge\frac{\varepsilon}{3}\right).
$$

Now we use the maximal inequality of Theorem A.10. To apply this inequality note that

$$
\max_{\tau_n(m/M)+1\le j\le\tau_n((m+1)/M)}\left|\sum_{k=\tau_n(m/M)+1}^{j}X_{nk}\right|=\max_{1\le j\le\tau_n((m+1)/M)}\left|\sum_{k=1}^{j}X_{nk}I_n(k)\right|
$$

with $I_n(k):=1_{\{\tau_n(m/M)+1\le k\le\tau_n((m+1)/M)\}}$, where m and M are dropped from the notation $I_n(k)$ for convenience. Because $\tau_n(t)$ is a stopping time w.r.t. $\left(\mathcal{F}_{n,k}^T\right)_{k\ge 0}$ for every $t\in[0,\infty)$, the random variable $I_n(k)$ is $\mathcal{F}_{n,k-1}^T$-measurable and, consequently, $(X_{nk}I_n(k))_{k\ge 1}$ is a square integrable martingale difference sequence w.r.t. $\left(\mathcal{F}_{n,k}^T\right)_{k\ge 0}$. For the associated square integrable martingale we have, for all $j\in\mathbb{N}$,

$$
E\left(\left(\sum_{k=1}^{j}X_{nk}I_n(k)\right)^2\right)=E\left(\sum_{k=\tau_n(m/M)+1}^{\tau_n((m+1)/M)}E\left(X_{nk}^2|\mathcal{F}_{n,k-1}^T\right)\right)\le\frac{m+1}{M}
$$

by definition of $\tau_n((m+1)/M)$ so that this martingale is uniformly integrable. Therefore

$$
P\left(\max_{\tau_n(m/M)+1\le j\le\tau_n((m+1)/M)}\left|\sum_{k=\tau_n(m/M)+1}^{j}X_{nk}\right|\ge\frac{\varepsilon}{3}\right)
$$

$$
\le\frac{6}{\varepsilon}E\left(\left|\sum_{k=1}^{\tau_n((m+1)/M)}X_{nk}I_n(k)\right|1_{\left\{\left|\sum_{k=1}^{\tau_n((m+1)/M)}X_{nk}I_n(k)\right|\ge\frac{\varepsilon}{6}\right\}}\right)
$$

$$
\le\frac{6}{\varepsilon}E\left(\left(\sum_{k=1}^{\tau_n((m+1)/M)}X_{nk}I_n(k)\right)^2\right)^{1/2}P\left(\left|\sum_{k=1}^{\tau_n((m+1)/M)}X_{nk}I_n(k)\right|\ge\frac{\varepsilon}{6}\right)^{1/2}
$$

by Theorem A.10 and the Cauchy-Schwarz inequality. Thus we find

$$
P\left(\sup_{\substack{0\le s,t\le\widetilde{T}\\|s-t|\le 1/M}}|T_n(s)-T_n(t)|\ge\varepsilon\right)
$$

$$
\le\frac{6}{\varepsilon}\sum_{m=0}^{\widetilde{T}M-1}E\left(\sum_{k=\tau_n(m/M)+1}^{\tau_n((m+1)/M)}E\left(X_{nk}^2|\mathcal{F}_{n,k-1}^T\right)\right)^{1/2}
$$

$$
\times P\left(\left|\sum_{k=1}^{\tau_n((m+1)/M)}X_{nk}I_n(k)\right|\ge\frac{\varepsilon}{6}\right)^{1/2}
$$

$$
\le\frac{6}{\varepsilon}\left(\sum_{m=0}^{\widetilde{T}M-1}E\left(\sum_{k=\tau_n(m/M)+1}^{\tau_n((m+1)/M)}E\left(X_{nk}^2|\mathcal{F}_{n,k-1}^T\right)\right)\right)^{1/2}
$$

$$
\times\left(\sum_{m=0}^{\widetilde{T}M-1}P\left(\left|\sum_{k=1}^{\tau_n((m+1)/M)}X_{nk}I_n(k)\right|\ge\frac{\varepsilon}{6}\right)\right)^{1/2}
$$

$$
\le\frac{6}{\varepsilon}\widetilde{T}^{1/2}\left(\sum_{m=0}^{\widetilde{T}M-1}P\left(\left|\sum_{k=1}^{\tau_n((m+1)/M)}X_{nk}I_n(k)\right|\ge\frac{\varepsilon}{6}\right)\right)^{1/2}
$$

because

$$
\sum_{m=0}^{\widetilde{T}M-1}E\left(\sum_{k=\tau_n(m/M)+1}^{\tau_n((m+1)/M)}E\left(X_{nk}^2|\mathcal{F}_{n,k-1}^T\right)\right)=E\left(\sum_{k=1}^{\tau_n(\widetilde{T})}E\left(X_{nk}^2|\mathcal{F}_{n,k-1}^T\right)\right)\le\widetilde{T}
$$

by definition of $\tau_n\left(\widetilde{T}\right)$.

The probabilities on the right-hand side of the last chain of inequalities will be handled by the martingale central limit theorem. Note that for all $t\in[0,\infty)$, $\varepsilon>0$ and $n\in\mathbb{N}$ with $\varepsilon\sqrt{n}>1$

$$
\sum_{k=1}^{\tau_n(t)}E\left(X_{nk}^2 1_{\{|X_{nk}|\ge\varepsilon\}}|\mathcal{F}_{n,k-1}^T\right)\le\sum_{k=1}^{nT}E\left(X_{nk}^2 1_{\{|X_{nk}|\ge\varepsilon\}}|\mathcal{F}_{k-1}\right)
$$

because $1_{\{|X_{nk}|\ge\varepsilon\}}=0$ for all $k\ge nT+1$. Therefore, (CLB_t) implies

$$
\sum_{k=1}^{\tau_n(t)}E\left(X_{nk}^2 1_{\{|X_{nk}|\ge\varepsilon\}}|\mathcal{F}_{n,k-1}^T\right)\to 0\quad\text{in probability as }n\to\infty
$$

for all $t \in [0, \infty)$ and $\varepsilon > 0$. Moreover, for all $t \in [0, \infty)$ and $n \in \mathbb{N}$,

$$\max_{1 \le k \le \tau_n(t)+1} E\left(X_{nk}^2 | \mathcal{F}_{n,k-1}^T\right) \le \max_{1 \le k \le nT} E\left(X_{nk}^2 | \mathcal{F}_{k-1}\right) \vee \frac{1}{n}$$

so that, by (CLB$_t$) and Proposition 6.7, for all $t \in [0, \infty)$,

$$\max_{1 \le k \le \tau_n(t)+1} E\left(X_{nk}^2 | \mathcal{F}_{n,k-1}^T\right) \to 0 \quad \text{in probability as } n \to \infty.$$

Since by definition of $\tau_n(t)$ we have

$$\sum_{k=1}^{\tau_n(t)} E\left(X_{nk}^2 | \mathcal{F}_{n,k-1}^T\right) \le t < \sum_{k=1}^{\tau_n(t)+1} E\left(X_{nk}^2 | \mathcal{F}_{n,k-1}^T\right)$$

it follows that

$$\sum_{k=1}^{\tau_n(t)} E\left(X_{nk}^2 | \mathcal{F}_{n,k-1}^T\right) \to t \quad \text{in probability as } n \to \infty.$$

Therefore, in probability as $n \to \infty$, for all $0 \le m \le \widetilde{T}M - 1$ and $M \in \mathbb{N}$,

$$\sum_{k=1}^{\tau_n((m+1)/M)} E\left((X_{nk}I_n(k))^2 \, 1_{\{|X_{nk}I_n(k)| \ge \varepsilon\}} | \mathcal{F}_{n,k-1}^T\right) \to 0$$

and

$$\sum_{k=1}^{\tau_n((m+1)/M)} E\left((X_{nk}I_n(k))^2 | \mathcal{F}_{n,k-1}^T\right) = \sum_{k=\tau_n(m/M)+1}^{\tau_n((m+1)/M)} E\left(X_{nk}^2 | \mathcal{F}_{n,k-1}^T\right)$$

$$= \sum_{k=1}^{\tau_n((m+1)/M)} E\left(X_{nk}^2 | \mathcal{F}_{n,k-1}^T\right) - \sum_{k=1}^{\tau_n(m/M)} E\left(X_{nk}^2 | \mathcal{F}_{n,k-1}^T\right) \to \frac{1}{M}.$$

The martingale central limit theorem in the form of Corollary 6.4 gives

$$\sum_{k=1}^{\tau_n((m+1)/M)} X_{nk}I_n(k) \overset{d}{\to} N_M \quad \text{as } n \to \infty$$

where $P^{NM} = N(0, 1/M)$ so that

$$\lim_{n \to \infty} P\left(\left|\sum_{k=1}^{\tau_n((m+1)/M)} X_{nk} I_n(k)\right| \geq \frac{\varepsilon}{6}\right) = 2\left(1 - \Phi\left(\frac{\varepsilon}{6}M^{1/2}\right)\right),$$

where Φ denotes the distribution function of the standard normal distribution. Hence

$$\limsup_{n \to \infty} P\left(\sup_{\substack{0 \leq s,t \leq \tilde{T} \\ |s-t| \leq 1/M}} |T_n(s) - T_n(t)| \geq \varepsilon\right) \leq \frac{12}{\varepsilon}\tilde{T}M^{1/2}\left(1 - \Phi\left(\frac{\varepsilon}{6}M^{1/2}\right)\right)^{1/2}.$$

The bound on the right-hand side clearly converges to zero as $M \to \infty$ and completes the proof.

Step 2. In the second part of the proof we will switch from the time scales $\tau_n(\cdot)$ to the time scales $[n \cdot]$ used in the definition of $S_{(n)}$. The potentialities of such a random change of time in martingale central limit theory are elucidated in [81, 82]. Note that

$$t < \sum_{k=1}^{\tau_n(t)+1} E\left(X_{nk}^2 | \mathcal{F}_{n,k-1}^T\right)$$

by definition of $\tau_n(t)$ so that $\tau_n(t) \to \infty$ almost surely as $t \to \infty$. Consequently,

$$\tau_n^{-1}(j) := \inf\{t \in [0, \infty) : \tau_n(t) \geq j\}$$

is almost surely well-defined for all $j \geq 0$. If $j \geq 0$ is fixed, then for all $t \in [0, \infty)$, by definition of $\tau_n^{-1}(j)$ and $\tau_n(t)$,

$$t < \tau_n^{-1}(j) \quad \Leftrightarrow \quad \tau_n(t) < j \quad \Leftrightarrow \quad \sum_{k=1}^{j} E\left(X_{nk}^2 | \mathcal{F}_{n,k-1}^T\right) > t,$$

which implies

$$\tau_n^{-1}(j) = \sum_{k=1}^{j} E\left(X_{nk}^2 | \mathcal{F}_{n,k-1}^T\right)$$

so that

$$j \in \left\{\tilde{j} \geq 0 : \sum_{k=1}^{\tilde{j}} E\left(X_{nk}^2 | \mathcal{F}_{n,k-1}^T\right) \leq \tau_n^{-1}(j)\right\}$$

and hence $\tau_n \left(\tau_n^{-1}(j) \right) \geq j$. Moreover, for all $j \geq 0$,

$$\sum_{k=1}^{\tau_n\left(\tau_n^{-1}(j)\right)} X_{nk} = \sum_{k=1}^{j} X_{nk} \quad \text{a.s.}$$

To see this, write

$$\sum_{k=1}^{\tau_n\left(\tau_n^{-1}(j)\right)} X_{nk} = \sum_{k=1}^{j} X_{nk} + \sum_{k=j+1}^{\tau_n\left(\tau_n^{-1}(j)\right)} X_{nk}$$

and note that

$$\sum_{k=1}^{j} E\left(X_{nk}^2 | \mathcal{F}_{n,k-1}^T \right) + \sum_{k=j+1}^{\tau_n\left(\tau_n^{-1}(j)\right)} E\left(X_{nk}^2 | \mathcal{F}_{n,k-1}^T \right)$$

$$= \sum_{k=1}^{\tau_n\left(\tau_n^{-1}(j)\right)} E\left(X_{nk}^2 | \mathcal{F}_{n,k-1}^T \right) \leq \tau_n^{-1}(j) = \sum_{k=1}^{j} E\left(X_{nk}^2 | \mathcal{F}_{n,k-1}^T \right)$$

which gives

$$\sum_{k=j+1}^{\tau_n\left(\tau_n^{-1}(j)\right)} E\left(X_{nk}^2 | \mathcal{F}_{n,k-1}^T \right) = 0 \quad \text{a.s.},$$

whence

$$\sum_{k=j+1}^{\tau_n\left(\tau_n^{-1}(j)\right)} X_{nk} = 0 \quad \text{a.s.}$$

because $X_{nk}^2 = 0$ almost surely on the event $\left\{ E\left(X_{nk}^2 | \mathcal{F}_{n,k-1}^T \right) = 0 \right\}$.

By monotonicity in t we get

$$\sup_{0 \leq t \leq T} \left| \sum_{k=1}^{[nt]} E\left(X_{nk}^2 | \mathcal{F}_{k-1} \right) - \eta^2(t) \right|$$

$$\leq \max_{0 \leq m \leq MT} \left| \sum_{k=1}^{[nm/M]} E\left(X_{nk}^2 | \mathcal{F}_{k-1} \right) - \eta^2\left(\frac{m}{M} \right) \right|$$

$$+ \sup_{\substack{0 \leq s, t \leq T \\ |s-t| \leq 1/M}} \left| \eta^2(s) - \eta^2(t) \right|$$

for every $M \in \mathbb{N}$, so that

$$\sup_{0 \le t \le T} \left| \sum_{k=1}^{[nt]} E\left(X_{nk}^2 | \mathcal{F}_{k-1}\right) - \eta^2(t) \right| \to 0 \quad \text{in probability as } n \to \infty$$

by (N_t) and continuity of the paths of the process η^2. For all $n \in \mathbb{N}$ and $0 \le t \le T$ we have $nt \le nT$ and therefore

$$\sum_{k=1}^{[nt]} E\left(X_{nk}^2 | \mathcal{F}_{n,k-1}^T\right) = \tau_n^{-1}([nt]) =: \eta_n(t)$$

so that

$$\sup_{0 \le t \le T} \left| \eta_n(t) - \eta^2(t) \right| \to 0 \quad \text{in probability as } n \to \infty.$$

Now we can show that

$$\lim_{\delta \downarrow 0} \limsup_{n \to \infty} P\left(\sup_{\substack{0 \le s,t \le T \\ |s-t| \le \delta}} \left| \sum_{k=1}^{[ns]} X_{nk} - \sum_{k=1}^{[nt]} X_{nk} \right| \ge \varepsilon \right) = 0.$$

To do this, observe again that $nt \le nT$ for all $t \in [0, T]$ so that with probability one

$$\sum_{k=1}^{[nt]} X_{nk} = \sum_{k=1}^{\tau_n\left(\tau_n^{-1}([nt])\right)} X_{nk} = \sum_{k=1}^{\tau_n(\eta_n(t))} X_{nk} = T_n(\eta_n(t)).$$

Therefore

$$P\left(\sup_{\substack{0 \le s,t \le T \\ |s-t| \le \delta}} \left| \sum_{k=1}^{[ns]} X_{nk} - \sum_{k=1}^{[nt]} X_{nk} \right| \ge \varepsilon \right)$$

$$= P\left(\sup_{\substack{0 \le s,t \le T \\ |s-t| \le \delta}} |T_n(\eta_n(s)) - T_n(\eta_n(t))| \ge \varepsilon \right).$$

For all $\tilde{T} \in \mathbb{N}$ and $0 < d \le 4$ on the event

$$A := \left\{ \sup_{\substack{0 \le s,t \le T \\ |s-t| \le \delta}} |T_n(\eta_n(s)) - T_n(\eta_n(t))| \ge \varepsilon \right\} \cap \left\{ \eta^2(T) \le \widetilde{T} \right\} \cap$$

$$\left\{ \sup_{0 \le t \le T} \left| \eta_n(t) - \eta^2(t) \right| \le \frac{d}{4} \right\} \cap \left\{ \sup_{\substack{0 \le s,t \le T \\ |s-t| \le \delta}} \left| \eta^2(s) - \eta^2(t) \right| \le \frac{d}{2} \right\}$$

we get, for all $s, t \in [0, T]$ with $|s - t| \le \delta$,

$$|\eta_n(s) - \eta_n(t)| \le 2 \sup_{0 \le t \le T} \left| \eta_n(t) - \eta^2(t) \right| + \sup_{\substack{0 \le s,t \le T \\ |s-t| \le \delta}} \left| \eta^2(s) - \eta^2(t) \right| \le d$$

and, recalling that the paths of η^2 are nondecreasing,

$$\eta_n(t) \le \eta^2(t) + \left| \eta_n(t) - \eta^2(t) \right| \le \widetilde{T} + \frac{d}{4} \le \widetilde{T} + 1$$

as well as $\eta_n(s) \le \widetilde{T} + 1$. Therefore

$$A \subset \left\{ \sup_{\substack{0 \le s,t \le \widetilde{T}+1 \\ |s-t| \le d}} |T_n(s) - T_n(t)| \ge \varepsilon \right\}$$

so that

$$P \left(\sup_{\substack{0 \le s,t \le T \\ |s-t| \le \delta}} \left| \sum_{k=1}^{[ns]} X_{nk} - \sum_{k=1}^{[nt]} X_{nk} \right| \ge \varepsilon \right)$$

$$\le P \left(\sup_{\substack{0 \le s,t \le \widetilde{T}+1 \\ |s-t| \le d}} |T_n(s) - T_n(t)| \ge \varepsilon \right) + P \left(\eta^2(T) > \widetilde{T} \right)$$

$$+ P \left(\sup_{0 \le t \le T} \left| \eta_n(t) - \eta^2(t) \right| > \frac{d}{4} \right) + P \left(\sup_{\substack{0 \le s,t \le T \\ |s-t| \le \delta}} \left| \eta^2(s) - \eta^2(t) \right| > \frac{d}{2} \right)$$

which yields, for all $\delta > 0$, $\widetilde{T} \in \mathbb{N}$ and $0 < d \le 4$,

$$\limsup_{n \to \infty} P\left(\sup_{\substack{0 \le s,t \le T \\ |s-t| \le \delta}} \left| \sum_{k=1}^{[ns]} X_{nk} - \sum_{k=1}^{[nt]} X_{nk} \right| \ge \varepsilon \right)$$

$$\le \limsup_{n \to \infty} P\left(\sup_{\substack{0 \le s,t \le \widetilde{T}+1 \\ |s-t| \le d}} |T_n(s) - T_n(t)| \ge \varepsilon \right)$$

$$+ P\left(\eta^2(T) > \widetilde{T} \right) + P\left(\sup_{\substack{0 \le s,t \le T \\ |s-t| \le \delta}} \left| \eta^2(s) - \eta^2(t) \right| > \frac{d}{2} \right).$$

By continuity of the paths of η^2 it follows for all $\widetilde{T} \in \mathbb{N}$ and $0 < d \le 4$ that

$$\lim_{\delta \downarrow 0} \limsup_{n \to \infty} P\left(\sup_{\substack{0 \le s,t \le T \\ |s-t| \le \delta}} \left| \sum_{k=1}^{[ns]} X_{nk} - \sum_{k=1}^{[nt]} X_{nk} \right| \ge \varepsilon \right)$$

$$\le \limsup_{n \to \infty} P\left(\sup_{\substack{0 \le s,t \le \widetilde{T}+1 \\ |s-t| \le d}} |T_n(s) - T_n(t)| \ge \varepsilon \right) + P\left(\eta^2(T) > \widetilde{T} \right).$$

The right-hand side of this inequality converges to zero as $d \downarrow 0$ followed by $\widetilde{T} \to \infty$, which concludes the proof of

$$\lim_{\delta \downarrow 0} \limsup_{n \to \infty} P\left(\sup_{\substack{0 \le s,t \le T \\ |s-t| \le \delta}} \left| \sum_{k=1}^{[ns]} X_{nk} - \sum_{k=1}^{[nt]} X_{nk} \right| \ge \varepsilon \right) = 0.$$

Because

$$\sup_{0 \le t \le T} \left| S_{(n)}(t) - \sum_{k=1}^{[nt]} X_{nk} \right| \le \max_{1 \le k \le nT} |X_{nk}| \to 0 \quad \text{in probability as } n \to \infty$$

the sequence $\left(S_{(n)}(t) \right)_{t \in [0,T]}$, $n \in \mathbb{N}$, satisfies

$$\lim_{\delta \downarrow 0} \limsup_{n \to \infty} P\left(\sup_{\substack{0 \le s,t \le T \\ |s-t| \le \delta}} \left| S_{(n)}(s) - S_{(n)}(t) \right| \ge \varepsilon \right) = 0,$$

as claimed. $\qquad\square$

Now we show that conditions (N_t) and (CLB_t) may be replaced by other sets of sufficient conditions which are functional versions of the conditions appearing in Sect. 6.3.

For a square integrable array $(X_{nk})_{n,k\in\mathbb{N}}$ of random variables we introduce

$(M_{2,t})$ $$E\left(\max_{1\le k\le[nt]} X_{nk}^2\right) \to 0 \quad \text{as } n \to \infty \text{ for all } t \in [0, \infty)$$

whereas the conditions

$(M_{1,t})$ $$E\left(\max_{1\le k\le[nt]} |X_{nk}|\right) \to 0 \quad \text{as } n \to \infty \text{ for all } t \in [0, \infty)$$

and

$(CLB_{1,t})$ $$\sum_{k=1}^{[nt]} E\left(|X_{nk}|\, 1_{\{|X_{nk}|\ge\varepsilon\}}|\mathcal{F}_{k-1}\right) \to 0 \quad \text{in probability as } n \to \infty$$

for every $\varepsilon > 0$ and all $t \in [0, \infty)$

only require integrable random variables.

The functional form of Raikov's condition

(R_t) $$\sum_{k=1}^{[nt]} X_{nk}^2 \to \eta^2(t) \quad \text{in probability as } n \to \infty \text{ for all } t \in [0, \infty)$$

and, for any $a > 0$, the conditions

$(T_{a,t})$ $$\sup_{0\le s\le t}\left|\sum_{k=1}^{[ns]}\left[X_{nk}1_{\{|X_{nk}|>a\}} + E\left(X_{nk}1_{\{|X_{nk}|\le a\}}|\mathcal{F}_{k-1}\right)\right]\right| \to 0$$

in probability as $n \to \infty$ for all $t \in [0, \infty)$

and

$(TR_{a,t})$ $$\sum_{k=1}^{[nt]}\left[X_{nk}1_{\{|X_{nk}|\le a\}} - E\left(X_{nk}1_{\{|X_{nk}|\le a\}}|\mathcal{F}_{k-1}\right)\right]^2 \to \eta^2(t)$$

in probability as $n \to \infty$ for all $t \in [0, \infty)$

are meaningful without any integrability assumption on the X_{nk}.

For these conditions we have the following analogue of 6.20. Here we assume for every $n \in \mathbb{N}$ that $(X_{nk})_{k\in\mathbb{N}}$ is adapted to $\mathbb{F}$ and that $(X_{nk})_{k\in\mathbb{N}}$ is a martingale difference sequence w.r.t. $\mathbb{F}$ for the implication $\overset{(*)}{\Longrightarrow}$.

7.2 *Conditions in the functional martingale central limit theorem*:

$$(M_{2,t}) \text{ and } (N_t) \;\Rightarrow\; (M_{1,t}) \text{ and } (R_t)$$

$$\Downarrow \qquad\qquad\qquad \Downarrow$$

$$(CLB_t) \text{ and } (N_t) \Rightarrow (CLB_{1,t}) \text{ and } (R_t) \overset{(*)}{\Rightarrow} (T_{a,t}) \text{ and } (TR_{a,t})$$

Note that the implications $(M_{2,t}) \Rightarrow (CLB_t)$ and $(M_{2,t}) \Rightarrow (M_{1,t})$ follow from Proposition 6.15 for $k_n = [nt]$. Moreover, $(CLB_t) \Rightarrow (CLB_{1,t})$ follows from Proposition 6.17, and $(M_{1,t}) \Rightarrow (CLB_{1,t})$ follows from Proposition 6.18 for $k_n = [nt]$. Under $(M_{2,t})$ and (N_t) as well as (CLB_t) and (N_t) Proposition 6.16 is applicable to derive (R_t) from (N_t), again with $k_n = [nt]$. Thus, the four implications in 7.2 without $(*)$ are true. To establish the implication with $(*)$ note that $(T_{a,t})$ follows from $(CLB_{1,t})$ and (R_t) for every $a > 0$ by Proposition 6.19. To derive $(TR_{a,t})$ for every $a > 0$ from $(CLB_{1,t})$ we use, for all $n \in \mathbb{N}$ and $t \in [0, \infty)$, the martingale difference property of the X_{nk} to obtain the inequality

$$\sup_{0 \le s \le t} \left| \sum_{k=1}^{[ns]} X_{nk} 1_{\{|X_{nk}|>a\}} + E\left(X_{nk} 1_{\{|X_{nk}|\le a\}} | \mathcal{F}_{k-1}\right) \right|$$

$$= \sup_{0 \le s \le t} \left| \sum_{k=1}^{[ns]} X_{nk} 1_{\{|X_{nk}|>a\}} - E\left(X_{nk} 1_{\{|X_{nk}|>a\}} | \mathcal{F}_{k-1}\right) \right|$$

$$\le \sum_{k=1}^{[nt]} |X_{nk}| 1_{\{|X_{nk}|>a\}} + \sum_{k=1}^{[nt]} E\left(|X_{nk}| 1_{\{|X_{nk}|>a\}} | \mathcal{F}_{k-1}\right) .$$

Here, the right-hand side converges to zero in probability as $n \to \infty$ by $(CLB_{1,t})$ and Lemma 6.5. Thus, all implications in 7.2 are proven.

The analogue of Proposition 6.21 reads as follows.

Proposition 7.3 *For every $n \in \mathbb{N}$, let $(X_{nk})_{k \in \mathbb{N}}$ be adapted to $\mathbb{F} = (\mathcal{F}_k)_{k \ge 0}$. Assume that there exists some $a > 0$ for which the conditions $(T_{a,t})$ and $(TR_{a,t})$ are satisfied. For all $k, n \in \mathbb{N}$ set*

$$X_{nk}(a) := X_{nk} 1_{\{|X_{nk}|\le a\}} - E\left(X_{nk} 1_{\{|X_{nk}|\le a\}} | \mathcal{F}_{k-1}\right)$$

and for all $n \in \mathbb{N}$ and $t \in [0, \infty)$

$$S_{(n,a)}(t) := \sum_{k=1}^{[nt]} X_{nk}(a) + (nt - [nt]) X_{n,[nt]+1}(a) .$$

Then for every $n \in \mathbb{N}$, $(X_{nk}(a))_{k \in \mathbb{N}}$ is a bounded martingale difference sequence w.r.t. $\mathbb{F}$ which satisfies $(M_{2,t})$ and (N_t) as well as

$$\sup_{0 \le s \le t} \left| S_{(n)}(s) - S_{(n,a)}(s) \right| \to 0 \quad \text{in probability as } n \to \infty \text{ for all } t \in [0, \infty).$$

Proof Let $a > 0$ be fixed such that $(T_{a,t})$ and $(TR_{a,t})$ hold. Then by $(TR_{a,t})$ for all $t \in [0, \infty)$

$$\sum_{k=1}^{[nt]} X_{nk}^2(a) \to \eta^2(t) \quad \text{in probability as } n \to \infty$$

from which, by monotonicity in t for all $t \in [0, \infty)$,

$$\sup_{0 \le s \le t} \left| \sum_{k=1}^{[ns]} X_{nk}^2(a) - \eta^2(s) \right| \to 0 \quad \text{in probability as } n \to \infty.$$

Put $Z_{(n)}(s) := \sum_{k=1}^{[ns]} X_{nk}^2(a)$ and let $f(s-0)$ denote the left-hand limit of $f : [0, \infty) \to \mathbb{R}$ at $s \in (0, \infty)$ provided the limit exists. By continuity of the paths of η^2 we obtain from the last display that

$$\sup_{0 \le s \le t} \left| Z_{(n)}(s-0) - \eta^2(s) \right| \to 0 \quad \text{in probability as } n \to \infty$$

so that for all $t \in [0, \infty)$

$$\max_{1 \le k \le [nt]} X_{nk}^2(a) \le \sup_{0 \le s \le t} \left| Z_{(n)}(s) - Z_{(n)}(s-0) \right| \to 0 \quad \text{in probability as } n \to \infty$$

which shows that the array $(X_{nk}(a))_{1 \le k \le [nt], n \in \mathbb{N}}$ satisfies condition (TM_a). Therefore, Proposition 6.21 implies that conditions $(M_{2,t})$ and (N_t) are satisfied for $(X_{nk}(a))_{k \in \mathbb{N}}$, $n \in \mathbb{N}$. For all $n \in \mathbb{N}$ and $0 \le s \le t < \infty$ we have

$$\left| S_{(n)}(s) - S_{(n,a)}(s) \right| \le \left| \sum_{k=1}^{[ns]} X_{nk} - \sum_{k=1}^{[ns]} X_{nk}(a) \right| + \left| X_{n,[ns]+1} - X_{n,[ns]+1}(a) \right|$$

so that

$$\sup_{0 \le s \le t} \left| S_{(n)}(s) - S_{(n,a)}(s) \right|$$

$$\le \sup_{0 \le s \le t} \left| \sum_{k=1}^{[ns]} X_{nk} - \sum_{k=1}^{[ns]} X_{nk}(a) \right| + \max_{1 \le k \le [nt]+1} |X_{nk} - X_{nk}(a)|$$

$$= \sup_{0 \le s \le t} \left| \sum_{k=1}^{[ns]} \left[X_{nk} 1_{\{|X_{nk}|>a\}} + E\left(X_{nk} 1_{\{|X_{nk}| \le a\}} | \mathcal{F}_{k-1} \right) \right] \right|$$

$$+ \max_{1 \le k \le [nt]+1} \left| X_{nk} 1_{\{|X_{nk}|>a\}} + E\left(X_{nk} 1_{\{|X_{nk}| \le a\}} | \mathcal{F}_{k-1} \right) \right|.$$

Both summands on the right-hand side converge to zero in probability as $n \to \infty$ by condition $(T_{a,t})$ (observe that $[nt] + 1 \leq [n(t+1)]$ and note that

$$X_{nk}1_{\{|X_{nk}|>a\}} + E\left(X_{nk}1_{\{|X_{nk}|\leq a\}}|\mathcal{F}_{k-1}\right)$$

is the jump of the process

$$\sum_{k=1}^{[ns]} X_{nk}1_{\{|X_{nk}|>a\}} + E\left(X_{nk}1_{\{|X_{nk}|\leq a\}}|\mathcal{F}_{k-1}\right), \quad s \in [0, \infty)$$

at time $s = k/n$ and that these processes converge to zero in probability uniformly on compact intervals by $(T_{a,t})$). This completes the proof of the proposition. □

In many applications martingale difference arrays are obtained from a single martingale difference sequence through renormalization. For this, let $(X_k)_{k\in\mathbb{N}}$ be a square integrable martingale difference sequence w.r.t. $\mathbb{F}$. For every $n \in \mathbb{N}$ and $t \in [0, \infty)$ we set

$$S_n(t) := \sum_{k=1}^{[nt]} X_k + (nt - [nt]) X_{[nt]+1}$$

so that $(S_n(t))_{t\in[0,\infty)}$ is a random process with paths in $C(\mathbb{R}_+)$. Its convergence in distribution requires renormalization. For this, let $(a_n)_{n\in\mathbb{N}}$ be a sequence of positive real numbers with $a_n \to \infty$ as $n \to \infty$. For a nonnegative stochastic process $(\eta(t))_{t\in[0,\infty)}$ with paths in $C(\mathbb{R}_+)$ and square integrable X_k we introduce the conditions

$(N_{a_n,t})$ $\dfrac{1}{a_n^2}\sum_{k=1}^{[nt]} E\left(X_k^2|\mathcal{F}_{k-1}\right) \to \eta^2(t)$ in probability as $n \to \infty$

           for all $t \in [0, \infty)$

and

$(CLB_{a_n,t})$ $\dfrac{1}{a_n^2}\sum_{k=1}^{[nt]} E\left(X_k^2 1_{\{|X_k|\geq\varepsilon a_n\}}|\mathcal{F}_{k-1}\right) \to 0$ in probability as $n \to \infty$

           for all $\varepsilon > 0$ and all $t \in [0, \infty)$.

Note that any process η^2 appearing in $(N_{a_n,t})$ is nonnegative with almost surely nondecreasing paths. The following result is a special case of Theorem 7.1.

Theorem 7.4 *Let $(X_k)_{k\in\mathbb{N}}$ be a square integrable martingale difference sequence w.r.t. $\mathbb{F} = (\mathcal{F}_k)_{k\geq0}$. Under $(N_{a_n,t})$ and $(CLB_{a_n,t})$,*

$$\frac{1}{a_n}S_n \to \left(W\left(\eta^2(t)\right)\right)_{t\in[0,\infty)} \quad \mathcal{F}_\infty\text{-stably as } n \to \infty,$$

where $W = (W(t))_{t\geq0}$ is a Brownian motion which is independent of $\mathcal{F}_\infty$.

The conditions $(N_{a_n,t})$ and $(CLB_{a_n,t})$ may be replaced by

$$(R_{a_n,t}) \qquad \frac{1}{a_n^2}\sum_{k=1}^{[nt]}X_k^2 \to \eta^2(t) \quad \text{in probability as } n \to \infty \text{ for all } t \in [0,\infty)$$

and

$$(M_{1,a_n,t}) \qquad \frac{1}{a_n}E\left(\max_{1\leq k\leq[nt]}|X_k|\right) \to 0 \quad \text{as } n \to \infty \text{ for all } t \in [0,\infty),$$

which are meaningful for all martingale difference sequences, i.e. without the assumption of square integrability.

Theorem 7.5 *Let $(X_k)_{k\in\mathbb{N}}$ be a martingale difference sequence w.r.t. $\mathbb{F} = (\mathcal{F}_k)_{k\geq0}$. Under $(R_{a_n,t})$ and $(M_{1,a_n,t})$,*

$$\frac{1}{a_n}S_n \to \left(W\left(\eta^2(t)\right)\right)_{t\in[0,\infty)} \quad \mathcal{F}_\infty\text{-stably as } n \to \infty,$$

where $W = (W(t))_{t\geq0}$ is a Brownian motion which is independent of $\mathcal{F}_\infty$.

Proof For all $n, k \in \mathbb{N}$ set $X_{nk} := X_k/a_n$. Because $(M_{1,a_n,t})$ and $(R_{a_n,t})$ are identical to $(M_{1,t})$ and (R_t) for the array $(X_{nk})_{k,n\in\mathbb{N}}$, it follows from 7.2 that for every $a > 0$ the conditions $(T_{a,t})$ and $(TR_{a,t})$ are satisfied for the array $(X_{nk})_{k,n\in\mathbb{N}}$. Now Proposition 7.3 and Theorem 7.1 imply for

$$X_{nk}(a) := X_{nk}1_{\{|X_{nk}|\leq a\}} - E\left(X_{nk}1_{\{|X_{nk}|\leq a\}}|\mathcal{F}_{k-1}\right)$$

and

$$S_{(n,a)}(t) := \sum_{k=1}^{[nt]}X_{nk}(a) + (nt - [nt])X_{n,[nt]+1}(a)$$

that

$$S_{(n,a)} \to \left(W\left(\eta^2(t)\right)\right)_{t\in[0,\infty)} \quad \mathcal{F}_\infty\text{-stably in } C(\mathbb{R}_+) \text{ as } n \to \infty.$$

Consequently, for all $0 < T < \infty$, by Theorem 3.18 (c) and continuity of the restriction map,

$$\left(S_{(n,a)}(t)\right)_{t\in[0,T]} \rightarrow \left(W\left(\eta^2(t)\right)\right)_{t\in[0,T]} \quad \mathcal{F}_\infty\text{-stably as } n\rightarrow\infty \text{ in } C\left([0,T]\right).$$

For the process

$$\frac{1}{a_n}S_n(t) = \frac{1}{a_n}\left(\sum_{k=1}^{[nt]}X_k + (nt-[nt])X_{[nt]+1}\right)$$

$$= \sum_{k=1}^{[nt]}X_{nk} + (nt-[nt])X_{n,[nt]+1} = S_{(n)}(t)$$

we have, also by Proposition 7.3,

$$\sup_{0\leq t\leq T}\left|\frac{1}{a_n}S_n(t) - S_{(n,a)}(t)\right| \rightarrow 0 \quad \text{in probability as } n\rightarrow\infty.$$

Theorem 3.18 (a) now implies

$$\left(\frac{1}{a_n}S_n(t)\right)_{t\in[0,T]} \rightarrow \left(W\left(\eta^2(t)\right)\right)_{t\in[0,T]} \quad \mathcal{F}_\infty\text{-stably as } n\rightarrow\infty \text{ in } C\left([0,T]\right),$$

and

$$\frac{1}{a_n}S_n \rightarrow \left(W\left(\eta^2(t)\right)\right)_{t\in\mathbb{R}_+} \quad \mathcal{F}_\infty\text{-stably as } n\rightarrow\infty \text{ in } C\left(\mathbb{R}_+\right)$$

follows from Corollary 3.23.   □

Remark 7.6 Let the sequence $(a_n)_{n\geq 1}$ be regularly varying, that is

$$\frac{a_{[n\lambda]}}{a_n} \rightarrow \Psi(\lambda) \quad \text{as } n\rightarrow\infty \text{ for all } \lambda\in(0,\infty)$$

and some positive function Ψ which is necessarily of the form $\Psi(\lambda) = \lambda^\rho$ for some $\rho\in\mathbb{R}_+$; see [10], Theorem 1.9.5. Assume $\rho > 0$.
(a) Condition

$$(N_{a_n}) \qquad \frac{1}{a_n^2}\sum_{k=1}^{n}E\left(X_k^2|\mathcal{F}_{k-1}\right) \rightarrow \eta^2 \quad \text{in probability as } n\rightarrow\infty$$

$$\text{for some real random variable } \eta\geq 0$$

from Sect. 6.4 implies condition $(\mathrm{N}_{a_n,t})$ with $\eta^2(t) = \Psi^2(t)\eta^2$ for all $t \in [0, \infty)$ (with $\Psi(0) := 0$) because

$$\frac{1}{a_n^2} \sum_{k=1}^{[nt]} E\left(X_k^2 | \mathcal{F}_{k-1}\right) = \left(\frac{a_{[nt]}}{a_n}\right)^2 \frac{1}{a_{[nt]}^2} \sum_{k=1}^{[nt]} E\left(X_k^2 | \mathcal{F}_{k-1}\right) \to \Psi^2(t)\eta^2$$

in probability as $n \to \infty$ for all $t \in (0, \infty)$.
(b) Condition

(CLB_{a_n}) $\displaystyle\frac{1}{a_n^2} \sum_{k=1}^{n} E\left(X_k^2 \mathbf{1}_{\{|X_k| \ge \varepsilon a_n\}} | \mathcal{F}_{k-1}\right) \to 0$ in probability as $n \to \infty$

for all $\varepsilon > 0$

implies $(\mathrm{CLB}_{a_n,t})$ because

$$\frac{1}{a_n^2} \sum_{k=1}^{[nt]} E\left(X_k^2 \mathbf{1}_{\{|X_k| \ge \varepsilon a_n\}} | \mathcal{F}_{k-1}\right)$$

$$= \left(\frac{a_{[nt]}}{a_n}\right)^2 \frac{1}{a_{[nt]}^2} \sum_{k=1}^{[nt]} E\left(X_k^2 \mathbf{1}_{\{|X_k| \ge (\varepsilon a_n/a_{[nt]})a_{[nt]}\}} | \mathcal{F}_{k-1}\right) \to 0$$

in probability as $n \to \infty$ for all $t \in (0, \infty)$, taking into account that $\left(a_{[nt]}/a_n\right)^2$ is bounded in n and $\varepsilon a_n/a_{[nt]} \to \varepsilon/\Psi(t) > 0$ as $n \to \infty$.
(c) Condition

(R_{a_n}) $\displaystyle\frac{1}{a_n^2} \sum_{k=1}^{n} X_k^2 \to \eta^2$ in probability as $n \to \infty$

for some real random variable $\eta \ge 0$

implies condition $(\mathrm{R}_{a_n,t})$ with $\eta^2(t) = \Psi^2(t)\eta^2$ for all $t \in [0, \infty)$ (with $\Psi(0) := 0$) by the same argument as in (a).
(d) Condition

(M_{1,a_n}) $\displaystyle\frac{1}{a_n} E\left(\max_{1 \le k \le n} |X_k|\right) \to 0$ as $n \to \infty$

implies $(\mathrm{M}_{1,a_n,t})$ because

$$\frac{1}{a_n} E\left(\max_{1 \le k \le [nt]} |X_k|\right) = \frac{a_{[nt]}}{a_n} \frac{1}{a_{[nt]}} E\left(\max_{1 \le k \le [nt]} |X_k|\right) \to 0 \quad \text{as } n \to \infty$$

for all $t \in (0, \infty)$ in view of the boundedness of $\left(a_{[nt]}/a_n\right)$.

Corollary 7.7 (Classical stable functional central limit theorem) *Let $(X_k)_{k \geq 1}$ be an independent sequence such that $(|X_k|)_{k \geq 1}$ is an identically distributed sequence, $X_1 \in \mathcal{L}^2(P)$ and $E X_k = 0$ for every $k \in \mathbb{N}$. Then*

$$\frac{1}{\sqrt{n}} S_n \to \sigma W \quad \mathcal{F}_\infty\text{-mixing as } n \to \infty \text{ in } C(\mathbb{R}_+),$$

where $\sigma^2 = \operatorname{Var} X_1$, $\mathcal{F}_\infty = \sigma(X_k, k \geq 1)$ and W is a Brownian motion which is independent of $\mathcal{F}_\infty$.

Proof Take $a_n = \sqrt{n}$, $\mathcal{F}_k = \sigma(X_1, \ldots, X_k)$, $\eta^2(t) = \sigma^2 t$ in Theorem 7.4, and use the scaling property of W. $\qquad\square$

Corollary 7.8 (Stationary martingale differences) *Let $(X_k)_{k \geq 1}$ be a stationary martingale difference sequence w.r.t. $\mathbb{F}$ with $X_1 \in \mathcal{L}^2(P)$. Then*

$$\frac{1}{\sqrt{n}} S_n \to E\left(X_1^2 | \mathcal{I}_X\right)^{1/2} W \quad \mathcal{F}_\infty\text{-stably as } n \to \infty \text{ in } C(\mathbb{R}_+),$$

where $\mathcal{I}_X$ is the invariant σ-field of the stationary process $(X_k)_{k \geq 1}$ and W is a Brownian motion which is independent of $\mathcal{F}_\infty$.

Proof The proof of Corollary 6.26 shows that $(X_k)_{k \geq 1}$ satisfies (R_{a_n}) and (M_{1,a_n}) for $a_n = \sqrt{n}$ and $\eta^2 = E\left(X_1^2 | \mathcal{I}_X\right)$. Therefore, according to Remark 7.6 (c) and (d), $(\mathrm{R}_{a_n,t})$ and $(\mathrm{M}_{1,a_n,t})$ are also satisfied. Theorem 7.5 implies

$$\frac{1}{\sqrt{n}} S_n \to \left(W\left(\eta^2(t)\right)\right)_{t \in [0,\infty)} \quad \mathcal{F}_\infty\text{-stably as } n \to \infty \text{ in } C(\mathbb{R}_+)$$

with $\eta^2(t) = E\left(X_1^2 | \mathcal{I}_X\right) t$ for all $t \in [0, \infty)$. But, by independence of $\sigma(W)$ and $\mathcal{F}_\infty$ and $\mathcal{F}_\infty$-measurability of $E\left(X_1^2 | \mathcal{I}_X\right)$, $\left(W\left(E\left(X_1^2 | \mathcal{I}_X\right) t\right)\right)_{t \in [0,\infty)}$ and $E\left(X_1^2 | \mathcal{I}_X\right)^{1/2} W$ have the same conditional distribution w.r.t $\mathcal{F}_\infty$, which yields the assertion. $\qquad\square$

Corollary 7.9 (Exchangeable processes) *Let $(Z_k)_{k \geq 1}$ be an exchangeable sequence of real random variables with $Z_1 \in \mathcal{L}^2(P)$ and let $X_k := Z_k - E(Z_1 | \mathcal{T}_Z)$. Then*

$$\frac{1}{\sqrt{n}} S_n \to \operatorname{Var}(Z_1 | \mathcal{T}_Z)^{1/2} W \quad \mathcal{F}_\infty\text{-stably as } \mathrm{n} \to \infty \text{ in } C(\mathbb{R}_+),$$

where $\mathcal{T}_Z$ is the tail-σ-field of the sequence $(Z_k)_{k \geq 1}$ and W is a Brownian motion which is independent of $\mathcal{F}_\infty = \sigma(Z_k; k \geq 1)$.

Proof Corollary 7.8; see also the proof of Corollary 6.27. $\qquad\square$

For arbitrary stationary sequences $(X_k)_{k \in \mathbb{N}}$, Corollary 7.8 combined with martingale approximations of the partial sums $\left(\sum_{k=1}^{n} X_k\right)_{n \in \mathbb{N}}$ yield, under suitable conditions, a stable functional central limit theorem (cf. e.g. [38, 66]). Recall that by Example 5.7, a distributional functional central limit theorem in the ergodic case is automatically mixing.

Chapter 8
A Stable Limit Theorem with Exponential Rate

In this chapter we establish a stable limit theorem for "explosive" processes with exponential rates. The increments of these processes are not asymptotically negligible and thus do not satisfy the conditional Lindeberg condition. A simple example is given by an independent sequence $(Z_n)_{n \geq 1}$ with $P^{Z_n} = N\left(0, 2^{n-1}\right)$, $X_0 := 0$, $X_n := \sum_{i=1}^{n} Z_i$ and rate $a_n := 2^{n/2}$. The subsequent limit theorem is suitable for such situations. In order to formulate this limit theorem we need the following observation.

Lemma 8.1 *Let $(Z_n)_{n \geq 0}$ be an independent and identically distributed sequence of real random variables and $t \in \mathbb{R}$ with $|t| > 1$. Then*

(i) $t^{-n} Z_n \to 0$ *a.s.,*

(ii) $\sum_{n=0}^{\infty} t^{-n} Z_n$ *converges a.s. in $\mathbb{R}$,*

(iii) $\sum_{n=0}^{\infty} |t|^{-n} |Z_n| < \infty$ *a.s.,*

(iv) $E \log^+ |Z_0| < \infty$

are equivalent assertions.

Proof (iii) $\Rightarrow$ (ii) $\Rightarrow$ (i) are obvious.

(i) $\Rightarrow$ (iv). We have $P\left(\limsup_{n \to \infty} \left\{\left|t^{-n} Z_n\right| > 1\right\}\right) = 0$, implying by the Borel-Cantelli lemma

$$\infty > \sum_{n=0}^{\infty} P\left(|t|^{-n} |Z_n| > 1\right) = \sum_{n=0}^{\infty} P\left(|Z_0| > |t|^n\right) = \sum_{n=0}^{\infty} P\left(\log^+ |Z_0| > n \log |t|\right),$$

hence (iv).

(iv) $\Rightarrow$ (iii). Choose $1 < s < |t|$. Then

$$\sum_{n=0}^{\infty} P\left(|Z_n| > s^n\right) = \sum_{n=0}^{\infty} P\left(\log^+ |Z_0| > n \log s\right) < \infty$$

© Springer International Publishing Switzerland 2015
E. Häusler and H. Luschgy, *Stable Convergence and Stable Limit Theorems*,
Probability Theory and Stochastic Modelling 74,
DOI 10.1007/978-3-319-18329-9_8

and thus again by the Borel-Cantelli lemma, $P\left(\liminf_{n\to\infty}\{|Z_n|\le s^n\}\right)=1$. This gives (iii). $\qquad\qquad\square$

In the sequel $\mathbb{F}=(\mathcal{F}_n)_{n\ge0}$ denotes a filtration in $\mathcal{F}$ and $\mathcal{F}_\infty:=\sigma\left(\bigcup_{n\in\mathbb{N}_0}\mathcal{F}_n\right)$. For a real process $X=(X_n)_{n\ge0}$ the increments ΔX_n are defined by $\Delta X_0=0$ and $\Delta X_n=X_n-X_{n-1}$ for $n\ge1$.

Theorem 8.2 *Let $X=(X_n)_{n\ge0}$ and $A=(A_n)_{n\ge0}$ be $\mathbb{F}$-adapted real processes, where A is nonnegative with $A_n>0$ for every $n\ge n_0$ and some $n_0\in\mathbb{N}$, let $(a_n)_{n\ge1}$ be a sequence in $(0,\infty)$ with $a_n\to\infty$, and let $G\in\mathcal{F}_\infty$ with $P(G)>0$. Assume that the following conditions are satisfied:*

(i) *There exists a nonnegative real random variable η with $P\left(G\cap\{\eta^2>0\}\right)>0$ and*

$$\frac{A_n}{a_n^2}\to\eta^2\quad\text{in } P_G\text{-probability as } n\to\infty,$$

(ii) *$(X_n/a_n)_{n\ge1}$ is bounded in $P_{G\cap\{\eta^2>0\}}$-probability,*

(iii) *there exists a $p\in(1,\infty)$ such that*

$$\lim_{n\to\infty}\frac{a_{n-r}^2}{a_n^2}=\frac{1}{p^r}\quad\text{for every } r\in\mathbb{N},$$

(iv) *there exists a probability distribution μ on $\mathcal{B}(\mathbb{R})$ with $\int\log^+|x|\,d\mu(x)<\infty$ such that*

$$E_P\left(\exp\left(it\frac{\Delta X_n}{A_n^{1/2}}\right)\Big|\mathcal{F}_{n-1}\right)\to\int\exp(itx)\,d\mu(x)\quad\text{in } P_{G\cap\{\eta^2>0\}}\text{-probability}$$

as $n\to\infty$ for every $t\in\mathbb{R}$.

Then

$$\frac{X_n}{A_n^{1/2}}\to\sum_{j=0}^{\infty}p^{-j/2}Z_j\quad\mathcal{F}_\infty\text{-mixing under } P_{G\cap\{\eta^2>0\}}$$

and

$$\frac{X_n}{a_n}\to\eta\sum_{j=0}^{\infty}p^{-j/2}Z_j\quad\mathcal{F}_\infty\text{-stably under } P_{G\cap\{\eta^2>0\}}$$

as $n\to\infty$, where $(Z_j)_{j\ge0}$ denotes an independent and identically distributed sequence of real random variables independent of $\mathcal{F}_\infty$ with $P^{Z_0}=\mu$.

Note that the almost sure convergence of the above series follows from Lemma 8.1. Condition (ii) means

$$\lim_{c \to \infty} \sup_{n \in \mathbb{N}} P_{G \cap \{\eta^2 > 0\}} \left(\frac{|X_n|}{a_n} > c \right) = 0$$

and is equivalent to the tightness of the sequence $\left(P_{G \cap \{\eta^2 > 0\}}^{X_n / a_n} \right)_{n \geq 1}$. Typical rates are $a_n = c p^{n/2}$ with $p \in (1, \infty)$ and $c \in (0, \infty)$.

If $\nu \in \mathcal{M}^1(\mathbb{R})$ denotes the distribution of $\sum_{j=0}^{\infty} p^{-j/2} Z_j$ under P, $\varphi : \Omega \times \mathbb{R} \to \mathbb{R}$, $\varphi(\omega, x) := \eta(\omega) x$ and $K(\omega, \cdot) := \nu^{\varphi(\omega, \cdot)}$, then the assertions of Theorem 8.2 may be read as

$$\frac{X_n}{A_n^{1/2}} \to \nu \quad \text{mixing under } P_{G \cap \{\eta^2 > 0\}}$$

and

$$\frac{X_n}{a_n} \to K \quad \text{stably under } P_{G \cap \{\eta^2 > 0\}} .$$

Of course, in this formulation one does not need the P-independence of $(Z_j)_{j \geq 0}$ and $\mathcal{F}_\infty$.

For measures μ which are not symmetric around zero the following variant of Theorem 8.2 turns out to be useful, for example, for the investigation of autoregressive processes in Chap. 9. If μ is symmetric around zero, both theorems coincide.

Theorem 8.3 *Replace condition* (iv) *in Theorem 8.2 by*

(v) *there exists a probability distribution μ on $\mathcal{B}(\mathbb{R})$ with $\int \log^+ |x| \, d\mu(x) < \infty$ such that*

$$E_P \left(\exp \left(it \frac{(-1)^n \Delta X_n}{A_n^{1/2}} \right) \Big| \mathcal{F}_{n-1} \right) \to \int \exp(itx) \, d\mu(x)$$

in $P_{G \cap \{\eta^2 > 0\}}$-probability as $n \to \infty$ for every $t \in \mathbb{R}$.

Then

$$\frac{(-1)^n X_n}{A_n^{1/2}} \to \sum_{j=0}^{\infty} (-1)^j \, p^{-j/2} Z_j \quad \mathcal{F}_\infty\text{-mixing under } P_{G \cap \{\eta^2 > 0\}}$$

and

$$\frac{(-1)^n X_n}{a_n} \to \eta \sum_{j=0}^{\infty} (-1)^j \, p^{-j/2} Z_j \quad \mathcal{F}_\infty\text{-stably under } P_{G \cap \{\eta^2 > 0\}} .$$

For the proofs, we need the following elementary result.

Lemma 8.4 *For complex numbers $b_0, \ldots, b_r, c_0, \ldots, c_r$ we have*

$$\prod_{j=0}^{r} c_j - \prod_{j=0}^{r} b_j = \sum_{j=0}^{r} d_j \left(c_j - b_j\right),$$

where

$$d_j := \prod_{k=0}^{j-1} c_k \prod_{k=j+1}^{r} b_k.$$

Proof For $-1 \le j \le r$ let $e_j := \prod_{k=0}^{j} c_k \prod_{k=j+1}^{r} b_k$. Then $d_j c_j = e_j$ and $d_j b_j = e_{j-1}$ for $0 \le j \le r$ and therefore

$$\sum_{j=0}^{r} d_j \left(c_j - b_j\right) = \sum_{j=0}^{r} \left(e_j - e_{j-1}\right) = e_r - e_{-1} = \prod_{k=0}^{r} c_k - \prod_{k=0}^{r} b_k. \qquad \square$$

Proof of Theorem 8.2 and Theorem 8.3. Let $Q := P_{G \cap \{\eta^2 > 0\}}$ and for $n \ge 0$ let

$$L_n := \frac{P\left(G \cap \{\eta^2 > 0\} \,|\, \mathcal{F}_n\right)}{P\left(G \cap \{\eta^2 > 0\}\right)}.$$

Note that $(L_n)_{n \ge 0}$ is the density process of Q with respect to P, that is, $L_n = dQ \,|\, \mathcal{F}_n / dP \,|\, \mathcal{F}_n$ for every $n \ge 0$.

We may assume without loss of generality that η^2 is $\mathcal{F}_\infty$-measurable. Then the martingale convergence theorem yields

$$L_n \to \frac{1_{G \cap \{\eta^2 > 0\}}}{P\left(G \cap \{\eta^2 > 0\}\right)} = \frac{dQ}{dP} \quad \text{in } \mathcal{L}^1(P) \text{ as } n \to \infty.$$

Also, $\left(Z_j\right)_{j \ge 1}$ and $\mathcal{F}_\infty$ are independent under Q. Furthermore, let

$$\psi(t) := \int \exp(itx) \, d\mu(x) = E_P \exp(itZ_0) = E_Q \exp(itZ_0),$$

where the last equation is a consequence of the independence of $\mathcal{F}_\infty$ and Z_0, and

$$\beta := \begin{cases} p^{1/2} \,, & \text{Theorem 8.2}, \\ -p^{1/2} \,, & \text{Theorem 8.3}. \end{cases}$$

Then by (iii), for every $r \in \mathbb{N}$, we have $a_{n-r}/a_n \to |\beta|^{-r}$ and for every $n \in \mathbb{N}$

$$b_n := (\text{sign}(\beta))^n = \begin{cases} 1 & , \text{ Theorem 8.2}, \\ (-1)^n & , \text{ Theorem 8.3}. \end{cases}$$

Step 1. For every $r \in \mathbb{N}_0$ we have

$$\sum_{j=0}^{r} \frac{b_{n-j} \Delta X_{n-j}}{\beta^j A_{n-j}^{1/2}} \to \sum_{j=0}^{r} \beta^{-j} Z_j \quad \mathcal{F}_\infty\text{-mixing under } Q \text{ as } n \to \infty.$$

By Corollary 3.19 with $\mathcal{G} = \mathcal{F}_\infty$ and $\mathcal{E} = \bigcup_{n \in \mathbb{N}_0} \mathcal{F}_n$ it is enough to show that

$$\int_F \exp\left(it \sum_{j=0}^{r} \frac{b_{n-j} \Delta X_{n-j}}{\beta^j A_{n-j}^{1/2}} \right) dQ \to Q(F) \prod_{j=0}^{r} \psi\left(\frac{t}{\beta^j} \right) \quad \text{as } n \to \infty$$

for every $t \in \mathbb{R}$, $F \in \mathcal{E}$ and $r \in \mathbb{N}_0$. Fixing $t \in \mathbb{R}$ and using the notation $B_{n,j} := \exp\left(itb_{n-j} \Delta X_{n-j}/\beta^j A_{n-j}^{1/2} \right)$, $C_j := \psi\left(t/\beta^j\right)$ and $g_n := \prod_{j=0}^{r} C_j - \prod_{j=0}^{r} B_{n,j}$ this means $\int_F g_n \, dQ \to 0$. Assume $F \in \mathcal{F}_{n_1}$ for some $n_1 \in \mathbb{N}_0$. For $0 \le j \le r$, let

$$D_{n,j} := \prod_{k=0}^{j-1} C_k \prod_{k=j+1}^{r} B_{n,k}.$$

Then $|D_{n,j}| \le 1$, $D_{n,j}$ is $\mathcal{F}_{n-j-1}$-measurable and for $n \ge (n_0 + r) \vee (n_1 + r + 1)$ and $0 \le j \le r$ the random variable $1_F L_{n-r-1}$ is $\mathcal{F}_{n-r-1}$-measurable and hence $\mathcal{F}_{n-j-1}$-measurable. In view of Lemma 8.4 and since $L_n \le 1/P\left(G \cap \{\eta^2 > 0\}\right)$ and $\left|C_j - E_P\left(B_{n,j}|\mathcal{F}_{n-j-1}\right)\right| \le 2$, we obtain for $n \ge (n_0 + r) \vee (n_1 + r + 1)$

$$\left| \int_F L_{n-r-1} g_n \, dP \right| = \left| \sum_{j=0}^{r} \int_F L_{n-r-1} D_{n,j} \left(C_j - E_P\left(B_{n,j}|\mathcal{F}_{n-j-1}\right)\right) dP \right|$$

$$\le \sum_{j=0}^{r} \int_F L_{n-r-1} \left|C_j - E_P\left(B_{n,j}|\mathcal{F}_{n-j-1}\right)\right| dP$$

$$\le \sum_{j=0}^{r} \int \left|C_j - E_P\left(B_{n,j}|\mathcal{F}_{n-j-1}\right)\right| dQ$$

$$+ 2\sum_{j=0}^{r} \int_{(G \cap \{\eta^2 > 0\})^c} L_{n-r-1} \, dP.$$

It follows from (iv) and (v), respectively, that

$$\int \left| C_j - E_P\left(B_{n,j}|\mathcal{F}_{n-j-1}\right)\right| dQ \to 0 \quad \text{as } n \to \infty.$$

Moreover,

$$\int_{(G\cap\{\eta^2>0\})^c} L_{n-r-1}\, dP \to Q\left(\left(G\cap\{\eta^2>0\}\right)^c\right) = 0$$

so that $\int_F L_{n-r-1}g_n\, dP \to 0$ as $n \to \infty$. Since $|g_n| \le 2$, we get

$$\left|\int_F g_n\, dQ - \int_F L_{n-r-1}g_n\, dP\right| \le 2\int \left|\frac{dQ}{dP} - L_{n-r-1}\right| dP \to 0 \quad \text{as } n \to \infty,$$

which gives the assertion.

Step 2. For any $r \in \mathbb{N}_0$, we have

$$\frac{X_n - X_{n-r-1}}{b_n A_n^{1/2}} \to \sum_{j=0}^{r} \beta^{-j} Z_j \quad \mathcal{F}_\infty\text{-mixing under } Q \text{ as } n \to \infty.$$

In fact, for $0 \le j \le r$ we obtain

$$\frac{b_{n-j}\Delta X_{n-j}}{\beta^j A_{n-j}^{1/2}} - \frac{a_{n-r-1}\Delta X_{n-j}}{a_n b_n A_{n-r-1}^{1/2}} = \frac{b_{n-j}\Delta X_{n-j}}{\beta^j A_{n-j}^{1/2}}\left(1 - \frac{a_{n-j}\beta^j}{a_n b_{n-j}b_n}\frac{A_{n-j}^{1/2}/a_{n-j}}{A_{n-r-1}^{1/2}/a_{n-r-1}}\right) \to 0$$

in Q-probability as $n \to \infty$, since by (i) and (iii) the second factor converges to zero in Q-probability and the first factor converges by Step 1 with $r = 0$ in distribution under Q. Consequently,

$$\sum_{j=0}^{r}\frac{b_{n-j}\Delta X_{n-j}}{\beta^j A_{n-j}^{1/2}} - \frac{a_{n-r-1}}{a_n b_n A_{n-r-1}^{1/2}}\sum_{j=0}^{r}\Delta X_{n-j} \to 0 \quad \text{in } Q\text{-probability as } n \to \infty.$$

Since $\sum_{j=0}^{r}\Delta X_{n-j} = X_n - X_{n-r-1}$, Step 1 and Theorem 3.18 (a) imply

$$\frac{a_{n-r-1}}{a_n}\frac{X_n - X_{n-r-1}}{b_n A_{n-r-1}^{1/2}} \to \sum_{j=0}^{r}\beta^{-j} Z_j \quad \mathcal{F}_\infty\text{-mixing under } Q.$$

Using (i) again we have

$$\frac{A_{n-r-1}^{1/2}/a_{n-r-1}}{A_n^{1/2}/a_n} \to 1 \quad \text{in } Q\text{-probability},$$

so that the assertion follows from Theorem 3.18 (b) and (c).

Step 3. For every $\varepsilon > 0$ we have

$$\lim_{r \to \infty} \limsup_{n \to \infty} Q\left(\left|\frac{X_n}{b_n A_n^{1/2}} - \frac{X_n - X_{n-r-1}}{b_n A_n^{1/2}}\right| > \varepsilon\right) = 0.$$

Indeed, for $r \in \mathbb{N}_0$, $n \geq n_0 \vee (r+2)$ and $\delta, \varepsilon > 0$ we obtain the estimate

$$Q\left(\frac{|X_{n-r-1}|}{A_n^{1/2}} > \varepsilon\right)$$

$$= Q\left(\frac{|X_{n-r-1}|}{A_n^{1/2}} > \varepsilon, \frac{A_n}{a_n^2} > \delta\right) + Q\left(\frac{|X_{n-r-1}|}{A_n^{1/2}} > \varepsilon, \frac{A_n}{a_n^2} \leq \delta\right)$$

$$\leq Q\left(|X_{n-r-1}| > \varepsilon\sqrt{\delta}a_n\right) + Q\left(\frac{A_n}{a_n^2} \leq \delta, \eta^2 > 2\delta\right) + Q\left(\eta^2 \leq 2\delta\right)$$

$$\leq \sup_{j \in \mathbb{N}} Q\left(\frac{|X_j|}{a_j} > \frac{\varepsilon\sqrt{\delta}a_n}{a_{n-r-1}}\right) + Q\left(\left|\frac{A_n}{a_n^2} - \eta^2\right| > \delta\right) + Q\left(\eta^2 \leq 2\delta\right).$$

Condition (iii) yields $a_n/a_{n-r-1} \geq p^{(r+1)/2}/2$ for $n \geq n_2(r)$, say. This implies in view of (i), (ii) and the subadditivity of limsup

$$\limsup_{r \to \infty} \limsup_{n \to \infty} Q\left(\frac{|X_{n-r-1}|}{A_n^{1/2}} > \varepsilon\right)$$

$$\leq \limsup_{r \to \infty} \left(\sup_{j \in \mathbb{N}} Q\left(\frac{|X_j|}{a_j} > \frac{1}{2}\varepsilon\sqrt{\delta}p^{(r+1)/2}\right) + Q\left(\eta^2 \leq 2\delta\right)\right)$$

$$= Q\left(\eta^2 \leq 2\delta\right).$$

We have $Q\left(\eta^2 \leq 2\delta\right) \to Q\left(\eta^2 = 0\right) = 0$ as $\delta \to 0$, hence the assertion.
Step 4. Since

$$\sum_{j=0}^{r} \beta^{-j} Z_j \to \sum_{j=0}^{\infty} \beta^{-j} Z_j \quad P\text{-a.s.}$$

and hence $\mathcal{F}_\infty$-mixing under Q as $r \to \infty$, we obtain

$$\frac{X_n}{b_n A_n^{1/2}} \to \sum_{j=0}^{\infty} \beta^{-j} Z_j \quad \mathcal{F}_\infty\text{-mixing under } Q$$

from Steps 2 and 3 and Theorem 3.21. By (i), $A_n^{1/2}/a_n \to \eta$ in Q-probability so that by Theorem 3.18

$$\frac{X_n}{b_n a_n} = \frac{X_n}{b_n A_n^{1/2}} \frac{A_n^{1/2}}{a_n} \to \eta \sum_{j=0}^{\infty} \beta^{-j} Z_j \quad \mathcal{F}_\infty\text{-stably under } Q. \qquad \square$$

In the situation of Theorem 8.2 or 8.3 with $\mu \neq \delta_0$ the conditional Lindeberg condition under P or only under $P_{G \cap \{\eta^2 > 0\}}$ (with rate a_n) cannot be satisfied for $\mathcal{L}^2(P)$-processes X. Otherwise, we have $\Delta X_n / a_n \to 0$ in $P_{G \cap \{\eta^2 > 0\}}$-probability (cf. Proposition 6.6) and hence $\Delta X_n / A_n^{1/2} \to 0$ in $P_{G \cap \{\eta^2 > 0\}}$-probability by condition (i), in contradiction to the mixing convergence $b_n \Delta X_n / A_n^{1/2} \to \mu$ under $P_{G \cap \{\eta^2 > 0\}}$, which has been shown in Step 1 of the above proof.

Corollary 8.5 (Stable central limit theorem) *Assume* $\mu = N(0, \sigma^2)$ *for some* $\sigma^2 \in [0, \infty)$ *in Theorem 8.2 (iv). Then*

$$\frac{X_n}{A_n^{1/2}} \to Z \quad \mathcal{F}_\infty\text{-mixing under } P_{G \cap \{\eta^2 > 0\}}$$

and

$$\frac{X_n}{a_n} \to \eta Z \quad \mathcal{F}_\infty\text{-stably under } P_{G \cap \{\eta^2 > 0\}},$$

where Z *is* P-independent of $\mathcal{F}_\infty$ *and* $P^Z = N(0, \sigma^2 p / (p-1))$.

Proof Since $\sum_{j=0}^{\infty} p^{-j} = p / (p-1)$, the assertion follows directly from Theorem 8.2. $\qquad \square$

The assertions of Corollary 8.5 may also be read as

$$\frac{X_n}{A_n^{1/2}} \to N\left(0, \frac{\sigma^2 p}{p-1}\right) \quad \text{mixing under } P_{G \cap \{\eta^2 > 0\}}$$

and

$$\frac{X_n}{a_n} \to N\left(0, \frac{\sigma^2 p}{p-1} \eta^2\right) \quad \text{stably under } P_{G \cap \{\eta^2 > 0\}}.$$

For $\mathcal{L}^2$-martingales X and $A = \langle X \rangle$ the above central limit theorem for $G = \Omega$ is a consequence of a limit theorem of Scott [88] (up to a non-trivial improvement concerning the rate a_n), where the quadratic characteristic $\langle X \rangle$ of X is given by $\langle X \rangle_0 = 0$ and $\langle X \rangle_n = \sum_{j=0}^{n} E\left((\Delta X_j)^2 | \mathcal{F}_{j-1}\right)$ for $n \geq 1$.

Remark 8.6 If in Theorem 8.2 or 8.3 the process X is an $\mathcal{L}^2$-martingale and $A = \langle X \rangle$, then condition (i) with $G = \Omega$ implies that $(X_n / a_n)_{n \geq 1}$ is bounded in P-probability and, in particular, condition (ii) holds. In fact, since $\left(X_0^2 + \langle X \rangle_n\right) / a_n^2 \to \eta^2$ in

P-probability, the sequence $\left((X_0^2 + \langle X \rangle_n) / a_n \right)_{n \geq 1}$ is bounded in P-probability. By the Lenglart inequality (see Theorem A.8 (a)) we have for every $n \geq 1$ and $b, c > 0$

$$P\left(\frac{|X_n|}{a_n} \geq b \right) = P\left(X_n^2 \geq b^2 a_n^2 \right) \leq \frac{c}{b^2} + P\left(X_0^2 + \langle X \rangle_n > c a_n^2 \right)$$

so that

$$\sup_{n \in \mathbb{N}} P\left(\frac{|X_n|}{a_n} \geq b \right) \leq \frac{c}{b^2} + \sup_{n \in \mathbb{N}} P\left(\frac{X_0^2 + \langle X \rangle_n}{a_n^2} > c \right).$$

This yields the assertion (cf. Proposition 6.9).

Exercise 8.1 Assume that $\mu = C(0, b)$ for some $b \in (0, \infty)$ in Theorem 8.2 (iv), where $C(0, b)$ denotes the *Cauchy-distribution* with scale parameter b (given by the λ-density $x \mapsto \frac{1}{\pi b (1 + x^2/b^2)}$, $x \in \mathbb{R}$). Show that

$$\frac{X_n}{A_n^{1/2}} \to Z \quad \mathcal{F}_\infty\text{-mixing under } P_{G \cap \{\eta^2 > 0\}} \text{ as } n \to \infty$$

and

$$\frac{X_n}{a_n} \to \eta Z \quad \mathcal{F}_\infty\text{-stably under } P_{G \cap \{\eta^2 > 0\}} \text{ as } n \to \infty,$$

where Z is P-independent of $\mathcal{F}_\infty$ and $P^Z = C\left(0, b\sqrt{p} / (\sqrt{p} - 1) \right)$.

An $\mathcal{L}^2$-martingale $X = (X_n)_{n \geq 0}$ is said to have $\mathbb{F}$-*conditional Gaussian incre-ments* if $P^{\Delta X_n | \mathcal{F}_{n-1}} = N(0, \Delta \langle X \rangle_n)$ for every $n \in \mathbb{N}$.

Corollary 8.7 *Let $X = (X_n)_{n \geq 0}$ be an $\mathcal{L}^2$-martingale with $\mathbb{F}$-conditional Gaussian increments and $\langle X \rangle_n > 0$ for every $n \geq n_0$ and some $n_0 \in \mathbb{N}$. Assume that conditions (i) and (iii) in Theorem 8.2 are satisfied with $G = \Omega$ and $A = \langle X \rangle$. Then*

$$\frac{X_n}{\langle X \rangle_n^{1/2}} \to N \quad \mathcal{F}_\infty\text{-mixing under } P_{\{\eta^2 > 0\}} \text{ as } n \to \infty$$

and

$$\frac{X_n}{a_n} \to \eta N \quad \mathcal{F}_\infty\text{-stably under } P_{\{\eta^2 > 0\}} \text{ as } n \to \infty,$$

where N is P-independent of $\mathcal{F}_\infty$ and $P^N = N(0, 1)$.

Proof Conditions (i) and (iii) of Theorem 8.2 are fulfilled by assumption and imply, as $n \to \infty$,

$$\frac{\langle X \rangle_n}{p a_{n-1}^2} = \frac{a_n^2}{p a_{n-1}^2} \frac{\langle X \rangle_n}{a_n^2} \to \eta^2 \quad \text{in } P\text{-probability}$$

which yields

$$\frac{\Delta \langle X \rangle_n}{a_{n-1}^2} \to p \eta^2 - \eta^2 = (p-1)\eta^2 \quad \text{in } P\text{-probability}.$$

This implies

$$\frac{\Delta \langle X \rangle_n}{a_n^2} = \frac{a_{n-1}^2}{a_n^2} \frac{\Delta \langle X \rangle_n}{a_{n-1}^2} \to \frac{p-1}{p} \eta^2 \quad \text{in } P\text{-probability}$$

and therefore

$$\frac{\Delta \langle X \rangle_n}{\langle X \rangle_n} \to \frac{p-1}{p} \quad \text{in } P_{\{\eta^2 > 0\}}\text{-probability}.$$

Furthermore, since $\langle X \rangle_n$ is $\mathcal{F}_{n-1}$-measurable,

$$P^{(\Delta X_n, \langle X \rangle_n)|\mathcal{F}_{n-1}} = P^{\Delta X_n|\mathcal{F}_{n-1}} \otimes \delta_{\langle X \rangle_n} = N\left(0, \Delta \langle X \rangle_n\right) \otimes \delta_{\langle X \rangle_n}$$

(see Lemma A.5 (a)) so that

$$E_P\left(\exp\left(it\frac{\Delta X_n}{\langle X \rangle_n^{1/2}}\right)\bigg|\mathcal{F}_{n-1}\right) = \int \exp\left(it\frac{x}{\langle X \rangle_n^{1/2}}\right) N\left(0, \Delta \langle X \rangle_n\right)(dx)$$

$$= \exp\left(-\frac{t^2 \Delta \langle X \rangle_n}{2 \langle X \rangle_n}\right) \to \exp\left(-\frac{t^2 (p-1)}{2p}\right) \quad \text{in } P_{\{\eta^2 > 0\}}\text{-probability}$$

as $n \to \infty$ for every $t \in \mathbb{R}$. The assertion follows from Corollary 8.5 with $G = \Omega$ and Remark 8.6.   □

Corollary 8.8 *In the situation of Theorem 8.2 with* $G = \Omega$ *replace condition* (iv) *by*

(vi) *there exist a probability distribution* μ *on* $\mathcal{B}(\mathbb{R})$ *with* $\int \log^+ |x| \, d\mu(x) < \infty$ *and a real* $\mathcal{F}_\infty$-*measurable discrete random variable* S *such that*

$$E_P\left(\exp\left(it\frac{\Delta X_n}{A_n^{1/2}}\right)\bigg|\mathcal{F}_{n-1}\right) \to \int \exp(itSx) \, d\mu(x) \quad \text{in } P_{\{\eta^2 > 0\}}\text{-probability}$$

as $n \to \infty$ *for every* $t \in \mathbb{R}$.

Then

$$\frac{X_n}{A_n^{1/2}} \to S \sum_{j=0}^{\infty} p^{-j/2} Z_j \quad \mathcal{F}_{\infty}\text{-stably under } P_{\{\eta^2 > 0\}}$$

and

$$\frac{X_n}{a_n} \to S\eta \sum_{j=0}^{\infty} p^{-j/2} Z_j \quad \mathcal{F}_{\infty}\text{-stably under } P_{\{\eta^2 > 0\}}$$

as $n \to \infty$.

Proof Let supp $\left(P^S\right) = \{s_k : k \geq 1\}$, $G_k := \{S = s_k\}$ and

$$I := \left\{ k \geq 1 : P\left(G_k \cap \left\{\eta^2 > 0\right\}\right) > 0 \right\}.$$

Then

$$E_P\left(\exp\left(it\frac{\Delta X_n}{A_n^{1/2}}\right)\Big| \mathcal{F}_{n-1}\right) \to \int \exp\left(its_k x\right) d\mu\left(x\right) \quad \text{in } P_{G_k \cap \{\eta^2 > 0\}}\text{-probability}$$

as $n \to \infty$ for every $k \in I$. Therefore, by Theorem 8.2

$$\frac{X_n}{A_n^{1/2}} \to s_k \sum_{j=0}^{\infty} p^{-j/2} Z_j \quad \mathcal{F}_{\infty}\text{-mixing under } P_{G_k \cap \{\eta^2 > 0\}}$$

and

$$\frac{X_n}{a_n} \to s_k\eta \sum_{j=1}^{\infty} p^{-j/2} Z_j \quad \mathcal{F}_{\infty}\text{-stably under } P_{G_k \cap \{\eta^2 > 0\}},$$

which can be read as

$$\frac{X_n}{A_n^{1/2}} \to S \sum_{j=0}^{\infty} p^{-j/2} Z_j \quad \mathcal{F}_{\infty}\text{-mixing under } P_{G_k \cap \{\eta^2 > 0\}}$$

and

$$\frac{X_n}{a_n} \to S\eta \sum_{j=1}^{\infty} p^{-j/2} Z_j \quad \mathcal{F}_{\infty}\text{-stably under } P_{G_k \cap \{\eta^2 > 0\}}$$

for every $k \in I$, as $n \to \infty$. Using $\sum_{k \in I} P_{\{\eta^2 > 0\}} (G_k) P_{G_k \cap \{\eta^2 > 0\}} = P_{\{\eta^2 > 0\}}$, the assertion follows from Proposition 3.24. $\square$

In just the same way, one deduces from Theorem 8.3 the

Corollary 8.9 *In the situation of Theorem 8.3 with $G = \Omega$ replace condition* (v) *by*

(vii) *there exist a probability distribution μ on $\mathcal{B}(\mathbb{R})$ with $\int \log^+ |x| \, d\mu(x) < \infty$ and a real $\mathcal{F}_\infty$-measurable discrete random variable S such that*

$$E_P \left(\exp \left(it \frac{(-1)^n \Delta X_n}{A_n^{1/2}} \right) \Big| \mathcal{F}_{n-1} \right) \to \int \exp(it S x) \, d\mu(x)$$

in $P_{\{\eta^2 > 0\}}$-probability as $n \to \infty$ for every $t \in \mathbb{R}$.

Then

$$\frac{(-1)^n X_n}{A_n^{1/2}} \to S \sum_{j=0}^{\infty} (-1)^j \, p^{-j/2} Z_j \quad \mathcal{F}_\infty\text{-stably under } P_{\{\eta^2 > 0\}}$$

and

$$\frac{(-1)^n X_n}{a_n} \to S \eta \sum_{j=0}^{\infty} (-1)^j p^{-j/2} Z_j \quad \mathcal{F}_\infty\text{-stably under } P_{\{\eta^2 > 0\}}$$

as $n \to \infty$.

The Corollaries 8.8 and 8.9 may possibly be extended to more general random variables S. But for our purposes the results are good enough.

The subsequent example provides an illustration of Corollary 8.7.

Example 8.10 (Explosive Gaussian autoregression of order one) Let $(Z_n)_{n \geq 1}$ be an independent and identically distributed sequence of $N(0, \sigma^2)$-distributed random variables with $\sigma^2 \in (0, \infty)$ and let $X_0 \in \mathcal{L}^2(P)$ be independent of $(Z_n)_{n \geq 1}$. Consider the autoregression defined by

$$X_n = \vartheta X_{n-1} + Z_n, \quad n \geq 1,$$

where $\vartheta \in \mathbb{R}$. The least squares estimator of ϑ on the basis of the observations $X_0, X_1, \ldots, X_n$ is given by

$$\widehat{\vartheta}_n = \frac{\sum_{j=1}^{n} X_j X_{j-1}}{\sum_{j=1}^{n} X_{j-1}^2}, \quad n \geq 2.$$

Note that $X_n^2 > 0$ for all $n \in \mathbb{N}$ because, by the independence of X_{n-1} and Z_n, the distribution of X_n is continuous. We assume $|\vartheta| > 1$ and derive a stable central limit theorem for $\widehat{\vartheta}_n$. Let $\mathcal{F}_n := \sigma(X_0, X_1, \ldots, X_n) = \sigma(X_0, Z_1, \ldots, Z_n)$, $\mathbb{F} :=$

$(\mathcal{F}_n)_{n\geq 0}$ and $M_n := \sum_{j=1}^{n} X_{j-1} Z_j / \sigma^2$ with $M_0 = 0$. Then M is an $\mathcal{L}^2$-martingale w.r.t. the filtration $\mathbb{F}$ with quadratic characteristic $\langle M \rangle_n = \sum_{j=1}^{n} X_{j-1}^2 / \sigma^2$. Since

$$\sum_{j=1}^{n} X_j X_{j-1} = \sum_{j=1}^{n} \left(\vartheta X_{j-1} + Z_j \right) X_{j-1} = \vartheta \sum_{j=1}^{n} X_{j-1}^2 + \sigma^2 M_n ,$$

we obtain $\widehat{\vartheta}_n - \vartheta = M_n / \langle M \rangle_n$ for all $n \geq 2$. By induction, $X_n = \vartheta^n X_0 + \sum_{j=1}^{n} \vartheta^{n-j} Z_j$ for all $n \geq 0$ so that by Lemma 8.1 (or the martingale convergence theorem)

$$\vartheta^{-n} X_n \to Y := X_0 + \sum_{j=1}^{\infty} \vartheta^{-j} Z_j \quad \text{a.s. as } n \to \infty$$

and clearly $P^{Y-X_0} = N\left(0, \sigma^2/(\vartheta^2 - 1)\right)$. In particular, P^Y is continuous. Let $a_n := |\vartheta|^n / \left(\vartheta^2 - 1\right)$ for all $n \in \mathbb{N}$. The discrete rule of de l'Hospital in Lemma 6.28 (b) yields

$$\frac{\sum_{j=1}^{n} X_{j-1}^2}{\sum_{j=1}^{n} \vartheta^{2(j-1)}} \to Y^2 \quad \text{a.s. as } n \to \infty .$$

Since $\sum_{j=1}^{n} \vartheta^{2(j-1)} = \left(\vartheta^{2n} - 1\right) / \left(\vartheta^2 - 1\right) \sim a_n^2 \left(\vartheta^2 - 1\right)$, we get

$$\frac{\langle M \rangle_n}{a_n^2} \to \frac{\left(\vartheta^2 - 1\right) Y^2}{\sigma^2} =: \eta^2 \quad \text{a.s. as } n \to \infty$$

and $P\left(\eta^2 > 0\right) = 1$. Furthermore, M obviously has $\mathbb{F}$-conditional Gaussian increments. Consequently, by Corollary 8.7

$$\left(\sum_{j=1}^{n} X_{j-1}^2 \right)^{1/2} \left(\widehat{\vartheta}_n - \vartheta \right) = \frac{\sigma M_n}{\langle M \rangle_n^{1/2}} \to \sigma N \quad \mathcal{F}_\infty\text{-mixing} ,$$

where N is independent of $\mathcal{F}_\infty$ (and thus of Y) and $P^N = N(0, 1)$, and using Theorem 3.18

$$a_n \left(\widehat{\vartheta}_n - \vartheta \right) = \frac{a_n^2}{\langle M \rangle_n} \frac{M_n}{a_n} \to \frac{N}{\eta} = \frac{\sigma N}{\left(\vartheta^2 - 1\right)^{1/2} |Y|} \quad \mathcal{F}_\infty\text{-stably as } n \to \infty .$$

By the symmetry around zero of P^U, we obtain, as $n \to \infty$,

$$\frac{\vartheta^n}{\vartheta^2 - 1} \left(\widehat{\vartheta}_n - \vartheta \right) \to \frac{\sigma N}{\left(\vartheta^2 - 1\right)^{1/2} |Y|} \quad \mathcal{F}_\infty\text{-stably}$$

or, what is that same in view of Lemma A.4 (c),

$$\frac{\vartheta^n}{\vartheta^2 - 1} \left(\widehat{\vartheta}_n - \vartheta \right) \rightarrow \frac{\sigma N}{\left(\vartheta^2 - 1 \right)^{1/2} Y} \quad \mathcal{F}_\infty\text{-stably} .$$

If $P^{X_0} = N\left(0, \tau^2\right)$ with $\tau \in [0, \infty)$, then using the independence of Y and N we get $P^{\sigma N/\left(\vartheta^2 - 1\right)^{1/2} Y} = C\left(0, b\right)$ with $b = \left(\tau^2 \left(\vartheta^2 - 1\right) / \sigma^2 + 1\right)^{-1/2}$ so that

$$\frac{\vartheta^n}{\vartheta^2 - 1} \left(\widehat{\vartheta}_n - \vartheta \right) \xrightarrow{d} C\left(0, b\right) .$$

In case $\tau^2 = 0$, that is, $X_0 = 0$, we obtain $b = 1$. This distributional convergence of the estimator is a classical result due to White [97]. The distributional convergence under random norming is contained in [5], Theorem 2.8 and [98]. General (non-normal) innovations Z_n are treated in the next chapter.   □

Further applications can be found in Chaps. 9 and 10.

Chapter 9
Autoregression of Order One

In this and the subsequent chapter we present concrete applications of previous stable limit theorems. Here we consider an *autoregressive process of order one* $X = (X_n)_{n \geq 0}$ generated recursively by

$$X_n = \vartheta X_{n-1} + Z_n \, , \ n \geq 1 \, ,$$

where $\vartheta \in \mathbb{R}$, $(Z_n)_{n \geq 1}$ is an independent and identically distributed sequence of real random variables and X_0 is a real random variable independent of $(Z_n)_{n \geq 1}$. We assume that P^{Z_1} is continuous. Then $X_n^2 > 0$ almost surely for all $n \geq 1$ since by independence of X_{n-1} and Z_n, P^{X_n} is continuous for $n \geq 1$. The usual *least squares estimator* for the parameter ϑ on the basis of the observations $X_0, \ldots, X_n$ is given by

$$\widehat{\vartheta}_n := \frac{\sum_{j=1}^{n} X_j X_{j-1}}{\sum_{j=1}^{n} X_{j-1}^2} \, , \ n \geq 2 \, ,$$

provided $Z_1 \in \mathcal{L}^1(P)$ and $E Z_1 = 0$. In the explosive case $|\vartheta| > 1$, the effect of the mean of Z_1 disappears asymptotically so that $\widehat{\vartheta}_n$ is also reasonable in that case if $E Z_1 \neq 0$. We prove stable limit theorems for $\widehat{\vartheta}_n$ under deterministic and random norming.

Let $\mathcal{F}_n := \sigma(X_0, X_1, \ldots, X_n) = \sigma(X_0, Z_1, \ldots, Z_n)$ for all $n \geq 0$ and $\mathbb{F} := (\mathcal{F}_n)_{n \geq 0}$. Define $\mathbb{F}$-adapted processes by

$$A_n := \sum_{j=1}^{n} X_{j-1}^2 \quad \text{with} \quad A_0 = 0$$

and

© Springer International Publishing Switzerland 2015
E. Häusler and H. Luschgy, *Stable Convergence and Stable Limit Theorems*,
Probability Theory and Stochastic Modelling 74,
DOI 10.1007/978-3-319-18329-9_9

$$B_n := \sum_{j=1}^{n} X_{j-1} Z_j \quad \text{with} \quad B_0 = 0.$$

Since $\sum_{j=1}^{n} X_j X_{j-1} = \sum_{j=1}^{n} \left(\vartheta X_{j-1} + Z_j \right) X_{j-1} = \vartheta A_n + B_n$, we obtain

$$\widehat{\vartheta}_n - \vartheta = B_n / A_n \quad \text{for all } n \geq 2.$$

Furthermore, by induction, we have $X_n = \vartheta^n X_0 + \sum_{j=1}^{n} \vartheta^{n-j} Z_j$ for all $n \geq 0$.

If $X_0, Z_1 \in \mathcal{L}^2$ and $E Z_1 = 0$ then $B = (B_n)_{n \geq 0}$ is an $\mathcal{L}^2$-martingale w.r.t. $\mathbb{F}$ with $\langle B \rangle = \sigma^2 A$. Therefore, in this setting, the strong law of large numbers for martingales of Theorem A.9 yields $\widehat{\vartheta}_n \to \vartheta$ almost surely, which says that $\widehat{\vartheta}_n$ is a strongly consistent estimator of ϑ (using $\sum_{j=1}^{\infty} Z_j^2 \leq 2 \left(1 + \vartheta^2 \right) \sum_{j=1}^{\infty} X_{j-1}^2$, so that $\langle B \rangle_\infty = \infty$ almost surely by Kolmogorov's strong law of large numbers).

The ergodic case

In the ergodic case $|\vartheta| < 1$ stable asymptotic normality of $\widehat{\vartheta}_n$ holds.

Theorem 9.1 *Assume* $|\vartheta| < 1$, $X_0, Z_1 \in \mathcal{L}^2$ *and* $E Z_1 = 0$. *Then*

$$\sqrt{n} \left(\widehat{\vartheta}_n - \vartheta \right) \to N \left(0, 1 - \vartheta^2 \right) \quad \text{mixing}$$

and

$$\left(\sum_{j=1}^{n} X_{j-1}^2 \right)^{1/2} \left(\widehat{\vartheta}_n - \vartheta \right) \to N \left(0, \sigma^2 \right) \quad \text{mixing}$$

as $n \to \infty$, *where* $\sigma^2 := \operatorname{Var} Z_1$.

Note that $\sigma^2 > 0$ by the continuity of P^{Z_1}. The above statements may also be read as

$$\sqrt{n} \left(\widehat{\vartheta}_n - \vartheta \right) \to \left(1 - \vartheta^2 \right)^{1/2} N \quad \mathcal{F}_\infty\text{-mixing}$$

and

$$\left(\sum_{j=1}^{n} X_{j-1}^2 \right)^{1/2} \left(\widehat{\vartheta}_n - \vartheta \right) \to \sigma N \quad \mathcal{F}_\infty\text{-mixing},$$

where N is a real random variable independent of $\mathcal{F}_\infty$ with $P^N = N(0, 1)$. Distributional convergence under deterministic norming was first investigated in [5], Theorem 4.3.

The main idea of the following proof is taken from [74], p. 174 and p. 186.

Proof of Theorem 9.1. The process $B = (B_n)_{n \geq 0}$ is a square integrable $\mathbb{F}$-martingale with quadratic characteristic $\langle B \rangle = \sigma^2 A$, where $A = (A_n)_{n \geq n}$. We apply the stable central limit theorem of Theorem 6.23.

Step 1. We rely on the fact that $X^2 = \left(X_n^2\right)_{n \geq 0}$ is uniformly integrable. To prove this, break Z_n into a sum $Z_n = V_n + W_n$, where

$$V_n = V_n(c) := Z_n 1_{\{|Z_n| \leq c\}} - E Z_n 1_{\{|Z_n| \leq c\}} \quad \text{and} \quad W_n = W_n(c) := Z_n - V_n$$

for some large truncation level $c \in (0, \infty)$. Define

$$G_n = G_n(c) := \sum_{j=1}^{n} \vartheta^{n-j} V_j \quad \text{and} \quad H_n := H_n(c) := \sum_{j=1}^{n} \vartheta^{n-j} W_j \, .$$

Then $X_n = \vartheta^n X_0 + G_n + H_n$ for $n \geq 0$. Observe that

$$|G_n| \leq \sum_{j=1}^{n} |\vartheta|^{n-j} |V_j| \leq 2c \sum_{j=1}^{n} |\vartheta|^{n-j} = 2c \sum_{i=0}^{n-1} |\vartheta|^i = 2c \frac{1 - |\vartheta|^n}{1 - |\vartheta|} \leq \frac{2c}{1 - |\vartheta|}$$

for every $n \geq 0$ so that $G = (G_n)_{n \geq 0}$ is uniformly bounded. Since the sequence $(W_n)_{n \geq 1}$ is independent and identically distributed with $E W_1 = E Z_1 = 0$, the process $H = (H_n)_{n \geq 0}$ satisfies

$$E H_n^2 = \sum_{j=1}^{n} \vartheta^{2(n-j)} E W_1^2 = E W_1^2 \frac{1 - \vartheta^{2n}}{1 - \vartheta^2} \leq \frac{E W_1^2}{1 - \vartheta^2}$$

for every $n \geq 0$. Using $W_1 = Z_1 1_{\{|Z_1| > c\}} + E Z_1 1_{\{|Z_1| \leq c\}}$ and $Z_1 \in \mathcal{L}^2(P)$, dominated convergence yields $E W_1(c)^2 \to (E Z_1)^2 = 0$ as $c \to \infty$. Let $\varepsilon > 0$. Choose $c \in (0, \infty)$ such that $\sup_{n \geq 0} E H_n(c)^2 \leq E W_1(c)^2 / \left(1 - \vartheta^2\right) \leq \varepsilon/2$ and then $a \geq 8c^2 / (1 - |\vartheta|)^2$. Since

$$\left\{ G_n^2 + H_n^2 > a \right\} \subset \left\{ G_n^2 \leq H_n^2, H_n^2 > a/2 \right\} \cup \left\{ G_n^2 \geq H_n^2, G_n^2 > a/2 \right\}$$

we obtain

$$\left(G_n^2 + H_n^2 \right) 1_{\{G_n^2 + H_n^2 > a\}} \leq 2 H_n^2 1_{\{H_n^2 > a/2\}} + 2 G_n^2 1_{\{G_n^2 > a/2\}} \leq 2 H_n^2$$

for every $n \geq 0$ and hence

$$\sup_{n \geq 0} E \left(G_n^2 + H_n^2 \right) 1_{\{G_n^2 + H_n^2 > a\}} \leq 2 \sup_{n \geq 0} E H_n^2 \leq \varepsilon \, .$$

This gives uniform integrability of $G^2 + H^2$, which implies uniform integrability of X^2 because $X_n^2 \leq 4\left(X_0^2 + G_n^2 + H_n^2\right)$. In particular, X^2 is $\mathcal{L}^1$-bounded.

Step 2. Now let us verify the assumptions of Theorem 6.23. We have for $n \geq 1$

$$\sum_{j=1}^{n} X_j^2 = \sum_{j=1}^{n} \left(\vartheta X_{j-1} + Z_j\right)^2 = \vartheta^2 A_n + 2\vartheta\, B_n + \sum_{j=1}^{n} Z_j^2$$

and thus by rearranging and dividing by n,

$$\frac{1 - \vartheta^2}{\sigma^2 n}\, \langle B \rangle_n = \frac{1}{n}\left(X_0^2 - X_n^2\right) + \frac{2\vartheta}{n}\, B_n + \frac{1}{n}\sum_{j=1}^{n} Z_j^2\,.$$

On the right-hand side, the first term converges in $\mathcal{L}^1$ to zero, because X^2 is $\mathcal{L}^1$-bounded. The middle term converges in $\mathcal{L}^2$ to zero, because

$$\frac{1}{n^2} E\, B_n^2 = \frac{1}{n^2} E\, \langle B \rangle_n = \frac{\sigma^2}{n^2}\sum_{j=1}^{n} E X_{j-1}^2 \leq \frac{\sigma^2 n}{n^2}\,\sup_{n \geq 0} E X_n^2 \to 0\,.$$

The third term converges almost surely to σ^2 by the Kolmogorov strong law of large numbers. Consequently,

$$\frac{\langle B \rangle_n}{n} \to \frac{\sigma^4}{1 - \vartheta^2}\ \text{in probability as } n \to \infty\,.$$

This is condition (N_{a_n}) with $a_n = \sqrt{n}$.

As concerns the conditional Lindeberg condition (CLB_{a_n}), we have for $\varepsilon > 0$ and $n \geq 1$

$$I_n(\varepsilon) := \frac{1}{n}\sum_{j=1}^{n} E\left(X_{j-1}^2 Z_j^2 1_{\{|X_{j-1} Z_j| \geq \varepsilon\sqrt{n}\}}\,|\,\mathcal{F}_{j-1}\right)$$

$$= \frac{1}{n}\sum_{j=1}^{n} X_{j-1}^2 E\left(Z_j^2 1_{\{|X_{j-1} Z_j| \geq \varepsilon\sqrt{n}\}}\,|\,\mathcal{F}_{j-1}\right)$$

$$\leq \frac{1}{n}\sum_{j=1}^{n} X_{j-1}^2 E\left(Z_j^2 1_{\{X_{j-1}^2 \geq \varepsilon\sqrt{n}\}} + Z_j^2 1_{\{Z_j^2 \geq \varepsilon\sqrt{n}\}}\,|\,\mathcal{F}_{j-1}\right)$$

$$= \frac{\sigma^2}{n}\sum_{j=1}^{n} X_{j-1}^2 1_{\{X_{j-1}^2 \geq \varepsilon\sqrt{n}\}} + \frac{1}{n\sigma^2}\,\langle B \rangle_n\, E Z_1^2 1_{\{Z_1^2 \geq \varepsilon\sqrt{n}\}}\,.$$

The first term converges in $\mathcal{L}^1$ to zero because X^2 is uniformly integrable by Step 1 and hence

$$\frac{1}{n}\sum_{j=1}^{n} EX_{j-1}^2 1_{\left\{X_{j-1}^2 \geq \varepsilon\sqrt{n}\right\}} \leq \sup_{j\geq 0} EX_j^2 1_{\left\{X_j^2 \geq \varepsilon\sqrt{n}\right\}} \to 0 \quad \text{as } n \to \infty.$$

The second term converges to zero in probability because $Z_1 \in \mathcal{L}^2$ and (N_{a_n}) holds. Consequently, $L_n(\varepsilon) \to 0$ in probability as $n \to \infty$.

Now Theorem 6.23 yields

$$\frac{B_n}{\sqrt{n}} \to N\left(0, \frac{\sigma^4}{1-\vartheta^2}\right) \quad \text{mixing}.$$

Using Theorem 3.7 (b), (c), this implies

$$\sqrt{n}\left(\widehat{\vartheta}_n - \vartheta\right) = \frac{\sqrt{n}B_n}{A_n} = \frac{B_n/\sqrt{n}}{\langle B\rangle_n/n\sigma^2} \to N\left(0, 1-\vartheta^2\right) \quad \text{mixing}$$

and

$$\left(\sum_{j=1}^{n} X_{j-1}^2\right)^{1/2}\left(\widehat{\vartheta}_n - \vartheta\right) = \frac{B_n}{A_n^{1/2}} = \frac{\sigma B_n}{\langle B\rangle_n^{1/2}} = \frac{\sigma B_n/\sqrt{n}}{(\langle B\rangle_n/n)^{1/2}} \to N\left(0, \sigma^2\right) \quad \text{mixing}$$

as $n \to \infty$. $\qquad\square$

The explosive case

In the explosive case $|\vartheta| > 1$ the asymptotic behavior of $\widehat{\vartheta}_n$ depends on the distribution of the innovations Z_n. Let sign $:= 1_{(0,\infty)} - 1_{(-\infty,0)}$.

Theorem 9.2 *Assume $|\vartheta| > 1$ and $E \log^+ |Z_1| < \infty$. Let $Y := X_0 + \sum_{j=1}^{\infty} \vartheta^{-j} Z_j$ (see Lemma 8.1) and let U be a real random variable independent of $\mathcal{F}_\infty$ with $P^U = P^{Y-X_0}$. Then*

$$\vartheta^n\left(\widehat{\vartheta}_n - \vartheta\right) \to \frac{\left(\vartheta^2 - 1\right) U}{Y} \quad \mathcal{F}_\infty\text{-stably},$$

$$(\text{sign}(\vartheta))^n \left(\sum_{j=1}^{n} X_{j-1}^2\right)^{1/2}\left(\widehat{\vartheta}_n - \vartheta\right) \to \text{sign}(Y)\left(\vartheta^2 - 1\right)^{1/2} U \quad \mathcal{F}_\infty\text{-stably}$$

and, if P^{Z_1} is symmetric around zero,

$$\left(\sum_{j=1}^{n} X_{j-1}^2\right)^{1/2}\left(\widehat{\vartheta}_n - \vartheta\right) \to \left(\vartheta^2 - 1\right)^{1/2} U \quad \mathcal{F}_\infty\text{-mixing}$$

as $n \to \infty$.

Under the stronger assumptions $X_0, Z_1 \in \mathcal{L}^2$ and $EZ_1 = 0$ distributional convergence was first investigated in [5], and stable convergence has been touched in [95]. Under the assumptions above, distributional convergence under deterministic norming has been stated in [54], Lemma 3.1 in case $X_0 = 0$ without proof.

In the special case of normal innovations, that is $P^{Z_1} = N\left(0, \sigma^2\right)$ with $\sigma^2 \in (0, \infty)$, Theorem 9.2 provides again the results of Example 8.10 (without assuming $X_0 \in \mathcal{L}^2$).

Proof We apply the stable limit Theorems 8.2 and 8.3, or more precisely, the Corollaries 8.8 and 8.9. We have $\vartheta^{-n} X_n \to Y$ almost surely as $n \to \infty$ by Lemma 8.1 so that the discrete rule of de l'Hospital, Lemma 6.28 (b), yields $A_n / \sum_{j=1}^{n} \vartheta^{2(j-1)} \to Y^2$ almost surely. We may assume that Y is $\mathcal{F}_\infty$-measurable. Let $a_n := |\vartheta|^n, n \geq 1$. Since $\sum_{j=1}^{n} \vartheta^{2(j-1)} = \left(\vartheta^{2n} - 1\right) / \left(\vartheta^2 - 1\right) \sim a_n^2 / \left(\vartheta^2 - 1\right)$ as $n \to \infty$, we get

$$\frac{A_n}{a_n^2} \to \frac{Y^2}{\vartheta^2 - 1} =: \eta^2 \quad \text{a.s.}$$

The distribution P^Y is continuous, hence $P\left(\eta^2 > 0\right) = 1$. This is condition (i) in Theorem 8.2 with $G = \Omega$. Condition (iii) of Theorem 8.2 holds with $p = \vartheta^2$. As for condition (ii) of Theorem 8.2 with respect to the process B, note first that

$$\frac{1}{a_n} |B_n| \leq \frac{1}{a_n} \sum_{j=1}^{n} |X_{j-1}| |Z_j| = \frac{\sum_{j=1}^{n} |X_{j-1}| |Z_j|}{\sum_{j=1}^{n} |\vartheta|^{j-1} |Z_j|} \frac{\sum_{j=1}^{n} |\vartheta|^{j-1} |Z_j|}{|\vartheta|^n}$$

and $|X_{n-1}| |Z_n| / |\vartheta|^{n-1} |Z_n| \to |Y|$ almost surely as $n \to \infty$. Since

$$\sum_{n=1}^{\infty} P\left(|\vartheta|^{n-1} |Z_n| > 1\right) = \sum_{n=1}^{\infty} P\left(|Z_1| > |\vartheta|^{-n+1}\right) = \infty,$$

the Borel-Cantelli lemma yields $P\left(\limsup_{n \to \infty} \{|\vartheta|^{n-1} |Z_n| > 1\}\right) = 1$ and therefore, $\sum_{n=1}^{\infty} |\vartheta|^{n-1} |Z_n| = \infty$ almost surely. Consequently, Lemma 6.28 (b) applies and gives

$$\frac{\sum_{j=1}^{n} |X_{j-1}| |Z_j|}{\sum_{j=1}^{n} |\vartheta|^{j-1} |Z_j|} \to |Y| \quad \text{a.s.}$$

Moreover, using Lemma 8.1,

$$\frac{1}{|\vartheta|^n} \sum_{j=1}^{n} |\vartheta|^{j-1} |Z_j| = \sum_{j=1}^{n} \frac{|Z_j|}{|\vartheta|^{n-j+1}} \overset{d}{=} \sum_{k=1}^{n} \frac{|Z_k|}{|\vartheta|^k} \to \sum_{k=1}^{\infty} \frac{|Z_k|}{|\vartheta|^k} < \infty \quad \text{a.s.}$$

This implies that $\left(\sum_{j=1}^{n} |X_{j-1} Z_j| / a_n\right)_{n \geq 1}$, as a product of an almost surely convergent sequence and a distributionally convergent sequence of real random variables, is bounded in probability and thus $(B_n/a_n)_{n \geq 1}$ is bounded in probability.

Let φ denote the Fourier transform of P^{Z_1}. Since A_n is $\mathcal{F}_{n-1}$-measurable, we obtain for all $t \in \mathbb{R}$ and $n \geq 2$

$$E\left(\exp\left(it \frac{\Delta B_n}{A_n^{1/2}}\right) \Big| \mathcal{F}_{n-1}\right) = E\left(\exp\left(it \frac{X_{n-1} Z_n}{A_n^{1/2}}\right) \Big| \mathcal{F}_{n-1}\right)$$

$$= \int \exp\left(it \frac{X_{n-1} z}{A_n^{1/2}}\right) dP^{Z_1}(z) = \varphi\left(\frac{t X_{n-1}}{A_n^{1/2}}\right).$$

If $\vartheta > 1$, then

$$\frac{X_{n-1}}{A_n^{1/2}} = \frac{X_{n-1}/\vartheta^{n-1}}{A_n^{1/2}/\vartheta^{n-1}}$$

$$\to \frac{Y}{\vartheta \eta} = \frac{Y(\vartheta^2 - 1)^{1/2}}{|Y| \vartheta} = \operatorname{sign}(Y) \frac{(\vartheta^2 - 1)^{1/2}}{\vartheta} \quad \text{a.s.}$$

and if $\vartheta < -1$,

$$\frac{(-1)^n X_{n-1}}{A_n^{1/2}} = \frac{(-1)^n X_{n-1}/\vartheta^{n-1}}{A_n^{1/2}/\vartheta^{n-1}} = \frac{a_n}{A_n^{1/2}} \frac{\vartheta^{n-1}}{(-1)^n a_n} \frac{X_{n-1}}{\vartheta^{n-1}}$$

$$\to \frac{Y}{\vartheta \eta} = \operatorname{sign}(Y) \frac{(\vartheta^2 - 1)^{1/2}}{\vartheta} \quad \text{a.s.}$$

Let $(W_n)_{n \geq 0}$ denote an independent and identically distributed sequence of real random variables independent of $\mathcal{F}_\infty$ with $P^{W_0} = P^{Z_1}$. In case $\vartheta > 1$, we obtain

$$E\left(\exp\left(it \frac{\Delta B_n}{A_n^{1/2}}\right) \Big| \mathcal{F}_{n-1}\right) = \varphi\left(\frac{t X_{n-1}}{A_n^{1/2}}\right) \to \varphi\left(t \operatorname{sign}(Y) \frac{(\vartheta^2 - 1)^{1/2}}{\vartheta}\right) \quad P\text{-a.s.}$$

as $n \to \infty$ for every $t \in \mathbb{R}$. This is condition (vi) in Corollary 8.8 with $\mu = P^{(\vartheta^2 - 1)^{1/2} Z_1/\vartheta}$ and $S = \operatorname{sign}(Y)$. From Corollary 8.8 follows

$$\left(\sum_{j=1}^{n} X_{j-1}^2\right)^{1/2} (\widehat{\vartheta}_n - \vartheta) = \frac{B_n}{A_n^{1/2}} \to \operatorname{sign}(Y) \frac{(\vartheta^2 - 1)^{1/2}}{\vartheta} \sum_{j=0}^{\infty} (\vartheta^2)^{-j/2} W_j$$

$\mathcal{F}_\infty$-stably. Since

$$\frac{\left(\vartheta^2 - 1\right)^{1/2}}{\vartheta} \sum_{j=0}^{\infty} \left(\vartheta^2\right)^{-j/2} W_j = \frac{\left(\vartheta^2 - 1\right)^{1/2}}{\vartheta} \sum_{j=0}^{\infty} \vartheta^{-j} W_j$$

$$\stackrel{d}{=} \left(\vartheta^2 - 1\right)^{1/2} \sum_{k=1}^{\infty} \vartheta^{-k} W_k \stackrel{d}{=} \left(\vartheta^2 - 1\right)^{1/2} U$$

(where distributional equality is always meant under P), this can be read as

$$\left(\sum_{j=1}^{n} X_{j-1}^2\right)^{1/2} \left(\widehat{\vartheta}_n - \vartheta\right) \to \text{sign}\,(Y)\left(\vartheta^2 - 1\right)^{1/2} U \quad \mathcal{F}_\infty\text{-stably}$$

(see Lemma A.5 (b)). In case $\vartheta < -1$, we obtain

$$E\left(\exp\left(it\frac{(-1)^n \Delta B_n}{A_n^{1/2}}\right)\Big|\mathcal{F}_{n-1}\right)$$

$$= \varphi\left(\frac{t\,(-1)^n\,X_{n-1}}{A_n^{1/2}}\right) \to \varphi\left(t\,\text{sign}\,(Y)\,\frac{\left(\vartheta^2 - 1\right)^{1/2}}{\vartheta}\right)$$

P-almost surely as $n \to \infty$ for every $t \in \mathbb{R}$ so that condition (vii) in Corollary 8.9 is satisfied with $\mu = P^{(\vartheta^2-1)^{1/2}Z_1/\vartheta}$ and $S = \text{sign}\,(Y)$. Thus Corollary 8.9 yields

$$(-1)^n \left(\sum_{j=1}^{n} X_{j-1}^2\right)^{1/2} \left(\widehat{\vartheta}_n - \vartheta\right)$$

$$= \frac{(-1)^n\,B_n}{A_n^{1/2}} \to \text{sign}\,(Y)\,\frac{\left(\vartheta^2 - 1\right)^{1/2}}{\vartheta} \sum_{j=0}^{\infty} (-1)^j\,(\vartheta^2)^{-j/2} W_j$$

$\mathcal{F}_\infty$-stably. Since

$$\frac{\left(\vartheta^2 - 1\right)^{1/2}}{\vartheta} \sum_{j=0}^{\infty} (-1)^j \left(\vartheta^2\right)^{-j/2} W_j = \frac{\left(\vartheta^2 - 1\right)^{1/2}}{\vartheta} \sum_{j=0}^{\infty} \vartheta^{-j} W_j \stackrel{d}{=} (\vartheta^2 - 1)^{1/2} U,$$

this reads as

$$(-1)^n \left(\sum_{j=1}^{n} X_{j-1}^2\right)^{1/2} \left(\widehat{\vartheta}_n - \vartheta\right) \to \text{sign}\,(Y)\left(\vartheta^2 - 1\right)^{1/2} U \quad \mathcal{F}_\infty\text{-stably}.$$

In both cases we thus obtain

$$(\text{sign}(\vartheta))^n \left(\sum_{j=1}^n X_{j-1}^2 \right)^{1/2} \left(\widehat{\vartheta}_n - \vartheta \right) \to \text{sign}(Y)(\vartheta^2 - 1)^{1/2} U \quad \mathcal{F}_\infty\text{-stably}.$$

As for the deterministic norming, we can conclude using Theorem 3.18

$$\vartheta^n \left(\widehat{\vartheta}_n - \vartheta \right) = \frac{\vartheta^n B_n}{A_n} = \frac{(\text{sign}(\vartheta))^n a_n B_n}{A_n} = \frac{a_n}{A_n^{1/2}} \frac{(\text{sign}(\vartheta))^n B_n}{A_n^{1/2}}$$

$$= \frac{a_n}{A_n^{1/2}} (\text{sign}(\vartheta))^n \left(\sum_{j=1}^n X_{j-1}^2 \right)^{1/2} \left(\widehat{\vartheta}_n - \vartheta \right)$$

$$\to \frac{\text{sign}(Y)(\vartheta^2 - 1)U}{|Y|} = \frac{(\vartheta^2 - 1)U}{Y} \quad \mathcal{F}_\infty\text{-stably}.$$

Now assume that P^{Z_1} is symmetric around zero. Then P^U is also symmetric around zero. Hence, by Lemma A.5 (c), $P^{\text{sign}(Y)(\vartheta^2-1)^{1/2}U|\mathcal{F}_\infty} = P^{(\vartheta^2-1)^{1/2}U}$ so that

$$(\text{sign}(\vartheta))^n \left(\sum_{j=1}^n X_{j-1}^2 \right)^{1/2} \left(\widehat{\vartheta}_n - \vartheta \right) \to (\vartheta^2 - 1)^{1/2} U \quad \mathcal{F}_\infty\text{-mixing}.$$

Thus, again by the symmetry of P^U,

$$\left(\sum_{j=1}^n X_{j-1}^2 \right)^{1/2} \left(\widehat{\vartheta}_n - \vartheta \right) \to (\vartheta^2 - 1)^{1/2} U \quad \mathcal{F}_\infty\text{-mixing}. \qquad \square$$

Exercise 9.1 Assume $|\vartheta| > 1$ and $P^{Z_1} = C(0, b)$ with scale parameter $b \in (0, \infty)$. Show that

$$\left(\sum_{j=1}^n X_{j-1}^2 \right)^{1/2} \left(\widehat{\vartheta}_n - \vartheta \right) \to C \left(0, \frac{b(\vartheta^2 - 1)^{1/2}}{|\vartheta| - 1} \right) \quad \text{mixing}$$

as $n \to \infty$. More generally, if $P^{Z_1} = S_\alpha(b)$, the *symmetric α-stable distribution* with Fourier transform $\int \exp(itx) \, dS_\alpha(b)(x) = e^{-b|t|^\alpha}$, $\alpha \in (0, 2)$, $b \in (0, \infty)$, then

$$\left(\sum_{j=1}^n X_{j-1}^2 \right)^{1/2} \left(\widehat{\vartheta}_n - \vartheta \right) \to S_\alpha \left(\frac{b \left(\vartheta^2 - 1 \right)^{\alpha/2}}{|\vartheta|^\alpha - 1} \right) \quad \text{mixing}.$$

(Note that $C\left(0, b\right) = S_1\left(b\right)$.)

Exercise 9.2 Assume that $|\vartheta| > 1$ and P^{Z_1} is symmetric around zero. Show that

$$|\vartheta|^n \left(\widehat{\vartheta}_n - \vartheta\right) \to \frac{(\vartheta^2 - 1)U}{Y} \quad \mathcal{F}_\infty\text{-stably}$$

with U and Y from Theorem 9.2.

The critical case

Theorem 9.3 *Assume* $|\vartheta| = 1$, $Z_1 \in \mathcal{L}^2$ *and* $EZ_1 = 0$. *Then*

$$n\left(\widehat{\vartheta}_n - \vartheta\right) \to \vartheta \frac{W_1^2 - 1}{2\int_0^1 W_t^2\, dt} \quad \mathcal{F}_\infty\text{-mixing}$$

and

$$\left(\sum_{j=1}^n X_{j-1}^2\right)^{1/2} \left(\widehat{\vartheta}_n - \vartheta\right) \to \vartheta\sigma \frac{W_1^2 - 1}{2\left(\int_0^1 W_t^2\, dt\right)^{1/2}} \quad \mathcal{F}_\infty\text{-mixing}$$

as $n \to \infty$, *where* $(W_t)_{t\in[0,1]}$ *denotes a Brownian motion independent of* $\mathcal{F}_\infty$.

Distributional convergence under deterministic norming for $\vartheta = 1$ has already been observed by [24, 73, 97]. One checks that the numerator and the denominator of the first limiting random variable are positively correlated so that they are not independent in both limiting random variables.

Proof of Theorem 9.3. Let $\vartheta = 1$. Then $X_n = X_0 + \sum_{i=1}^n Z_i$ and hence, for $n \geq 1$,

$$A_n = \sum_{j=1}^n \left(X_0 + \sum_{i=1}^{j-1} Z_i\right)^2 = nX_0^2 + 2X_0 \sum_{j=1}^n \sum_{i=1}^{j-1} Z_i + \sum_{j=1}^n \left(\sum_{l=1}^{j-1} Z_i\right)^2$$

$$= nX_0^2 + 2X_0 \sum_{i=1}^{n-1} (n-i) Z_i + \sum_{j=1}^n \left(\sum_{i=1}^{j-1} Z_i\right)^2$$

and

$$B_n = \sum_{j=1}^n \left(X_0 + \sum_{i=1}^{j-1} Z_i\right) Z_j = X_0 \sum_{j=1}^n Z_j + \sum_{j=1}^n \left(\sum_{i=1}^{j-1} Z_i\right) Z_j$$

$$= X_0 \sum_{j=1}^n Z_j + \frac{1}{2}\left(\sum_{j=1}^n Z_j\right)^2 - \frac{1}{2}\sum_{j=1}^n Z_j^2.$$

For $n \in \mathbb{N}$, let $X^n = (X_t^n)_{t \in [0,1]}$ denote the normalized (path-continuous) partial sum process based on (Z_n) from Example 3.14 and let $Y_t^n := \left(\sum_{j=1}^{[nt]} Z_j \right) / \sigma \sqrt{n}, t \in$ [0, 1]. The map $C\,([0, 1]) \to \mathbb{R}^2, x \mapsto \left(\frac{1}{2} x\,(1)^2, \left(\int_0^1 x\,(t)^2\,dt \right)^{1/2} \right)$ is continuous so that by Example 3.14 (or Corollary 7.7) and Theorem 3.18 (c)

$$\left(\frac{1}{2} (X_1^n)^2, \left(\int_0^1 (X_t^n)^2\,dt \right)^{1/2} \right) \to \left(\frac{1}{2} W_1^2, \left(\int_0^1 W_t^2\,dt \right)^{1/2} \right) \quad \mathcal{F}_\infty\text{-mixing}.$$

We have

$$\left| \left(\int_0^1 (X_t^n)^2\,dt \right)^{1/2} - \left(\int_0^1 (Y_t^n)^2\,dt \right)^{1/2} \right| \le \left(\int_0^1 |X_t^n - Y_t^n|^2\,dt \right)^{1/2}$$

$$\le \|X^n - Y^n\|_{\sup} \le \frac{1}{\sigma \sqrt{n}} \max_{1 \le j \le n} |Z_j| \to 0 \quad \text{in probability}$$

and moreover,

$$\int_0^1 (Y_t^n)^2\,dt = \frac{1}{\sigma^2 n} \sum_{j=1}^n \int_{(j-1)/n}^{j/n} \left(\sum_{i=1}^{j-1} Z_i \right)^2\,dt = \frac{1}{\sigma^2 n^2} \sum_{j=1}^n \left(\sum_{i=1}^{j-1} Z_i \right)^2.$$

Using Theorem 3.18 (a), (c) this implies

$$\left(\frac{1}{2\sigma^2 n} \left(\sum_{j=1}^n Z_j \right)^2, \frac{1}{\sigma^2 n^2} \sum_{j=1}^n \left(\sum_{i=1}^{j-1} Z_i \right)^2 \right) \to \left(\frac{1}{2} W_1^2, \int_0^1 W_t^2\,dt \right) \quad \mathcal{F}_\infty\text{-mixing}.$$

Since $\left(\sum_{j=1}^n Z_j \right) / n \to 0$ almost surely, $\left(\sum_{j=1}^n Z_j^2 \right) / 2\sigma^2 n \to 1/2$ almost surely by the Kolmogorov strong law of large numbers and $\left(\sum_{i=1}^n i Z_i \right) / n^2 \to 0$ almost surely by the Kolmogorov criterion (or Theorem A.9 with $B_n = n^2$, $p = 2$), we obtain in view of Theorem 3.18 (b), (c)

$$\left(\frac{B_n}{\sigma^2 n}, \frac{A_n}{\sigma^2 n^2} \right) \to \left(\frac{1}{2} W_1^2 - \frac{1}{2}, \int_0^1 W_t^2\,dt \right) \quad \mathcal{F}_\infty\text{-mixing}.$$

Consequently, by Theorem 3.18 (c), using $P\left(\int_0^1 W_t^2\,dt > 0 \right) = 1$,

$$n\,(\widehat{\vartheta}_n - \vartheta) = \frac{n B_n}{A_n} = \frac{B_n / \sigma^2 n}{A_n / \sigma^2 n^2} \to \frac{W_1^2 - 1}{2 \int_0^1 W_t^2\,dt} \quad \mathcal{F}_\infty\text{-mixing}$$

and

$$\left(\sum_{j=1}^{n} X_{j-1}^2\right)^{1/2} (\widehat{\vartheta}_n - \vartheta) = \frac{B_n}{A_n^{1/2}} = \sigma \frac{B_n/\sigma^2 n}{\sqrt{A_n/\sigma^2 n^2}} \to \sigma \frac{W_1^2 - 1}{2\left(\int_0^1 W_t^2\, dt\right)^{1/2}}$$

$\mathcal{F}_\infty$-mixing as $n \to \infty$.

In case $\vartheta = -1$, let $\widetilde{Z}_n := (-1)^n Z_n$. Then $X_n = (-1)^n \left(X_0 + \sum_{j=1}^{n} \widetilde{Z}_j\right)$ and hence, for all $n \geq 1$,

$$A_n = \sum_{j=1}^{n} \left((-1)^{j-1} \left(X_0 + \sum_{i=1}^{j-1} \widetilde{Z}_i\right)\right)^2$$

$$= n X_0^2 + 2 X_0 \sum_{i=1}^{n-1} (n-i)\, \widetilde{Z}_i + \sum_{j=1}^{n} \left(\sum_{i=1}^{j-1} \widetilde{Z}_i\right)^2$$

and

$$B_n = \sum_{j=1}^{n} \left((-1)^{j-1} \left(X_0 + \sum_{i=1}^{j-1} \widetilde{Z}_i\right)\right) Z_j = -\sum_{j=1}^{n} \left(X_0 + \sum_{i=1}^{j-1} \widetilde{Z}_i\right) \widetilde{Z}_j$$

$$= -X_0 \sum_{j=1}^{n} \widetilde{Z}_j - \frac{1}{2} \left(\sum_{j=1}^{n} \widetilde{Z}_j\right)^2 + \frac{1}{2} \sum_{j=1}^{n} \widetilde{Z}_j^2.$$

One may apply Corollary 7.7 to the normalized partial sum process based on $\left(\widetilde{Z}_n\right)_{n\geq 1}$. One simply has to observe that now $\left(\sum_{j=1}^{n} \widetilde{Z}_j\right)/n \to 0$ almost surely by the Kolmogorov criterion (or Theorem A.9). $\square$

We see that in the case $|\vartheta| \leq 1$ the limiting distributions of $\widehat{\vartheta}_n$ under deterministic and random norming do not depend on the distribution P^{Z_1} (and X_0) while in the explosive case $|\vartheta| > 1$ they do.

Notice that in case $|\vartheta| = 1$ there occurs a singularity in the sense that A_n/a_n^2 does not converge in probability (with $a_n = n$) in contrast to the case $|\vartheta| \neq 1$. This coincides with the fact that the observation process X is a martingale if $\vartheta = 1$ and $\left((-1)^n X_n\right)_{n\geq 0}$ is a martingale if $\vartheta = -1$ (see [63], [89], Chap. 5).

Remark 9.4 The preceding result provides a counterexample to Theorem 6.23 of the type of Example 6.12 for arrays: In condition (N_{a_n}) convergence in probability cannot be replaced by mixing convergence. Assume the setting of Theorem 9.3 with $X_0 = 0$, $Z_1 \in \mathcal{L}^p$ for some $p > 2$ and $\vartheta = 1$. Then $B = (B_n)_{n\geq 0}$, where

$$B_n = \sum_{j=1}^{n} \left(\sum_{i=1}^{j-1} Z_i \right) Z_j \quad \text{with} \quad B_0 = B_1 = 0$$

is a square integrable martingale with quadratic characteristic

$$\langle B \rangle_n = \sigma^2 A_n = \sigma^2 \sum_{j=1}^{n} \left(\sum_{i=1}^{j-1} Z_i \right)^2 .$$

The proof of Theorem 9.3 shows that

$$\frac{1}{n^2} \langle B \rangle_n = \frac{\sigma^4 A_n}{\sigma^2 n^2} \to \sigma^4 \int_0^1 W_t^2 \, dt \quad \mathcal{F}_\infty\text{-mixing} .$$

Hence, condition (N_{a_n}) with $a_n = n$ holds with mixing convergence instead of convergence in probability. Moreover, the conditional Lyapunov condition $(\text{CLY}_{a_n,p})$ is satisfied for B which implies (CLB_{a_n}) by Remark 6.25. In fact, we have

$$\frac{1}{n^p} \sum_{j=1}^{n} E\left(\left| \left(\sum_{i=1}^{j-1} Z_i \right) Z_j \right|^p \Big| \mathcal{F}_{j-1} \right) = \frac{1}{n^p} \sum_{j=1}^{n} \left| \sum_{i=1}^{j-1} Z_i \right|^p E |Z_1|^p .$$

Let $b := (p-1)/p$. Then $b > 1/2$ and hence, for example, the strong law of large numbers of Theorem A.9 (or Example 4.2) for the martingale $\left(\sum_{i=1}^{n} Z_i \right)_{n \geq 0}$ yields

$$\frac{\left| \sum_{i=1}^{n-1} Z_i \right|^p}{n^{pb}} \to 0 \quad \text{a.s.}$$

so that by the discrete rule of de l'Hospital in Lemma 6.28 (b)

$$\frac{\sum_{j=1}^{n} \left| \sum_{i=1}^{j-1} Z_i \right|^p}{\sum_{j=1}^{n} j^{pb}} \to 0 \quad \text{a.s.}$$

Since $\sum_{j=1}^{n} j^{pb} \sim n^{pb+1}/(pb+1) = n^p/p$, we obtain

$$\frac{1}{n^p} \sum_{j=1}^{n} E\left(\left| \left(\sum_{i=1}^{j-1} Z_i \right) Z_j \right|^p \Big| \mathcal{F}_{j-1} \right) \to 0 \quad \text{a.s.}$$

On the other hand, again, for example, by the proof of Theorem 9.3,

$$\frac{B_n}{n} \to \frac{\sigma^2}{2} \left(W_1^2 - 1 \right) \quad \mathcal{F}_\infty\text{-mixing}.$$

The distribution of the limiting random variable is not symmetric around zero and hence is not a variance mixture of centered Gaussian distributions.   □

Exercise 9.3 Assume $\vartheta = 1$, $Z_1 \in \mathcal{L}^2$ and $E Z_1 = 0$. Show that

$$n^{-3/2} \sum_{j=1}^{n} X_j \to \sigma \int_0^1 W_t \, dt \quad \mathcal{F}_\infty\text{-mixing as } n \to \infty,$$

where $(W_t)_{t\in[0,1]}$ denotes a Brownian motion independent of $\mathcal{F}_\infty$.

Exercise 9.4 (cf. [55]) Assume $|\vartheta| \leq 1$, $X_0, Z_1 \in \mathcal{L}^2$ and $E Z_1 = 0$, and let $\gamma > 0$ be fixed. For every $c \in \mathbb{N}$, set

$$\tau_c := \min \left\{ n \in \mathbb{N} : \sum_{j=1}^{n} X_{j-1}^2 \geq c\gamma \right\}.$$

Show that τ_c is almost surely finite for every $c \in \mathbb{N}$ and

$$\left(\sum_{j=1}^{\tau_c} X_{j-1}^2 \right)^{1/2} \left(\widehat{\vartheta}_{\tau_c} - \vartheta \right) \to \sigma \sqrt{\gamma} N \quad \mathcal{F}_\infty\text{-mixing as } c \to \infty$$

as well as

$$c^{1/2} \left(\widehat{\vartheta}_{\tau_c} - \vartheta \right) \to \frac{\sigma}{\sqrt{\gamma}} N \quad \mathcal{F}_\infty\text{-mixing as } c \to \infty,$$

where $P^N = N(0, 1)$ and N is independent of $\mathcal{F}_\infty$.

Hint: Apply Corollary 6.4. The proof of $X_n^2 / \sum_{j=1}^{n} X_{j-1}^2 \to 0$ almost surely as $n \to \infty$ is a crucial step.

Exercise 9.4 shows that sequential sampling with random sample size τ_c leads to the same normal limit for $\widehat{\vartheta}_{\tau_c}$ as $c \to \infty$ for the whole range $-1 \leq \vartheta \leq 1$ of the autoregression parameter, in contrast to the result of Theorem 9.3.

Chapter 10
Galton-Watson Branching Processes

Let $(Y_{nj})_{n,j \in \mathbb{N}}$ be independent and identically distributed random variables with values in $\mathbb{N}_0$, and let X_0 be some random variable with values in $\mathbb{N}$ which is independent of $(Y_{nj})_{n,j \in \mathbb{N}}$, where all these random variables are defined on the same probability space $(\Omega, \mathcal{F}, P)$. For every $n \in \mathbb{N}$ we set

$$X_n := \sum_{j=1}^{X_{n-1}} Y_{nj} .$$

The process $X = (X_n)_{n \geq 0}$ is the *Galton-Watson branching process*.

The process X can be interpreted as follows: In a population of particles (which may represent people, cells, neutrons, etc., depending on the field of application) each particle j of the $(n-1)$-th generation produces a random number Y_{nj} (which may be 0) of identical particles in the n-th generation, called the *offspring* of j, and it does so independently of all other particles from the $(n-1)$-th and all earlier generations. The *offspring distribution*, i.e. the distribution of Y_{nj}, is the same for all particles in all generations. Then X_n is the total number of particles in the n-th generation, with X_0 being the (random) number of particles in the 0-th generation. Note that excluding the value 0 of X_0 is not an essential restriction because by definition of X_n we would have $X_n = 0$ for all $n \in \mathbb{N}$ on the event $\{X_0 = 0\}$ so that $(X_n)_{n \geq 0}$ would be trivial on $\{X_0 = 0\}$.

For every $k \in \mathbb{N}_0$ set $p_k := P(Y_{11} = k)$. To exclude trivial cases, we always assume $p_0 < 1$ (if $p_0 = 1$, then $X_n = 0$ almost surely for all $n \in \mathbb{N}$) and $p_1 < 1$ (if $p_1 = 1$, then $X_n = X_0$ almost surely for all $n \in \mathbb{N}$). Clearly, if $X_n = 0$ for some $n \in \mathbb{N}$, then $X_m = 0$ for all $m \geq n$, and the population is said to be *extinct* at time n.

One of the main features of the process X is the fact that with probability one either $X_n = 0$ for all large n or $\lim_{n \to \infty} X_n = \infty$, that is, $P(\{\lim_{n \to \infty} X_n = 0\} \cup \{\lim_{n \to \infty} X_n = \infty\}) = 1$; see e.g. [64], Satz 9.1. Whether the *probability of extinction* $\rho := P(\lim_{n \to \infty} X_n = 0)$ equals 1 or is strictly less than 1 is completely

© Springer International Publishing Switzerland 2015

E. Häusler and H. Luschgy, *Stable Convergence and Stable Limit Theorems*,
Probability Theory and Stochastic Modelling 74,
DOI 10.1007/978-3-319-18329-9_10

determined by the *offspring mean* $\alpha := E(Y_{11})$. If $\alpha \leq 1$, then $\rho = 1$, and if $\alpha > 1$, then $\rho < 1$; see e.g. [64], Korollar 9.5. Observe that $\alpha > 0$ because $p_0 < 1$.

We are interested here in stable limit theorems motivated by asymptotic statistical inference about $\alpha > 1$ for $n \to \infty$. This is only meaningful on the event

$$M_+ := \left\{ \lim_{n \to \infty} X_n = \infty \right\}$$

because on the complementary event of extinction $\{\lim_{n \to \infty} X_n = 0\}$ the number of available data about the process X stays finite as n gets large. Therefore, we will restrict ourselves to the case $\alpha > 1$ in which $P(M_+) > 0$ under suitable moment conditions and in which the process X is called *supercritical*. In the sequel we will discuss several different estimators of α in the supercritical case and derive stable limit theorems for these estimators under deterministic and random norming. For this, we have to collect a few more basic facts about the process X. We always assume $Y_{11} \in \mathcal{L}^2(P)$ with $\sigma^2 := \mathrm{Var}(Y_{11}) > 0$ and $X_0 \in \mathcal{L}^2(P)$ as well as $\alpha > 1$.

Let $\mathcal{F}_0 = \sigma(X_0)$ and $\mathcal{F}_n = \sigma(X_0, Y_{ij}; 1 \leq i \leq n, j \in \mathbb{N})$ for all $n \in \mathbb{N}$. Clearly, $\mathbb{F} = (\mathcal{F}_n)_{n \geq 0}$ is a filtration and X is $\mathbb{F}$-adapted. As usual, $\mathcal{F}_\infty = \sigma\left(\bigcup_{n=0}^\infty \mathcal{F}_n \right)$. There exists a nonnegative $M_\infty \in \mathcal{L}^2(\mathcal{F}_\infty, P)$ with

$$M_n := \alpha^{-n} X_n \to M_\infty \quad \text{a.s. and in } \mathcal{L}^2 \text{ as } n \to \infty.$$

This is a consequence of the fact that $(M_n)_{n \geq 0}$ is an $\mathcal{L}^2$-bounded martingale w.r.t. $\mathbb{F}$ and the martingale convergence theorem; see e.g. [64], Lemma 9.3 and Satz 9.4. Moreover, $\{\lim_{n \to \infty} X_n = 0\} = \{M_\infty = 0\}$ almost surely so that

$$M_+ = \{M_\infty > 0\} \quad \text{a.s.}$$

and $P(M_+) > 0$; see e.g. [64], Satz 9.4 and the remark following it in combination with Satz 9.6 and our assumption $Y_{11} \in \mathcal{L}^2(P)$.

A moment estimator

The first estimator which we will consider here is a simple moment estimator. It appears in [44]. For all $n \in \mathbb{N}$ we have

$$E(X_n | \mathcal{F}_{n-1}) = \sum_{j=1}^{X_{n-1}} E(Y_{nj} | \mathcal{F}_{n-1}) = \alpha X_{n-1}$$

because $E(Y_{nj} | \mathcal{F}_{n-1}) = E(Y_{nj}) = \alpha$ by independence of Y_{nj} and $\mathcal{F}_{n-1}$. Consequently, $E(X_n) = \alpha E(X_{n-1})$ for all $n \in \mathbb{N}$, whence $E(X_n) = \alpha^n E(X_0)$ and

$$\alpha = \frac{E(X_n)^{1/n}}{E(X_0)^{1/n}}.$$

Ignoring the denominator because $E\left(X_0\right)^{1/n} \to 1$ as $n \to \infty$, the principle of moments yields the (unconditional) moment estimator

$$\widehat{\alpha}_n^{(M)} := X_n^{1/n} \, .$$

On M_+ we have $M_\infty > 0$ so that $\alpha^{-n} X_n \to M_\infty$ almost surely implies

$$\log X_n - n \log \alpha \to \log M_\infty \quad \text{a.s. as } n \to \infty \, .$$

This yields $\frac{1}{n} \log X_n - \log \alpha \to 0$ almost surely so that $\widehat{\alpha}_n^{(M)} \to \alpha$ almost surely on M_+. Thus, $\widehat{\alpha}_n^{(M)}$ is a strongly consistent estimator for α on M_+. On the other hand, on M_+ we get

$$n\left(\log \widehat{\alpha}_n^{(M)} - \log \alpha\right) = \log X_n - n \log \alpha \to \log M_\infty \quad \text{a.s. as } n \to \infty$$

and, by the mean value theorem,

$$n\left(\log \widehat{\alpha}_n^{(M)} - \log \alpha\right) = \frac{n}{\xi_n}\left(\widehat{\alpha}_n^{(M)} - \alpha\right)$$

for some ξ_n between $\widehat{\alpha}_n^{(M)}$ and α. Therefore, $\xi_n \to \alpha$ almost surely as $n \to \infty$ and hence

$$n\left(\widehat{\alpha}_n^{(M)} - \alpha\right) \to \alpha \log M_\infty \quad \text{a.s. on } M_+ \, .$$

This exhibits a rather unusual asymptotic behavior of the estimator $\widehat{\alpha}_n^{(M)}$.

A conditional moment estimator

To motivate the second estimator we apply the principle of moments condition-ally to

$$\alpha = \frac{E\left(X_n | \mathcal{F}_{n-1}\right)}{X_{n-1}} \, ,$$

provided that $X_{n-1} \geq 1$. Replacing the conditional moment $E\left(X_n | \mathcal{F}_{n-1}\right)$ by X_n, we arrive at the estimator

$$\widehat{\alpha}_n^{(LN)} := \frac{X_n}{X_{n-1}} \, .$$

Note that $X_n = 0$ for some $n \in \mathbb{N}$ implies $X_m = 0$ for all $m \geq n$ so that we have $X_n \geq 1$ for all $\in \mathbb{N}_0$ on M_+ and hence

$$\widehat{\alpha}_n^{(LN)} = \alpha \frac{M_n}{M_{n-1}} \to \alpha \frac{M_\infty}{M_\infty} = \alpha \quad \text{a.s. on } M_+ \text{ as } n \to \infty \, ,$$

which says that $\widehat{\alpha}_n^{(LN)}$ is a strongly consistent estimator for α on M_+. This is the *Lotka-Nagaev estimator* considered in [67]. A stable limit theorem for $\widehat{\alpha}_n^{(LN)}$ will be derived here from the following stability result.

Theorem 10.1 *Under the above assumptions,*

$$\frac{1}{\alpha^{(n-1)/2}} \sum_{j=1}^{X_{n-1}} \left(Y_{nj} - \alpha\right) \to \sigma M_\infty^{1/2} N \quad \mathcal{F}_\infty\text{-stably as } n \to \infty,$$

where $P^N = N(0, 1)$ *and* N *is* P-*independent of* $\mathcal{F}_\infty$.

Proof For all $n \in \mathbb{N}$ and $j \geq 0$ set

$$\widetilde{\mathcal{F}}_{nj} := \sigma \left(X_0, Y_{mk}, 1 \leq m \leq n - 1, k \in \mathbb{N}; Y_{n1}, \dots, Y_{nj}\right)$$

so that $\widetilde{\mathcal{F}}_{10} = \sigma(X_0)$ and $\widetilde{\mathcal{F}}_{n0} = \sigma(X_0, Y_{mk}, 1 \leq m \leq n-1, k \in \mathbb{N}) = \mathcal{F}_{n-1}$ for all $n \geq 2$. The array $(\widetilde{\mathcal{F}}_{nj})_{j \geq 0, n \in \mathbb{N}}$ is clearly nondecreasing in j and n so that it satisfies the nesting condition. For every $n \in \mathbb{N}$ the $\mathbb{N}_0$-valued random variable X_{n-1} is measurable w.r.t. $\widetilde{\mathcal{F}}_{n0}$ and therefore a stopping time w.r.t. $(\widetilde{\mathcal{F}}_{nj})_{j \geq 0}$. Moreover, for every $n \in \mathbb{N}$, by independence of Y_{nj} and $\widetilde{\mathcal{F}}_{n,j-1}$ and $E(Y_{nj}) = \alpha$, the sequence $(Y_{nj} - \alpha)_{j \in \mathbb{N}}$ is a martingale difference sequence w.r.t. $(\widetilde{\mathcal{F}}_{nj})_{j \geq 0}$. Therefore, $(\alpha^{-(n-1)/2} (Y_{nj} - \alpha))_{j \in \mathbb{N}}$ is a martingale difference sequence w.r.t. $(\widetilde{\mathcal{F}}_{nj})_{j \geq 0}$ as well. By independence of Y_{nj} and $\widetilde{\mathcal{F}}_{n,j-1}$ again we have

$$\sum_{j=1}^{X_{n-1}} E\left(\left(\frac{Y_{nj} - \alpha}{\alpha^{(n-1)/2}}\right)^2 \Big| \widetilde{\mathcal{F}}_{n,j-1}\right) = \frac{1}{\alpha^{n-1}} \sum_{j=1}^{X_{n-1}} E\left((Y_{nj} - \alpha)^2\right)$$

$$= \sigma^2 \frac{X_{n-1}}{\alpha^{n-1}} \to \sigma^2 M_\infty \quad \text{a.s. as } n \to \infty$$

so that condition (N_{τ_n}) is satisfied with the finite stopping time $\tau_n = X_{n-1}$ and $\eta^2 = \sigma^2 M_\infty$. Moreover, again by independence of Y_{nj} and $\widetilde{\mathcal{F}}_{n,j-1}$,

$$\sum_{j=1}^{X_{n-1}} E\left(\left(\frac{Y_{nj} - \alpha}{\alpha^{(n-1)/2}}\right)^2 1_{\{|Y_{nj} - \alpha| \geq \varepsilon \alpha^{(n-1)/2}\}} \Big| \widetilde{\mathcal{F}}_{n,j-1}\right)$$

$$= \frac{1}{\alpha^{n-1}} \sum_{j=1}^{X_{n-1}} E\left((Y_{nj} - \alpha)^2 1_{\{|Y_{nj} - \alpha| \geq \varepsilon \alpha^{(n-1)/2}\}}\right)$$

$$= \frac{X_{n-1}}{\alpha^{n-1}} E\left((Y_{11} - \alpha)^2 1_{\{|Y_{11} - \alpha| \geq \varepsilon \alpha^{(n-1)/2}\}}\right) \to 0 \quad \text{a.s. as } n \to \infty$$

so that condition (CLB_{τ_n}) is satisfied with the finite stopping time $\tau_n = X_{n-1}$. Observe that $\mathcal{F}_\infty = \sigma\left(\bigcup_{n \in \mathbb{N}} \bigcup_{j \geq 0} \widetilde{\mathcal{F}}_{nj}\right)$. Therefore, the assertion follows from Corollary 6.4 and Remark 6.2 (d). $\square$

Corollary 10.2 *Under the above assumptions,*

$$X_{n-1}^{1/2}\left(\widehat{\alpha}_n^{(LN)} - \alpha\right) \to \sigma N \quad \mathcal{F}_\infty\text{-mixing under } P_{M_+} \text{ as } n \to \infty$$

and

$$\alpha^{(n-1)/2}\left(\widehat{\alpha}_n^{(LN)} - \alpha\right) \to \sigma M_\infty^{-1/2} N \quad \mathcal{F}_\infty\text{-stably under } P_{M_+} \text{ as } n \to \infty,$$

where $P^N = N(0, 1)$ *and* N *is* P*-independent of* $\mathcal{F}_\infty$.

Proof On M_+ we have almost surely

$$\sum_{j=1}^{X_{n-1}} (Y_{nj} - \alpha) = X_n - \alpha X_{n-1} = X_{n-1}\left(\widehat{\alpha}_n^{(LN)} - \alpha\right)$$

so that

$$X_{n-1}^{1/2}\left(\widehat{\alpha}_n^{(LN)} - \alpha\right) = \frac{\alpha^{(n-1)/2}}{X_{n-1}^{1/2}} \frac{1}{\alpha^{(n-1)/2}} \sum_{j=1}^{X_{n-1}} (Y_{nj} - \alpha) \ .$$

Consequently, the first assertion follows from Theorem 10.1 and $\alpha^{(n-1)/2}/X_{n-1}^{1/2} \to M_\infty^{-1/2} P_{M_+}$-almost surely as $n \to \infty$ via Theorem 3.18 (b) and (c) (use $g(x, y) = xy$).

On M_+ we also get almost surely

$$\alpha^{(n-1)/2}\left(\widehat{\alpha}_n^{(LN)} - \alpha\right) = \frac{\alpha^{(n-1)/2}}{X_{n-1}^{1/2}} X_{n-1}^{1/2}\left(\widehat{\alpha}_n^{(LN)} - \alpha\right)$$

so that the second assertion follows from the first one and $\alpha^{(n-1)/2}/X_{n-1}^{1/2} \to M_\infty^{-1/2}$ P_{M_+}-almost surely as $n \to \infty$, again via Theorem 3.18 (b) and (c). $\square$

A conditional least squares estimator

The third estimator is a conditional least squares estimator which is defined as the minimizer of the sum of squares

$$\sum_{i=1}^{n} (X_i - E(X_i|\mathcal{F}_{i-1}))^2 = \sum_{i=1}^{n} (X_i - \alpha X_{i-1})^2$$

and is given by

$$\widehat{\alpha}_n^{(LS)} := \frac{\sum_{i=1}^n X_i X_{i-1}}{\sum_{i=1}^n X_{i-1}^2}.$$

Since $\sum_{i=1}^n X_{i-1}^2 \geq X_0^2 \geq 1$, $\widehat{\alpha}_n^{(LS)}$ is well-defined. On M_+ we have

$$\frac{X_{i-1} X_i}{X_{i-1}^2} = \widehat{\alpha}_i^{(LN)} \to \alpha \quad \text{a.s. as } i \to \infty,$$

and the Toeplitz Lemma 6.28 (b) implies $\widehat{\alpha}_n^{(LS)} \to \alpha$ almost surely on M_+ as $n \to \infty$ so that $\widehat{\alpha}_n^{(LS)}$ is strongly consistent on M_+. To obtain stable limit theorems for $\widehat{\alpha}_n^{(LS)}$, we introduce the process $U^{(LS)} = \left(U_n^{(LS)} \right)_{n \geq 0}$ with $U_0^{(LS)} := 0$ and

$$U_n^{(LS)} := \sum_{i=1}^n \left(X_{i-1} X_i - \alpha X_{i-1}^2 \right) \quad \text{for } n \geq 1$$

which is an $\mathbb{F}$-martingale because $E(X_i | \mathcal{F}_{i-1}) = \alpha X_{i-1}$. If $E\left(X_0^4 \right) < \infty$ and $E\left(Y_{11}^4 \right) < \infty$, then $U^{(LS)}$ is square integrable with quadratic characteristic

$$\begin{aligned}
\left\langle U^{(LS)} \right\rangle_n &= \sum_{i=1}^n E\left(\left(\Delta U_i^{(LS)} \right)^2 \Big| \mathcal{F}_{i-1} \right) \\
&= \sum_{i=1}^n E\left(\left[X_{i-1} X_i - \alpha X_{i-1}^2 \right]^2 \Big| \mathcal{F}_{i-1} \right) \\
&= \sum_{i=1}^n X_{i-1}^2 E\left(\left[X_i - \alpha X_{i-1} \right]^2 \Big| \mathcal{F}_{i-1} \right) \\
&= \sum_{i=1}^n X_{i-1}^2 \left[E\left(X_i^2 | \mathcal{F}_{i-1} \right) - 2\alpha X_{i-1} E\left(X_i | \mathcal{F}_{i-1} \right) + \alpha^2 X_{i-1}^2 \right] \\
&= \sigma^2 \sum_{i=1}^n X_{i-1}^3
\end{aligned}$$

because $E\left(X_i^2 | \mathcal{F}_{i-1} \right) = \sigma^2 X_{i-1} + \alpha^2 X_{i-1}^2$ and $E(X_i | \mathcal{F}_{i-1}) = \alpha X_{i-1}$. The following application of Theorem 8.2 and Corollary 8.5 is crucial.

Theorem 10.3 *If $E\left(X_0^4 \right) < \infty$ and $E\left(Y_{11}^4 \right) < \infty$, then*

$$\frac{U_n^{(LS)}}{\left\langle U^{(LS)} \right\rangle_n^{1/2}} \to N \quad \mathcal{F}_\infty\text{-mixing under } P_{M_+} \text{ as } n \to \infty,$$

where N is P-independent of $\mathcal{F}_\infty$ with $P^N = N(0,1)$.

Proof Here, we are in the setting of Remark 8.6 so that condition (ii) in Theorem 8.2 follows from conditions (i), (iii) and (iv). Consequently, we only have to verify these conditions.

We verify condition (i) with $G = \Omega$, $a_n = \alpha^{3n/2}$ and $\eta = \sigma M_\infty^{3/2}/\left(\alpha^3 - 1\right)^{1/2}$. For this, note that $\alpha^{-3(i-1)} X_{i-1}^3 \to M_\infty^3$ almost surely as $i \to \infty$, so that the Toeplitz Lemma 6.28 (b) implies

$$\frac{\sum_{i=1}^n X_{i-1}^3}{\sum_{i=1}^n \alpha^{3(i-1)}} \to M_\infty^3 \quad \text{a.s. as } n \to \infty.$$

Because

$$\sum_{i=1}^n \alpha^{3(i-1)} = \frac{\alpha^{3n} - 1}{\alpha^3 - 1} \sim \frac{1}{\alpha^3 - 1} \alpha^{3n}$$

we get

$$\frac{\left\langle U^{(LS)}\right\rangle_n}{a_n^2} = \frac{\left\langle U^{(LS)}\right\rangle_n}{\alpha^{3n}} = \sigma^2 \frac{\sum_{i=1}^n X_{i-1}^3}{\alpha^{3n}} \to \frac{\sigma^2}{\alpha^3 - 1} M_\infty^3 \quad \text{a.s. as } n \to \infty,$$

which implies (i).

For all $n, r \in \mathbb{N}$ we have $a_{n-r}^2/a_n^2 = 1/\alpha^{3r}$ which is (ii) with $p = \alpha^3$.

Finally, we will verify (iv) for $\mu = N(0, b)$ with $b = \left(\alpha^3 - 1\right)/\alpha^3$, which means that Corollary 8.5 applies and yields $P^Z = N(0,1)$ because $bp/(p-1) = 1$ in the present case. For the proof of (iv) we write for every $t \in \mathbb{R}$ and $n \in \mathbb{N}$, using measurability of X_{n-1} and $\left\langle U^{(LS)}\right\rangle_n$ w.r.t. $\mathcal{F}_{n-1}$,

$$E_P\left(\exp\left(it\frac{\Delta U_n^{(LS)}}{\left\langle U^{(LS)}\right\rangle_n^{1/2}}\right)\bigg|\mathcal{F}_{n-1}\right) = E_P\left(\exp\left(it\frac{X_{n-1}X_n - \alpha X_{n-1}^2}{\left\langle U^{(LS)}\right\rangle_n^{1/2}}\right)\bigg|\mathcal{F}_{n-1}\right)$$

$$= \exp\left(it\frac{(-\alpha)X_{n-1}^2}{\left\langle U^{(LS)}\right\rangle_n^{1/2}}\right) E_P\left(\exp\left(it\frac{X_{n-1}}{\left\langle U^{(LS)}\right\rangle_n^{1/2}}\sum_{j=1}^{X_{n-1}} Y_{nj}\right)\bigg|\mathcal{F}_{n-1}\right)$$

$$= \exp\left(i\frac{(-\alpha)tX_{n-1}}{\left\langle U^{(LS)}\right\rangle_n^{1/2}}\right)^{X_{n-1}} \zeta\left(\frac{tX_{n-1}}{\left\langle U^{(LS)}\right\rangle_n^{1/2}}\right)^{X_{n-1}},$$

where ζ denotes the characteristic function of Y_{11} and we used independence of $\sigma\left(Y_{nj} : j \in \mathbb{N}\right)$ and $\mathcal{F}_{n-1}$. Employing the characteristic function

$$\phi(u) = \exp\left(i\left(\frac{-\alpha}{\sigma}\right)u\right)\zeta\left(\frac{u}{\sigma}\right), \quad u \in \mathbb{R},$$

of the normalized random variable $(Y_{11} - \alpha)/\sigma$, we get

$$E_P\left(\exp\left(it\frac{\Delta U_n^{(LS)}}{\langle U^{(LS)}\rangle_n^{1/2}}\right)\bigg|\mathcal{F}_{n-1}\right) = \phi\left(\frac{\sigma t X_{n-1}}{\langle U^{(LS)}\rangle_n^{1/2}}\right)^{X_{n-1}}.$$

Note that on M_+ we have

$$\frac{\sigma t X_{n-1}}{\langle U^{(LS)}\rangle_n^{1/2}} = \frac{\sigma t X_{n-1}^{3/2}}{\langle U^{(LS)}\rangle_n^{1/2}}\frac{1}{X_{n-1}^{1/2}}$$

with

$$\frac{\sigma t X_{n-1}^{3/2}}{\langle U^{(LS)}\rangle_n^{1/2}} = \sigma t\left(\frac{a_n^2}{\langle U^{(LS)}\rangle_n}\right)^{1/2}\left(\alpha^{-(n-1)}X_{n-1}\right)^{3/2}\alpha^{-3/2} \to t\left(\frac{\alpha^3-1}{\alpha^3}\right)^{1/2}$$

almost surely as $n \to \infty$. The classical central limit theorem for sums of independent and identically distributed random variables yields

$$\phi\left(\frac{x}{\sqrt{n}}\right)^n \to \exp\left(-\frac{1}{2}x^2\right) \quad \text{as } n \to \infty$$

uniformly in $x \in \mathbb{R}$ on compact intervals. Setting $x = \sigma t X_{n-1}^{3/2}/\langle U^{(LS)}\rangle_n^{1/2}$ and $n = X_{n-1}$ and combining the last two facts, we obtain

$$\phi\left(\frac{\sigma t X_{n-1}}{\langle U^{(LS)}\rangle_n^{1/2}}\right)^{X_{n-1}} \to \exp\left(-\frac{1}{2}t^2\frac{\alpha^3-1}{\alpha^3}\right) \quad \text{a.s. on } M_+ \text{ as } n \to \infty,$$

which implies condition (iv) with $b = (\alpha^3 - 1)/\alpha^3$ and concludes the proof. $\square$

Corollary 10.4 *Under the assumptions of Theorem 10.3,*

$$\frac{\sum_{i=1}^n X_{i-1}^2}{\left(\sum_{i=1}^n X_{i-1}^3\right)^{1/2}}\left(\widehat{\alpha}_n^{(LS)} - \alpha\right) \to \sigma N \quad \mathcal{F}_\infty\text{-mixing under } P_{M_+} \text{ as } n \to \infty$$

and

$$\frac{(\alpha^3-1)^{1/2}}{\alpha^2-1}\alpha^{n/2}\left(\widehat{\alpha}_n^{(LS)} - \alpha\right) \to \sigma M_\infty^{-1/2}N \quad \mathcal{F}_\infty\text{-stably under } P_{M_+} \text{ as } n \to \infty,$$

where $P^N = N(0,1)$ and N is P-independent of $\mathcal{F}_\infty$.

Proof For all $n \in \mathbb{N}$, we have

$$\widehat{\alpha}_n^{(LS)} = \frac{\sum_{i=1}^n \left(X_{i-1} X_i - \alpha X_{i-1}^2 \right)}{\sum_{i=1}^n X_{i-1}^2} + \alpha = \frac{U_n^{(LS)}}{\sum_{i=1}^n X_{i-1}^2} + \alpha$$

so that

$$\frac{\sum_{i=1}^n X_{i-1}^2}{\left(\sum_{i=1}^n X_{i-1}^3 \right)^{1/2}} \left(\widehat{\alpha}_n^{(LS)} - \alpha \right) = \sigma \frac{U_n^{(LS)}}{\left(U^{(LS)} \right)_n^{1/2}} .$$

Thus the first statement is immediate from Theorem 10.3. The second statement follows from the first and

$$\frac{\left(\alpha^3 - 1 \right)^{1/2}}{\alpha^2 - 1} \alpha^{n/2} \frac{\left(\sum_{i=1}^n X_{i-1}^3 \right)^{1/2}}{\sum_{i=1}^n X_{i-1}^2} \to M_\infty^{-1/2} \quad \text{a.s. on } M_+ \text{ as } n \to \infty .$$

For this, use the asymptotic almost sure behavior of $\sum_{i=1}^n X_{i-1}^3$ as $n \to \infty$ established before and

$$\alpha^{-2n} \sum_{i=1}^n X_{i-1}^2 \to \frac{1}{\alpha^2 - 1} M_\infty^2 \quad \text{a.s. on } M_+ \text{ as } n \to \infty$$

which follows from $\alpha^{-(i-1)} X_{i-1} \to M_\infty$ almost surely as $i \to \infty$ and the Toeplitz Lemma 6.28 (b).   □

A weighted conditional least squares estimator

To obtain a fourth estimator for α we observe that the conditional variance

$$\text{Var}\,(X_i | \mathcal{F}_{i-1}) = E \left[(X_i - E\,(X_i | \mathcal{F}_{i-1}))^2 \, \big| \mathcal{F}_{i-1} \right]$$

$$= E \left(X_i^2 | \mathcal{F}_{i-1} \right) - E\,(X_i | \mathcal{F}_{i-1})^2 = \sigma^2 X_{i-1}$$

of X_i given $\mathcal{F}_{i-1}$ strongly depends on i. It is therefore reasonable to stabilize this conditional variance of the summand $X_i - E\,(X_i | \mathcal{F}_{i-1}) = X_i - \alpha X_{i-1}$ in the conditional least squares approach, that is, to consider the minimizer of the weighted sum of squares

$$\sum_{i=1}^n \frac{(X_i - E\,(X_i | \mathcal{F}_{i-1}))^2}{\text{Var}\,(X_i | \mathcal{F}_{i-1})} = \sum_{i=1}^n \frac{(X_i - \alpha X_{i-1})^2}{\sigma X_{i-1}} ,$$

which is given by

$$\widehat{\alpha}_n^{(H)} := \frac{\sum_{i=1}^{n} X_i}{\sum_{i=1}^{n} X_{i-1}} .$$

Since $\sum_{i=1}^{n} X_{i-1} \geq X_0 \geq 1$, $\widehat{\alpha}_n^{(H)}$ is well-defined. On M_+ we have

$$\frac{X_i}{X_{i-1}} = \widehat{\alpha}_i^{(LN)} \to \alpha \quad \text{a.s. as } i \to \infty ,$$

and the Toeplitz Lemma 6.28 (b) implies $\widehat{\alpha}_n^{(H)} \to \alpha$ almost surely on M_+ so that $\widehat{\alpha}_n^{(H)}$ is strongly consistent on M_+. This is the *Harris estimator* introduced in [42]; see also [43].

To derive stable limit theorems for $\widehat{\alpha}_n^{(H)}$ we introduce the process $U^{(H)} = \left(U^{(H)}\right)_{n \geq 0}$ with $U_0^{(H)} := 0$ and

$$U_n^{(H)} := \sum_{i=1}^{n} (X_i - \alpha X_{i-1}) \quad \text{for } n \geq 1 .$$

Under our original moment assumptions $X_0, Y_{11} \in \mathcal{L}^2(P)$ the process $U^{(H)}$ is an $\mathcal{L}^2$-martingale w.r.t. $\mathbb{F}$ with quadratic characteristic

$$\left\langle U^{(H)} \right\rangle_n = \sum_{i=1}^{n} E\left(\left(\Delta U_i^{(H)} \right)^2 \Big| \mathcal{F}_{i-1} \right) = \sigma^2 \sum_{i=1}^{n} X_{i-1} = \sum_{i=1}^{n} \text{Var}\left(X_i | \mathcal{F}_{i-1} \right) .$$

Again, an application of Theorem 8.2 and Corollary 8.5 is crucial.

Theorem 10.5 *If $X_0, Y_{11} \in \mathcal{L}^2(P)$, then*

$$\frac{U_n^{(H)}}{\left\langle U^{(H)} \right\rangle_n^{1/2}} \to N \quad \mathcal{F}_\infty\text{-mixing under } P_{M_+} \text{ as } n \to \infty ,$$

where N is P-independent of $\mathcal{F}_\infty$ with $P^N = N(0, 1)$.

Proof We are again in the setting of Remark 8.6 so that we have to verify conditions (i), (iii) and (iv) of Theorem 8.2.

First, we will show that condition (i) holds with $G = \Omega$, $a_n = \alpha^{n/2}$ and $\eta = \sigma M_\infty / \left(\alpha^2 - 1 \right)^{1/2}$. As in the proof of Theorem 10.3, $\alpha^{-(i-1)} X_{i-1} \to M_\infty$ almost surely as $i \to \infty$ and the Toeplitz Lemma 6.28 (b) imply

$$\frac{\sum_{i=1}^{n} X_{i-1}}{\sum_{i=1}^{n} \alpha^{i-1}} \to M_\infty \quad \text{a.s. as } n \to \infty .$$

Because

$$\sum_{i=1}^{n} \alpha^{i-1} = \frac{\alpha^n - 1}{\alpha - 1} \sim \frac{1}{\alpha - 1}\alpha^n$$

we obtain

$$\frac{\langle U^{(H)}\rangle_n}{a_n^2} = \frac{\langle U^{(H)}\rangle_n}{\alpha^n} = \sigma^2 \frac{\sum_{i=1}^{n} X_{i-1}}{\alpha^n} \to \frac{\sigma^2}{\alpha - 1}M_\infty \quad \text{a.s. as } n \to \infty,$$

which gives (i).

For all $n, r \in \mathbb{N}$ we have $a_{n-r}^2/a_n^2 = 1/\alpha^r$ which is (ii) with $p = \alpha$.

Finally, we verify (iv) for $\mu = N(0, b)$ with $b = (\alpha - 1)/\alpha$, which means that Corollary 8.5 applies and yields $P^Z = N(0, 1)$ because $bp/(p - 1) = 1$. For the proof of (iv), as in the proof of Theorem 10.3 let ζ denote the characteristic function of the random variable Y_{11} and ϕ that of the normalized random variable $(Y_{11} - \alpha)/\sigma$. Then for every $t \in \mathbb{R}$ and $n \in \mathbb{N}$, by the same reasoning as in the proof of Theorem 10.3,

$$E_P\left(\exp\left(it\frac{\Delta U_n^{(H)}}{\langle U^{(H)}\rangle_n^{1/2}}\right)\Big|\mathcal{F}_{n-1}\right) = E_P\left(\exp\left(it\frac{X_n - \alpha X_{n-1}}{\langle U^{(H)}\rangle_n^{1/2}}\right)\Big|\mathcal{F}_{n-1}\right)$$

$$= \exp\left(it\frac{(-\alpha)X_{n-1}}{\langle U^{(H)}\rangle_n^{1/2}}\right)E_P\left(\exp\left(i\frac{t}{\langle U^{(H)}\rangle_n^{1/2}}\sum_{j=1}^{X_{n-1}}Y_{nj}\right)\Big|\mathcal{F}_{n-1}\right)$$

$$= \exp\left(it\frac{(-\alpha)}{\langle U^{(H)}\rangle_n^{1/2}}\right)^{X_{n-1}}\zeta\left(\frac{t}{\langle U^{(H)}\rangle_n^{1/2}}\right)^{X_{n-1}} = \phi\left(\frac{\sigma t}{\langle U^{(H)}\rangle_n^{1/2}}\right)^{X_{n-1}}.$$

On M_+ we have

$$\frac{\sigma t}{\langle U^{(H)}\rangle_n^{1/2}} = \frac{\sigma t X_{n-1}^{1/2}}{\langle U^{(H)}\rangle_n^{1/2}}\frac{1}{X_{n-1}^{1/2}}$$

with

$$\frac{\sigma t X_{n-1}^{1/2}}{\langle U^{(H)}\rangle_n^{1/2}} = \sigma t\left(\frac{a_n^2}{\langle U^{(H)}\rangle_n}\right)^{1/2}\left(\alpha^{-(n-1)}X_{n-1}\right)^{1/2}\alpha^{-1/2} \to t\left(\frac{\alpha - 1}{\alpha}\right)^{1/2}$$

almost surely as $n \to \infty$. Using again, as in the proof of Theorem 10.3,

$$\phi\left(\frac{x}{\sqrt{n}}\right)^n \to \exp\left(-\frac{1}{2}x^2\right) \quad \text{as } n \to \infty$$

uniformly in $x \in \mathbb{R}$ on compact intervals, now with $x = \sigma t X_{n-1}^{1/2} / (U^{(H)})_{|n}^{1/2}$ and $n = X_{n-1}$, we arrive at

$$
\phi\left(\frac{\sigma t}{(U^{(H)})_{|n}^{1/2}}\right)^{X_{n-1}} \to \exp\left(-\frac{1}{2}t^2\frac{\alpha - 1}{\alpha}\right) \quad \text{a.s. on } M_+ \text{ as } n \to \infty,
$$

which implies condition (iv) with $b = (\alpha - 1)/\alpha$ and concludes the proof.   □

Corollary 10.6 *Under the assumptions of Theorem 10.5,*

$$
\left(\sum_{i=1}^n X_{i-1}\right)^{1/2} \left(\widehat{\alpha}_n^{(H)} - \alpha\right) \to \sigma N \quad \mathcal{F}_\infty\text{-mixing under } P_{M_+} \text{ as } n \to \infty
$$

and

$$
\frac{\alpha^{n/2}}{(\alpha - 1)^{1/2}}\left(\widehat{\alpha}_n^{(H)} - \alpha\right) \to \sigma M_\infty^{-1/2} N \quad \mathcal{F}_\infty\text{-stably under } P_{M_+} \text{ as } n \to \infty,
$$

where $P^N = N(0, 1)$ and N is P-independent of $\mathcal{F}_\infty$.

Proof For all $n \in \mathbb{N}$,

$$
\widehat{\alpha}_n^{(H)} = \frac{\sum_{i=1}^n (X_i - \alpha X_{i-1})}{\sum_{i=1}^n X_{i-1}} + \alpha = \frac{U_n^{(H)}}{\sum_{i=1}^n X_{i-1}} + \alpha
$$

so that

$$
\left(\sum_{i=1}^n X_{i-1}\right)^{1/2}\left(\widehat{\alpha}_n^{(H)} - \alpha\right) = \sigma \frac{U_n^{(H)}}{(U^{(H)})_{|n}^{1/2}}.
$$

Thus the first statement follows immediately from Theorem 10.5. The second statement follows from the first and

$$
\frac{\alpha^{n/2}}{(\alpha - 1)^{1/2}}\left(\sum_{i=1}^n X_{i-1}\right)^{-1/2} = \frac{1}{(\alpha - 1)^{1/2}}\sigma\left(\frac{a_n^2}{(U^{(H)})_n}\right)^{1/2} \to M_\infty^{-1/2}
$$

almost surely on M_+ as $n \to \infty$.   □

The above stable central limit theorem for the Harris estimator and the stable central limit theorem of Corollary 10.2 for the Lotka-Nagaev estimator are due to Dion [25].

The moment estimator $\widehat{\alpha}_n^{(M)}$ converges to α at a linear rate and is therefore clearly inferior asymptotically to the other three estimators, all of which converge exponentially fast. As Corollaries 10.2, 10.4 and 10.6 show, the order of the rate of convergence is the same for all three of these estimators, namely $\alpha^{n/2}$. As the limits of all three estimators are the same, we compare these estimators as in [44] in a somewhat informal way by comparing the squares of the normalizing factors since an estimator with a bigger normalizing factor is obviously preferable to a competitor with a smaller one because, for example, it leads to shorter asymptotic confidence intervals. As mentioned in [44], this is a concept of asymptotic efficiency in an obvious, though not albeit standard sense.

Denoting the random normalizers of the three estimators $\widehat{\alpha}_n^{(LN)}$, $\widehat{\alpha}_n^{(LS)}$ and $\widehat{\alpha}_n^{(H)}$ by $N_{n,LN}$, $N_{n,LS}$ and $N_{n,H}$, respectively, and employing the asymptotic behavior of X_n as well as of $\sum_{i=1}^n X_{i-1}^k$ for $k = 1, 2, 3$, which was established in the previous proofs, we get almost surely as $n \to \infty$ for all $\alpha \in (1, \infty)$

$$\frac{N_{n,LN}^2}{N_{n,LS}^2} = \frac{X_{n-1} \sum_{i=1}^n X_{i-1}^3}{\left(\sum_{i=1}^n X_{i-1}^2\right)^2} \to \frac{\left(\alpha^2 - 1\right)^2}{\alpha\left(\alpha^3 - 1\right)} = \frac{\alpha^3 + \alpha^2 - \alpha - 1}{\alpha^3 + \alpha^2 + \alpha} < 1,$$

$$\frac{N_{n,LN}^2}{N_{n,H}^2} = \frac{X_{n-1}}{\sum_{i=1}^n X_{i-1}} \to \frac{\alpha - 1}{\alpha} < 1$$

and

$$\frac{N_{n,LS}^2}{N_{n,H}^2} = \frac{\left(\sum_{i=1}^n X_{i-1}^2\right)^2}{\left(\sum_{i=1}^n X_{i-1}^3\right)\left(\sum_{i=1}^n X_{i-1}\right)} \to \frac{\left(\alpha^3 - 1\right)\left(\alpha - 1\right)}{\left(\alpha^2 - 1\right)^2} = \frac{\alpha^2 + \alpha + 1}{\alpha^2 + 2\alpha + 1} < 1.$$

These results show that the Harris estimator is asymptotically the best one, which is not really surprising because this estimator can be viewed as a nonparametric maximum likelihood estimator; see [40], Sect. 2.4. Of course, the results are the same if the deterministic normalizers from Corollaries 10.2, 10.4 and 10.6 are considered.

Exercise 10.1 Let X be a supercritical Galton-Watson branching process with $X_0, Y_{11} \in \mathcal{L}^2(P)$ and $\mathrm{Var}(Y_{11}) > 0$, and assume $p_k < 1$ for all $k \in \mathbb{N}_0$. If the complete family tree $\left(Y_{ij}\right)_{1 \le i \le n, 1 \le j \le X_{i-1}}$ up to generation $n \in \mathbb{N}$ of X is observable, then

$$\widehat{p}_{k,n} := \frac{1}{Z_n} \sum_{i=1}^n \sum_{j=1}^{X_{i-1}} 1_{\{Y_{ij}=k\}}$$

with $Z_n := \sum_{i=1}^n X_{i-1}$ is the nonparametric maximum likelihood estimator of p_k for every $k \in \mathbb{N}_0$; see [40, 42]. For every $k \in \mathbb{N}_0$, show that

$$\widehat{p}_{k,n} \to p_k \quad \text{a.s. as } n \to \infty \text{ on } M_+$$

and

$$\frac{\alpha^{n/2}}{(\alpha-1)^{1/2}}\left(\widehat{p}_{k,n}-p_k\right) \to (p_k(1-p_k))^{1/2}M_{\infty}^{-1/2}N \quad \mathcal{F}_{\infty}\text{-stably under } P_{M_+},$$

where $P^N = N(0,1)$ and N and $\mathcal{F}_{\infty}$ are P-independent.

Exercise 10.2 In the situation of Exercise 10.1, assume that after the $(n-1)$-th generation of X only the complete next generation $\left(Y_{nj}\right)_{1\leq j\leq X_{n-1}}$ is observable. For the estimator

$$\widetilde{p}_{k,n} := \frac{1}{X_{n-1}} \sum_{j=1}^{X_{n-1}} 1_{\{Y_{nj}=k\}}$$

of p_k show that for all $k \in \mathbb{N}_0$

$$\widetilde{p}_{k,n} \to p_k \quad \text{a.s. as } n \to \infty \text{ on } M_+$$

and

$$X_{n-1}^{1/2}\left(\widetilde{p}_{k,n}-p_k\right) \to (p_k(1-p_k))^{1/2}N \quad \mathcal{F}_{\infty}\text{-mixing under } P_{M_+} \text{ as } n \to \infty,$$

where $P^N = N(0,1)$ and N and $\mathcal{F}_{\infty}$ are P-independent.

Hint: The strong consistency of $\widetilde{p}_{k,n}$ on M_+ can be derived from the strong consistency of $\widehat{p}_{k,n}$ on M_+ appearing in Exercise 10.1.

Appendix A

Here we collect some basic facts about the weak topology on $\mathcal{M}^1(\mathcal{X})$, conditional distributions and martingales.

A.1 Weak Topology and Conditional Distributions

Let $\mathcal{X}$ be a separable metrizable topological space equipped with its Borel-σ-field $\mathcal{B}(\mathcal{X})$ and $\mathcal{M}^1(\mathcal{X})$ the set of all probability measures on $\mathcal{B}(\mathcal{X})$ equipped with the weak topology. Let d be a metric on $\mathcal{X}$ that induces the topology and let $U_b(\mathcal{X}, d)$ denote the subspace of $C_b(\mathcal{X})$ consisting of all d-uniformly continuous, bounded real functions.

Theorem A.1 (Portmanteau theorem) *Let $(\nu_\alpha)_\alpha$ be a net in $\mathcal{M}^1(\mathcal{X})$ and $\nu \in \mathcal{M}^1(\mathcal{X})$. Let β be the system of all finite intersections of open balls in $\mathcal{X}$. The following statements are equivalent:*

(i) $\nu_\alpha \to \nu$ *weakly*,

(ii) $\lim_\alpha \int h \, d\nu_\alpha = \int h \, d\nu$ *for every* $h \in U_b(\mathcal{X}, d)$,

(iii) $\liminf_\alpha \nu_\alpha(O) \geq \nu(O)$ *for every open subset* $O \subset \mathcal{X}$,

(iv) $\limsup_\alpha \nu_\alpha(C) \leq \nu(C)$ *for every closed subset* $C \subset \mathcal{X}$,

(v) $\lim_\alpha \nu_\alpha(B) = \nu(B)$ *for every* $B \in \mathcal{B}(\mathcal{X})$ *satisfying* $\nu(\partial B) = 0$,

(vi) $\lim_\alpha \nu_\alpha(B) = \nu(B)$ *for every* $B \in \beta$ *satisfying* $\nu(\partial B) = 0$.

Proof For the equivalences (i)–(v) see [69], Theorem II.6.1.

  (v) $\Rightarrow$ (vi) is obvious.

  (vi) $\Rightarrow$ (iii). Let $\beta_1 := \{B \in \beta : \nu(\partial B) = 0\}$ and let β_2 denote the system of all finite unions of sets from β_1. Using that β_1 is closed under finite intersections since

© Springer International Publishing Switzerland 2015
E. Häusler and H. Luschgy, *Stable Convergence and Stable Limit Theorems*,
Probability Theory and Stochastic Modelling 74,
DOI 10.1007/978-3-319-18329-9

$\partial \left(\bigcap_{i=1}^{k} B_i \right) \subset \bigcup_{i=1}^{k} \partial B_i$, the inclusion-exclusion formula yields $\lim_\alpha \nu_\alpha (G) = \nu (G)$ for every $G \in \beta_2$. Moreover, we observe that β_1 is a base for the topology on $\mathcal{X}$. In fact, if $O \subset \mathcal{X}$ is any open subset and $x \in O$, then there exists an $r > 0$ such that $B(x, r) := \{d(x, \cdot) < r\} \subset O$. Since $\partial B(x, s) \subset \{d(x, \cdot) = s\}$, $s > 0$, these boundaries are pairwise disjoint and thus $R := \{s > 0 : \nu (\partial B(x, s)) > 0\}$ is countable. Hence, $(0, r] \cap R^c \neq \emptyset$, and for $s \in (0, r] \cap R^c$ we obtain $x \in B(x, s) \subset O$ and $B(x, s) \in \beta_1$. So β_1 is a base. The space $\mathcal{X}$ having a countable base is strongly Lindelöf, that is, every open cover of any open subset of $\mathcal{X}$ has a countable subcover. Consequently, for every open set $O \subset \mathcal{X}$, there exists a sequence (G_n) in β_2 such that $G_n \uparrow O$. One obtains

$$\liminf_\alpha \nu_\alpha (O) \geq \lim_\alpha \nu_\alpha (G_n) = \nu (G_n) \quad \text{for every } n \in \mathbb{N}$$

and $\lim_{n \to \infty} \nu (G_n) = \nu (O)$ which yields $\liminf_\alpha \nu_\alpha (O) \geq \nu (O)$. $\qquad \square$

Lemma A.2 *We have*

$$\mathcal{B} \left(\mathcal{M}^1 (\mathcal{X}) \right) = \sigma \left(\nu \mapsto \int h d\nu, h \in C_b (\mathcal{X}) \right) = \sigma (\nu \mapsto \nu (B), B \in \mathcal{B} (\mathcal{X})).$$

Proof Let $g_B (\nu) = \nu (B)$ and $g_h (\nu) = \int h d\nu$. A base β of the weak topology on $\mathcal{M}^1 (\mathcal{X})$ belonging to $\sigma (g_h, h \in C_b (\mathcal{X}))$ is given by the collection of finite intersections of sets of the form $\{\{g_h \in U\} : h \in C_b (\mathcal{X}), U \subset \mathbb{R} \text{ open}\}$. The space $\mathcal{M}^1 (\mathcal{X})$ being separable and metrizable and thus having a countable base is strongly Lindelöf. Consequently, every open subset of $\mathcal{M}^1 (\mathcal{X})$ is a countable union of sets from β. This implies $\mathcal{B} \left(\mathcal{M}^1 (\mathcal{X}) \right) \subset \sigma (g_h, h \in C_b (\mathcal{X}))$.

The inclusion $\sigma (g_h, h \in C_b (\mathcal{X})) \subset \sigma (g_B, B \in \mathcal{B} (\mathcal{X}))$ follows from the usual approximation of h by $\mathcal{B} (\mathcal{X})$-simple functions.

The system $\mathcal{D} := \left\{ B \in \mathcal{B} (\mathcal{X}) : g_B \text{ is } \mathcal{B} \left(\mathcal{M}^1 (\mathcal{X}) \right)\text{-measurable} \right\}$ is a Dynkin-system which contains every open subset of $\mathcal{X}$ by the Portmanteau theorem. Thus $\mathcal{D} = \mathcal{B} (\mathcal{X})$ and we deduce $\sigma (g_B, B \in \mathcal{B} (\mathcal{X})) \subset \mathcal{B} \left(\mathcal{M}^1 (\mathcal{X}) \right)$. $\qquad \square$

Let $(\Omega, \mathcal{F}, P)$ be a probability space, $\mathcal{G} \subset \mathcal{F}$ a sub-σ-field and $X : (\Omega, \mathcal{F}) \to (\mathcal{X}, \mathcal{B} (\mathcal{X}))$ a random variable. The *distribution* of X is denoted by P^X. The *conditional distribution* $P^{X|\mathcal{G}}$ of X given $\mathcal{G}$ is the P-almost surely unique Markov kernel in $\mathcal{K}^1 (\mathcal{G}, \mathcal{X})$ such that

$$P^{X|\mathcal{G}} (\cdot, B) = P (X \in B|\mathcal{G}) \; P\text{-a.s. for every } B \in \mathcal{B} (\mathcal{X}) .$$

It is characterized by the Radon-Nikodym equations

$$\int_G P^{X|\mathcal{G}} (\omega, B) \, dP (\omega) = P \left(X^{-1} (B) \cap G \right) \text{ for every } G \in \mathcal{G}, B \in \mathcal{B} (\mathcal{X}) ,$$

or, what is the same, by measure uniqueness, $P \otimes P^{X|\mathcal{G}} = P \otimes \delta_X$ on $\mathcal{G} \otimes \mathcal{B} (\mathcal{X})$.

For Borel-measurable functions $f : \mathcal{X} \to \mathbb{R}$ such that $f(X) \in \mathcal{L}^1(P)$ we have

$$E\left(f(X) \mid \mathcal{G}\right) = \int f(x) \, P^{X\mid\mathcal{G}}(dx)$$

provided $P^{X\mid\mathcal{G}}$ exists ([26], Theorem 10.2.5).

Theorem A.3 *Assume that $\mathcal{X}$ is polish. Then the conditional distribution $P^{X\mid\mathcal{G}}$ exists.*

Proof [26], Theorem 10.2.2. $\square$

In the sequel we assume that the conditional distribution $P^{X\mid\mathcal{G}}$ exists.

Lemma A.4 (a) *If X is $\mathcal{G}$-measurable, then $P^{X\mid\mathcal{G}} = \delta_X$.*
(b) $P^{X\mid\mathcal{G}} = P^X$ *if and only if $\sigma(X)$ and $\mathcal{G}$ are independent.*
(c) *Let $\mathcal{Y}$ be a further separable metrizable topological space and $g : \mathcal{X} \to \mathcal{Y}$ be Borel-measurable. Then $P^{g(X)\mid\mathcal{G}} = \left(P^{X\mid\mathcal{G}}\right)^g$.*
(d) *Let Q be a probability distribution on $\mathcal{F}$ with $Q \ll P$ and dQ/dP be $\mathcal{G}$-measurable. Then $Q^{X\mid\mathcal{G}} = P^{X\mid\mathcal{G}}$ Q-almost surely.*
In particular, $Q \otimes P^{X\mid\mathcal{G}} = Q \otimes \delta_X$ on $\mathcal{G} \otimes \mathcal{B}(\mathcal{X})$ and $QP^{X\mid\mathcal{G}} = Q\delta_X = Q^X$.

Proof (a) We have $\delta_X \in \mathcal{K}^1(\mathcal{G}, \mathcal{X})$ and δ_X clearly satisfies the Radon-Nikodym equations for $P^{X\mid\mathcal{G}}$.
   (b) We have $P^{X\mid\mathcal{G}} = P^X$ if and only if $P^X(B)P(G) = P\left(X^{-1}(B) \cap G\right)$ for every $G \in \mathcal{G}, B \in \mathcal{B}(\mathcal{X})$, that is, the independence of $\sigma(X)$ and $\mathcal{G}$.
   (c) For every $G \in \mathcal{G}, C \in \mathcal{B}(\mathcal{Y})$ we have

$$\int_G \left(P^{X\mid\mathcal{G}}\right)^g (\omega, C) \, dP(\omega) = \int_G P^{X\mid\mathcal{G}}\left(\omega, g^{-1}(C)\right) dP(\omega)$$

$$= P\left(\{g(X) \in C\} \cap G\right) = \int_G P^{g(X)\mid\mathcal{G}}(\omega, C) \, dP(\omega) \; .$$

   (d) Let $f := dQ/dP$. For every $G \in \mathcal{G}$ and $B \in \mathcal{B}(\mathcal{X})$ we obtain

$$\int_G P^{X\mid\mathcal{G}}(\omega, B) \, dQ(\omega) = \int_G P^{X\mid\mathcal{G}}(\omega, B) f(\omega) \, dP(\omega)$$

$$= \int_G E_P\left(1_B(X) f \mid \mathcal{G}\right) dP = \int_G 1_B(X) f \, dP = Q\left(X^{-1}(B) \cap G\right) . \quad \square$$

Now let $\mathcal{Y}$ be a further separable metrizable topological space and $Y : (\Omega, \mathcal{F}) \to (\mathcal{Y}, \mathcal{B}(\mathcal{Y}))$ a random variable. Note that $\mathcal{B}(\mathcal{X} \times \mathcal{Y}) = \mathcal{B}(\mathcal{X}) \otimes \mathcal{B}(\mathcal{Y})$ ([26], Proposition 4.1.7). For $K_1 \in \mathcal{K}^1(\mathcal{F}, \mathcal{X})$ and $K_2 \in \mathcal{K}^1(\mathcal{F}, \mathcal{Y})$ define the *product kernel* $K_1 \otimes K_2 \in \mathcal{K}^1(\mathcal{F}, \mathcal{X} \times \mathcal{Y})$ by $K_1 \otimes K_2(\omega, \cdot) := K_1(\omega, \cdot) \otimes K_2(\omega, \cdot)$.

Lemma A.5 *Let Y be $\mathcal{G}$-measurable.*

(a) $P^{(X,Y)|\mathcal{G}} = P^{X|\mathcal{G}} \otimes \delta_Y$.

(b) *Let $\widetilde{X} : (\Omega, \mathcal{F}) \to (\mathcal{X}, \mathcal{B}(\mathcal{X}))$ be a random variable with $P^{\widetilde{X}} = P^X$. If $\sigma(X), \mathcal{G}$ are independent and $\sigma(\widetilde{X}), \mathcal{G}$ are independent, then $P^{(X,Y)|\mathcal{G}} = P^{(\widetilde{X},Y)|\mathcal{G}}$.*

(c) *Let $\mathcal{X} = \mathcal{Y} = \mathbb{R}$. If $\sigma(X)$ and $\mathcal{G}$ are independent and P^X is symmetric around zero, then $P^{X|Y||\mathcal{G}} = P^{XY|\mathcal{G}}$.*

In particular, if $|Y| = 1$ P-almost surely, then $P^{XY|\mathcal{G}} = P^{X|\mathcal{G}} = P^X$.

Proof (a) Let $K := P^{X|\mathcal{G}} \otimes \delta_Y$. Then $K \in \mathcal{K}^1(\mathcal{G}, \mathcal{X} \times \mathcal{Y})$ and for every $G \in \mathcal{G}$, $B \in \mathcal{B}(\mathcal{X}), C \in \mathcal{B}(\mathcal{Y})$

$$\int_G K(\omega, B \times C)\, dP(\omega) = \int_G P^{X|\mathcal{G}}(\omega, B)\, \delta_{Y(\omega)}(C)\, dP(\omega)$$

$$= \int_{G \cap Y^{-1}(C)} P^{X|\mathcal{G}}(\omega, B)\, dP(\omega) = P\left(X^{-1}(B) \cap Y^{-1}(C) \cap G\right)$$

$$= P\left((X, Y)^{-1}(B \times C) \cap G\right).$$

Measure uniqueness yields the assertion.

(b) By (a) and Lemma A.4 (b), $P^{(X,Y)|\mathcal{G}} = P^X \otimes \delta_Y = P^{\widetilde{X}} \otimes \delta_Y = P^{(\widetilde{X},Y)|\mathcal{G}}$.

(c) Let $g, h : \mathbb{R}^2 \to \mathbb{R}$ be defined by $g(x, y) := xy$ and $h(x, y) := x|y|$. Then by (a) and Lemma A.4 (b) and (c) for every $B \in \mathcal{B}(\mathbb{R})$

$$P^{X|Y||\mathcal{G}}(\cdot, B) = P^{(X,Y)|\mathcal{G}}\left(\cdot, h^{-1}(B)\right) = P^X \otimes \delta_Y\left(h^{-1}(B)\right)$$

$$= \int P^X(\{x \in \mathbb{R} : x|y| \in B\})\, d\delta_Y(y)$$

$$= \int_{(0,\infty)} P^X(\{x \in \mathbb{R} : xy \in B\})\, d\delta_Y(y)$$

$$+ \int_{\{0\}} P^X(\{x \in \mathbb{R} : 0 \in B\})\, d\delta_Y(y)$$

$$+ \int_{(-\infty,0)} P^X(\{x \in \mathbb{R} : -xy \in B\})\, d\delta_Y(y)$$

$$= P^X \otimes \delta_Y\left(g^{-1}(B)\right) = P^{XY|\mathcal{G}}(\cdot, B).$$

The assertion follows from Lemma 2.1 (b). $\qquad\qquad\qquad\qquad\qquad\qquad\square$

Let μ be a probability distribution on $\mathcal{F} \otimes \mathcal{B}(\mathcal{X})$ with $\mu^{\pi_1} = P$, $\pi_1 : \Omega \times \mathcal{X} \to \Omega$ being the projection. Then a Markov kernel $K \in \mathcal{K}^1(\mathcal{F}, \mathcal{X})$ is called a *disintegration* of μ w.r.t. π_1 if $P \otimes K = \mu$. The kernel K is then P-almost surely unique.

Theorem A.6 *Assume that $\mathcal{X}$ is polish. Then a disintegration w.r.t. π_1 exists for every probability distribution μ on $\mathcal{F} \otimes \mathcal{B}(\mathcal{X})$ with $\mu^{\pi_1} = P$.*

Proof By Theorem A.3, the conditional distribution $\widetilde{K} := \mu^{\pi_2|\mathcal{F}\otimes\{\emptyset,\mathcal{X}\}} \in \mathcal{K}^1(\mathcal{F}\otimes\{\emptyset,\mathcal{X}\},\mathcal{X})$ exists, $\pi_2 : \Omega\times\mathcal{X} \rightarrow \mathcal{X}$ being the projection. Since $\mathcal{F}\otimes\{\emptyset,\mathcal{X}\} = \sigma(\pi_1)$, there exists a $K \in \mathcal{K}^1(\mathcal{F},\mathcal{X})$ such that $K(\pi_1(\omega,x),B) = \widetilde{K}((\omega,x),B)$ for every $\omega \in \Omega$, $x \in \mathcal{X}$, $B \in \mathcal{B}(\mathcal{X})$. We obtain for $F \in \mathcal{F}$ and $B \in \mathcal{B}(\mathcal{X})$

$$P\otimes K(F\times B) = \int_F K(\omega,B)\,dP(\omega) = \int_{F\times\mathcal{X}} K(\pi_1(\omega,x),B)\,d\mu(\omega,x)$$

$$= \int_{F\times\mathcal{X}} \widetilde{K}((\omega,x),B)\,d\mu(\omega,x) = \mu\left(\pi_2^{-1}(B)\cap(F\times\mathcal{X})\right)$$

$$= \mu(F\times B).$$

Measure uniqueness yields $P\otimes K = \mu$.   □

In view of the Radon-Nikodym equations for $P^{X|\mathcal{G}}$ we see that $P^{X|\mathcal{G}}$ is the disintegration of $P\otimes\delta_X|\mathcal{G}\otimes\mathcal{B}(\mathcal{X})$ so that $P^{X|\mathcal{G}} = E(\delta_X|\mathcal{G})$ in the sense of Definition 2.4.

Lemma A.7 *Assume that $\mathcal{X}$ is polish. Let $K \in \mathcal{K}^1(\mathcal{F},\mathcal{X})$ and let $\mathcal{G}_1 \subset \mathcal{G}_2 \subset \mathcal{F}$ be sub-σ-fields.*

(a) $E(E(K|\mathcal{G}_2)|\mathcal{G}_1) = E(K|\mathcal{G}_1)$.

(b) $E\left(P^{X|\mathcal{G}_2}|\mathcal{G}_1\right) = P^{X|\mathcal{G}_1}$.

(c) $E(K|\mathcal{G}) = \int_{\mathcal{M}^1(\mathcal{X})} \nu\,P^{K|\mathcal{G}}(d\nu)$, *where on the right-hand side K is regarded as an $\left(\mathcal{M}^1(\mathcal{X}),\mathcal{B}(\mathcal{M}^1(\mathcal{X}))\right)$-valued random variable (see Lemma A.2). In particular, $E(K(\cdot,B)|\mathcal{G}) = E(K|\mathcal{G})(\cdot,B)$ for every $B \in \mathcal{B}(\mathcal{X})$.*

The conditional distribution $P^{K|\mathcal{G}}$ in part (c) exists by Theorem A.3 because $\mathcal{M}^1(\mathcal{X})$ is polish.

Proof (a) Let $H := E(K|\mathcal{G}_2)$ and $J := E(K|\mathcal{G}_1)$. Since $P\otimes H = P\otimes K$ on $\mathcal{G}_2\otimes\mathcal{B}(\mathcal{X})$ and $P\otimes J = P\otimes K$ on $\mathcal{G}_1\otimes\mathcal{B}(\mathcal{X})$, we get $P\otimes J = P\otimes H$ on $\mathcal{G}_1\otimes\mathcal{B}(\mathcal{X})$ so that $J = E(H|\mathcal{G}_1)$.

(b) Using (a), we obtain

$$E\left(P^{X|\mathcal{G}_2}|\mathcal{G}_1\right) = E(E(\delta_X|\mathcal{G}_2)|\mathcal{G}_1) = E(\delta_X|\mathcal{G}_1) = P^{X|\mathcal{G}_1}.$$

(c) The right-hand side, denoted by H, satisfies

$$H(\omega,B) = \int \nu(B)\,P^{K|\mathcal{G}}(\omega,d\nu)$$

so that $H \in \mathcal{K}^1 (\mathcal{G}, \mathcal{X})$. We obtain for every $G \in \mathcal{G}, B \in \mathcal{B}(\mathcal{X})$

$$
\begin{aligned}
P \otimes H (G \times B) &= \int_G \int \nu (B) \, P^{K|\mathcal{G}} (\omega, d\nu) \, dP (\omega) \\
&= \int 1_G (\omega) \nu (B) \, dP \otimes P^{K|\mathcal{G}} (\omega, \nu) \\
&= \int 1_G (\omega) \nu (B) \, dP \otimes \delta_K (\omega, \nu) \\
&= \int_G K (\omega, B) \, dP (\omega) \\
&= P \otimes K (G \times B)
\end{aligned}
$$

and measure uniqueness yields $P \otimes H = P \otimes K$ on $\mathcal{G} \otimes \mathcal{B}(\mathcal{X})$. This implies $H = E (K|\mathcal{G})$. Furthermore, for $B \in \mathcal{B}(\mathcal{X})$ let $g_B : \mathcal{M}^1 (\mathcal{X}) \to \mathbb{R}$, $g_B (\nu) := \nu (B)$. Then

$$
E (K (\cdot, B) |\mathcal{G}) = E (g_B (K) |\mathcal{G}) = \int g_B (\nu) \, P^{K|\mathcal{G}} (d\nu) = H (\cdot, B) . \qquad \square
$$

A.2 Martingales

Let $I = [\alpha, \beta] \cap \mathbb{Z}$ be an integer interval, where $\alpha \in \mathbb{Z}$, $\beta \in \mathbb{Z} \cup \{\infty\}$ and $\alpha < \beta$ (like $\{0, \ldots, k\}$, $\mathbb{N}$ or $\mathbb{N}_0$), let $\mathbb{F} = (\mathcal{F}_n)_{n \in I}$ be a *filtration* in $\mathcal{F}$, that is, a nondecreasing family of sub-σ-fields of $\mathcal{F}$, and let $X = (X_n)_{n \in I}$ be a real process defined on $(\Omega, \mathcal{F}, P)$. The *increments* (or differences) of X are given by $\Delta X_n = X_n - X_{n-1}$ for $n \in I$, $n \geq \alpha + 1$ so that $X_n = X_\alpha + \sum_{j=1}^n \Delta X_j$, $n \in I$. The process $[X]$ defined by $[X]_n := \sum_{j=\alpha+1}^n (\Delta X_j)^2$ for $n \in I$, $n \geq \alpha + 1$ with $[X]_\alpha = 0$ is called the *quadratic variation of* X. The process X is called $\mathbb{F}$-*adapted* if X_n is $\mathcal{F}_n$-measurable for every $n \in I$. If X is an $\mathbb{F}$-*martingale*, i.e. X is integrable, $\mathbb{F}$-adapted and $E (X_n|\mathcal{F}_{n-1}) = X_{n-1}$ for every $n \in I, n \geq \alpha+1$, then $E (\Delta X_n|\mathcal{F}_{n-1}) = 0, n \in I$, $n \geq \alpha + 1$. Conversely, if $Z = (Z_n)_{n \in I, n \geq \alpha+1}$ is an $\mathbb{F}$-*martingale increment* (or *martingale difference*) *sequence*, that is, Z is integrable, $\mathbb{F}$-adapted and $E (Z_n|\mathcal{F}_{n-1}) = 0$, $n \in I, n \geq \alpha+1$, then for any random variable $Z_\alpha \in \mathcal{L}^1 (\mathcal{F}_\alpha, P)$ the process X defined by $X_n := Z_\alpha + \sum_{j=\alpha+1}^n Z_j$ is an $\mathbb{F}$-martingale.

If X is integrable and $\mathbb{F}$-adapted, then its $\mathbb{F}$-*compensator* A is defined by

$$
A_n := \sum_{j=\alpha+1}^n E \left(\Delta X_j|\mathcal{F}_{j-1}\right) \quad \text{with} \quad A_\alpha = 0 .
$$

The compensated process $X - A$ is an $\mathbb{F}$-martingale. Furthermore, X is an $\mathbb{F}$-*submartingale*, that is $E(X_n|\mathcal{F}_{n-1}) \geq X_{n-1}$ for every $n \in I$, $n \geq \alpha + 1$, if and only if its $\mathbb{F}$-compensator is (almost surely) nondecreasing.

For square integrable martingales X, the process $\langle X \rangle$ defined by

$$\langle X \rangle_n := \sum_{j=\alpha+1}^{n} E\left((\Delta X_j)^2 | \mathcal{F}_{j-1}\right) \quad \text{with} \quad \langle X \rangle_\alpha = 0$$

is called the *quadratic $\mathbb{F}$-characteristic* of X and is the $\mathbb{F}$-compensator of the non-negative $\mathbb{F}$-submartingale $X^2 = (X_n^2)_{n \in I}$ and of $[X]$.

Theorem A.8 (Lenglart's inequalities) *Let X be a nonnegative $\mathbb{F}$-submartingale with $\mathbb{F}$-compensator A.*
(a) *For every $a, b > 0$,*

$$P\left(\sup_{n \in I} X_n \geq a\right) \leq \frac{b}{a} + P\left(X_\alpha + A_\beta > b\right),$$

where $A_\beta := \lim_{n \to \infty} A_n$ if $\beta = \infty$.
(b) *If X is nondecreasing, then for every $a, b > 0$,*

$$P\left(X_\alpha + A_\beta \geq a\right) \leq \frac{1}{a}\left(b + E\sup_{\substack{n \in I \\ n \geq \alpha+1}} \Delta X_n\right) + P\left(X_\beta > b\right),$$

where $X_\beta := \lim_{n \to \infty} X_n$ if $\beta = \infty$.

Proof [64], Satz 3.9. $\square$

A process $(B_n)_{n \in I}$ is said to be $\mathbb{F}$-*predictable* if B_α is $\mathcal{F}_\alpha$-measurable and B_n is $\mathcal{F}_{n-1}$-measurable for every $n \in I$, $n \geq \alpha + 1$. In this sense $\mathbb{F}$-compensators are $\mathbb{F}$-predictable.

Theorem A.9 (Strong law of large numbers; Chow) *Assume $\beta = \infty$. Let X be an $\mathbb{F}$-martingale, $p \in (0, 2]$ and let B be an $\mathbb{F}$-predictable, nonnegative, nondecreasing process. If*

$$\sum_{j=\alpha+1}^{\infty} \frac{E\left(|\Delta X_j|^p | \mathcal{F}_{j-1}\right)}{(1 + B_j)^p} < \infty \quad a.s.,$$

then $X_n/B_n \to 0$ almost surely on $\{B_\infty = \infty\}$ as $n \to \infty$. In particular, if X is square integrable and $a > 1/2$, then $X_n/\langle X \rangle_n^a \to 0$ almost surely on $\{\langle X \rangle_\infty = \infty\}$.

Proof [64], Satz 5.4 (a) and Korollar 5.5. $\square$

Theorem A.10 (Brown's inequality) *Assume* $\beta = \infty$. *Let* $X = (X_k)_{k \in I}$ *be a uniformly integrable* $\mathbb{F}$-*martingale. Then for all almost surely finite* $\mathbb{F}$-*stopping times* $\tau : \Omega \to I \cup \{\infty\}$ *(that is,* $\tau(\Omega) \subset I$ *almost surely) and all* $\varepsilon > 0$,

$$P\left(\max_{\alpha \leq k \leq \tau} |X_k| \geq \varepsilon\right) \leq \frac{2}{\varepsilon} E\left(|X_\tau| 1_{\{|X_\tau| \geq \varepsilon/2\}}\right) .$$

Proof Setting $M_m := \max_{\alpha \leq k \leq m} |X_{k \wedge \tau}|$ and observing that $(|X_{\tau \wedge k}|)_{k \in I}$ is a non-negative submartingale, by Doob's maximal inequality we obtain for all $m \in I$ and $\varepsilon > 0$

$$\begin{aligned}
2\varepsilon P\left(M_m \geq 2\varepsilon\right) &\leq E\left(|X_{m \wedge \tau}| 1_{\{M_m \geq 2\varepsilon\}}\right) \\
&= E\left(|X_{m \wedge \tau}| 1_{\{M_m \geq 2\varepsilon, |X_{m \wedge \tau}| \geq \varepsilon\}}\right) + E\left(|X_{m \wedge \tau}| 1_{\{M_m \geq 2\varepsilon, |X_{m \wedge \tau}| < \varepsilon\}}\right) \\
&\leq E\left(|X_{m \wedge \tau}| 1_{\{|X_{m \wedge \tau}| \geq \varepsilon\}}\right) + \varepsilon P\left(M_m \geq 2\varepsilon\right)
\end{aligned}$$

so that

$$P\left(M_m \geq 2\varepsilon\right) \leq \frac{1}{\varepsilon} E\left(|X_{m \wedge \tau}| 1_{\{|X_{m \wedge \tau}| \geq \varepsilon\}}\right) .$$

Using uniform integrability of the sequence $(X_{m \wedge \tau})_{m \in I}$ and letting m tend to infinity implies the assertion. $\square$

In [12] a sharper result is derived from Doob's upcrossing inequality, but Theorem A.10 is all that is needed in tightness proofs like that of Theorem 7.1. The 2ε-trick to obtain Theorem A.10 from Doob's maximal inequality may be found for example in [23], p. 18, or [96], Lemma 2.

Appendix B

Solutions of Exercises

2.1. The system $\mathcal{D} := \{B \in \mathcal{B}(\mathcal{X}) : K(\cdot, B) \text{ is } \mathcal{F}\text{-measurable}\}$ is a Dynkin-system. A standard argument yields the assertion.

2.2. If $K_\alpha \to K$ weakly, then by Theorem 2.3, $QK_\alpha \to QK$ weakly (in $\mathcal{M}^1(\mathcal{X})$) for every probability distribution Q on $\mathcal{F}$ such that $Q \equiv P$. Conversely, if Q is a probability distribution on $\mathcal{F}$ with $Q \ll P$ and $\overline{Q} := (Q + P)/2$, then $\overline{Q} \equiv P$ so that

$$\frac{1}{2}\int h\,dQK_\alpha + \frac{1}{2}\int h\,dPK_\alpha = \int h\,d\overline{Q}K_\alpha$$

$$\to \int h\,d\overline{Q}K = \frac{1}{2}\int h\,dQK + \frac{1}{2}\int h\,dPK$$

and

$$\int h\,dPK_\alpha \to \int h\,dPK$$

for every $h \in C_b(\mathcal{X})$. Consequently, $\int h\,dQK_\alpha \to \int h\,dQK$ and hence, $QK_\alpha \to QK$ weakly. It follows from Theorem 2.3 that $K_\alpha \to K$ weakly.

2.3. Assume $Q \ll P$ and let $g := dQ/dP$. Then $gf \in \mathcal{L}^1(P)$ for every $f \in \mathcal{L}^1(Q)$. The topology $\tau(Q)$ which is generated by the functions

$$K \mapsto \int f \otimes h\,dQ \otimes K = \int gf \otimes h\,dP \otimes K, \quad f \in \mathcal{L}^1(Q), h \in C_b(\mathcal{X}),$$

is thus coarser than $\tau(P)$.

2.5. Check the proof of Theorem 2.7.

© Springer International Publishing Switzerland 2015
E. Häusler and H. Luschgy, *Stable Convergence and Stable Limit Theorems*,
Probability Theory and Stochastic Modelling 74,
DOI 10.1007/978-3-319-18329-9

2.6. If $\{P^K : K \in \Gamma\}$ is tight in $\mathcal{M}^1(\mathcal{M}^1(\mathcal{X}))$, then for every $n \in \mathbb{N}$, there exists a weakly compact set $M_n \subset \mathcal{M}^1(\mathcal{X})$ such that $\sup_{K \in \Gamma} P^K(M_n^c) = \sup_{K \in \Gamma} P(K \notin M_n) \leq 2^{-n-1}$. Since M_n is tight for every $n \in \mathbb{N}$, there exist compact sets $A_n \subset \mathcal{X}$ such that $\sup_{\nu \in M_n} \nu(A_n^c) \leq 2^{-n-1}$. This implies for every $K \in \Gamma, n \in \mathbb{N}$,

$$
PK(A_n^c) = \int K(\cdot, A_n^c)\, dP = \int_{\{K \notin M_n\}} K(\cdot, A_n^c)\, dP + \int_{\{K \in M_n\}} K(\cdot, A_n^c)\, dP
$$

$$
\leq 2^{-n-1} + 2^{-n-1} = 2^{-n},
$$

and hence, $P\Gamma$ is tight.

Conversely, assume that $P\Gamma$ is tight in $\mathcal{M}^1(\mathcal{X})$. Then for every $n \in \mathbb{N}$, there exists a compact set $A_n \subset \mathcal{X}$ such that $\sup_{K \in \Gamma} PK(A_n^c) \leq 2^{-2n}$. Now for $m \in \mathbb{N}$ introduce the set

$$
M_m := \left\{\nu \in \mathcal{M}^1(\mathcal{X}) : \nu(A_n^c) \leq 2^{-n} \text{ for every } n > m\right\}.
$$

Clearly M_m is tight and, by the Portmanteau theorem, M_m is weakly closed so that M_m is a weakly compact subset of $\mathcal{M}^1(\mathcal{X})$. Using the Markov inequality, we obtain for every $K \in \Gamma, m \in \mathbb{N}$,

$$
P^K(M_m^c) = P(K \notin M_m) = P\left(\bigcup_{n>m}\{K(\cdot, A_n^c) > 2^{-n}\}\right)
$$

$$
\leq \sum_{n>m} P(K(\cdot, A_n^c) > 2^{-n}) \leq \sum_{n>m} 2^n PK(A_n^c) \leq \sum_{n>m} 2^{-n} \to 0
$$

as $m \to \infty$, which shows that $P\Gamma$ is tight in $\mathcal{M}^1(\mathcal{M}^1(\mathcal{X}))$.

2.7. Recall that $\mathcal{B}(\mathcal{X} \times \mathcal{Y}) = \mathcal{B}(\mathcal{X}) \otimes \mathcal{B}(\mathcal{Y})$. Let $w_0(\mathcal{X} \times \mathcal{Y})$ denote the topology on $\mathcal{M}^1(\mathcal{X} \times \mathcal{Y})$ generated by the maps $\mu \mapsto \int h \otimes k\, d\mu, h \in C_b(\mathcal{X}), k \in C_b(\mathcal{Y})$. In order to show that $w_0(\mathcal{X} \times \mathcal{Y})$ coincides with the weak topology on $\mathcal{M}^1(\mathcal{X} \times \mathcal{Y})$ we have to show that the map $\mu \mapsto \int g\, d\mu$ is $w_0(\mathcal{X} \times \mathcal{Y})$-continuous for every $g \in C_b(\mathcal{X} \times \mathcal{Y})$. Let $(\mu_\alpha)_\alpha$ be a net in $\mathcal{M}^1(\mathcal{X} \times \mathcal{Y})$ and $\mu \in \mathcal{M}^1(\mathcal{X} \times \mathcal{Y})$ such that $\mu_\alpha \to \mu$ with respect to $w_0(\mathcal{X} \times \mathcal{Y})$. Let $d_\mathcal{X}$ and $d_\mathcal{Y}$ be metrics inducing the topologies on $\mathcal{X}$ and $\mathcal{Y}$, respectively. Let $O \subset \mathcal{X}, U \subset \mathcal{Y}$ be open subsets and for $n \in \mathbb{N}$, let $h_{O,n}(x) := 1 \wedge n \inf_{z \in O^c} d_\mathcal{X}(x, z)$ and $k_{U,n}(y) := 1 \wedge n \inf_{z \in U^c} d_\mathcal{Y}(y, z)$. Then $h_{O,n} \in C_b(\mathcal{X}), k_{U,n} \in C_b(\mathcal{Y}), h_{O,n} \uparrow 1_O$ and $k_{U,n} \uparrow 1_U$ so that $h_{O,n} \otimes k_{U,n} \uparrow 1_O \otimes 1_U = 1_{O \times U}$. We obtain

$$
\liminf_\alpha \mu_\alpha(O \times U) \geq \lim_\alpha \int h_{O,n} \otimes k_{U,n}\, d\mu_\alpha = \int h_{O,n} \otimes k_{U,n}\, d\mu \quad \text{for every } n \in \mathbb{N}
$$

and by monotone convergence, $\lim_{n \to \infty} \int h_{O,n} \otimes k_{U,n}\, d\mu = \mu(O \times U)$ which yields $\liminf_\alpha \mu_\alpha(O \times U) \geq \mu(O \times U)$.

Analogously, if $V \subset \mathcal{X}$ and $W \subset \mathcal{Y}$ are closed subsets and $\tilde{h}_n := 1 - h_{V^c,n}$, $\tilde{k}_n := 1 - k_{W^c,n}$, then $\tilde{h}_n \downarrow 1_V$ and $\tilde{k}_n \downarrow 1_W$ so that

$$\limsup_\alpha \mu_\alpha (V \times W) \leq \lim_\alpha \int \tilde{h}_n \otimes \tilde{k}_n \, d\mu_\alpha = \int \tilde{h}_n \otimes \tilde{K}_n \, d\mu \quad \text{for every } n \in \mathbb{N}$$

and thus $\limsup_\alpha \mu_\alpha (V \times W) \leq \mu (V \times W)$.

Let $\beta := \{O \times U : O \subset \mathcal{X} \text{ open}, U \subset \mathcal{Y} \text{ open}\}$ and $\beta_1 := \{G \in \beta : \mu (\partial G) = 0\}$. Then for $G = O \times U \in \beta_1$, using $\overline{G} = \overline{O} \times \overline{U}$ and $\partial G = \overline{G} \setminus G$,

$$\mu (G) \leq \liminf_\alpha \mu_\alpha (G) \leq \limsup_\alpha \mu_\alpha (G) \leq \limsup_\alpha (\overline{G}) \leq \mu (\overline{G}) = \mu (G)$$

so that $\lim_\alpha \mu_\alpha (G) = \mu (G)$.

The metric $d ((x, y), (x_1, y_1)) := d_{\mathcal{X}} (x, x_1) \vee d_{\mathcal{Y}} (y, y_1)$ induces the product topology and the corresponding open balls satisfy

$$B ((x, y), r) := \{d ((x, y), \cdot) < r\} = \{d_{\mathcal{X}} (x, \cdot) < r\} \times \{d_{\mathcal{Y}} (y, \cdot) < r\} .$$

Hence $B ((x, y), r) \in \beta$. Furthermore, β is closed under finite intersections since $\bigcap_{i=1}^k G_i = \bigcap_{i=1}^k (O_i \times U_i) = \left(\bigcap_{i=1}^k O_i \right) \times \left(\bigcap_{i=1}^k U_i \right)$. Now the Portmanteau theorem yields $\mu_\alpha \to \mu$ weakly, that is, $\lim_\alpha \int g \, d\mu_\alpha = \int g \, d\mu$ for every $g \in C_b (\mathcal{X} \times \mathcal{Y})$. This completes the proof of the first assertion.

The second assertion is an immediate consequence of the first one and Theorem 2.3.

2.8. For every $F \in \mathcal{F}, h \in C_b (\mathcal{X}), k \in C_b (\mathcal{Y})$, setting $f := 1_F \int k (y) K (\cdot, dy)$, we have

$$\left| \int 1_F \otimes h \otimes k \, dP \otimes (H_\alpha \otimes K_\alpha) - \int 1_F \otimes h \otimes k \, dP \otimes (H \otimes K) \right|$$

$$\leq \left| \int 1_F \otimes h \otimes k \, dP \otimes (H_\alpha \otimes K_\alpha) - \int 1_F \otimes h \otimes k \, dP \otimes (H_\alpha \otimes K) \right|$$

$$+ \left| \int 1_F \otimes h \otimes k \, dP \otimes (H_\alpha \otimes K) - \int 1_F \otimes h \otimes k \, dP \otimes (H \otimes K) \right|$$

$$\leq \int \left| 1_F \int h (x) H_\alpha (\cdot, dx) \right| \left| \int k (y) K_\alpha (\cdot, dy) - \int k (y) K (\cdot, dy) \right| dP$$

$$+ \left| \int f \otimes h \, dP \otimes H_\alpha - \int f \otimes h \, dP \otimes H \right|$$

$$\leq \|h\|_{\sup} \int \left| \int k (y) K_\alpha (\cdot, dy) - \int k (y) K (\cdot, dy) \right| dP$$

$$+ \left| \int f \otimes h \, dP \otimes H_\alpha - \int f \otimes h \, dP \otimes H \right|$$

which yields

$$\lim_\alpha \left| \int 1_F \otimes h \otimes k \, dP \otimes (H_\alpha \otimes K_\alpha) - \int 1_F \otimes h \otimes k \, dP \otimes (H \otimes K) \right| = 0.$$

The assertion follows from Exercise 2.7.

3.1. Let $(F_n)_{n\geq 1}$ be nonincreasing. Then for every $h \in C_b(\mathcal{X})$,

$$\int h \, dP_{F_n}^{X_n} = \int_F h(X_n) \, dP \frac{1}{P(F_n)} + \int_{F_n \cap F^c} h(X_n) \, dP \frac{1}{P(F_n)}$$

$$= \int h \, dP_F^{X_n} \frac{P(F)}{P(F_n)} + \int_{F_n \cap F^c} h(X_n) \, dP \frac{1}{P(F_n)}.$$

Since $P(F)/P(F_n) \to 1$ and

$$\left| \int_{F_n \cap F^c} h(X_n) \, dP \right| \leq \|h\|_{\sup} P(F_n \cap F^c) \to 0,$$

we obtain

$$\lim_{n\to\infty} \int h \, dP_{F_n}^{X_n} = \lim_{n\to\infty} \int h \, dP_{F_n}^{X_n} = \int h \, d\nu$$

so that $P_{F_n}^{X_n} \to \nu$ weakly.

Now let $(F_n)_{n\geq 1}$ be nondecreasing. Then for $h \in C_b(\mathcal{X})$ and n sufficiently large

$$\int h \, dP_{F_n}^{X_n} = \int_F h(X_n) \, dP \frac{1}{P(F_n)} - \int_{F \cap F_n^c} h(X_n) \, dP \frac{1}{P(F_n)}$$

$$= \int h \, dP_F^{X_n} \frac{P(F)}{P(F_n)} - \int_{F \cap F_n^c} h(X_n) \, dP \frac{1}{P(F_n)}.$$

Since $P\left(F \cap F_n^c\right) \to 0$, we obtain as above $\lim_{n\to\infty} \int h \, dP_{F_n}^{X_n} = \int h \, d\nu$.

3.2. If $1_{F_n} \to K$ $\mathcal{G}$-stably, $G \in \mathcal{G}$ and $h \in C_b(\mathbb{R})$ satisfies $h(0) = 0$ and $h(1) = 1$, then by Theorem 3.2,

$$P(F_n \cap G) = E 1_G h\left(1_{F_n}\right) \to \int 1_G \otimes h \, dP \otimes K = \int_G \alpha \, dP.$$

Conversely, assume $\lim_{n\to\infty} P(F_n \cap G) = \int_G \alpha \, dP$ for every $G \in \mathcal{G}$. Then for $G \in \mathcal{G}$ with $P(G) > 0$, using $P_G K = \left(\int \alpha \, dP_G\right) \delta_1 + \left(\int (1 - \alpha) \, dP_G\right) \delta_0$, we get

$$P_G^{1_{F_n}}(\{1\}) = P_G(F_n) \to \int \alpha \, dP_G = P_G K(\{1\})$$

and

$$P_G^{1_{F_n}}(\{0\}) = 1 - P_G(F_n) \to \int (1 - \alpha)\, dP_G = P_G K(\{0\}) \, .$$

This yields weak convergence $P_G^{1_{F_n}} \to P_G K$ in $\mathcal{M}^1(\mathbb{R})$. The assertion $1_{F_n} \to K$ $\mathcal{G}$-stably follows from Theorem 3.2.

3.3. Clearly, the X_n are identically distributed with $P^{X_n} = (\delta_0 + \delta_1)/2$. Let $Q := 2t\, dt$. Then dQ/dP is $\mathcal{G}$-measurable and $Q^{X_n} = \left(\frac{3}{4} - a_n\right)\delta_0 + \left(a_n + \frac{1}{4}\right)\delta_1$. Choosing $h \in C_b(\mathbb{R})$ such that $h(0) = 0$ and $h(1) = 1$ (e.g. $h(t) = (t \wedge 1) \vee 0$) we get $\int h\, dQ^{X_n} = a_n + \frac{1}{4}$. Thus, if (a_n) is not convergent, (Q^{X_n}) is not weakly convergent. Consequently, by Theorem 3.2, (X_n) does not converge $\mathcal{G}$-stably. Alternatively, one can argue that the assertion follows immediately from Example 1.2 and Theorem 3.7.

3.4. (i) $\Rightarrow$ (ii) Assume $X_n \to K$ $\mathcal{G}$-stably for some $K \in \mathcal{K}^1(\mathcal{G})$. Then for $f \in \mathcal{L}^1(\mathcal{G}, P)$ and $h \in C_b(\mathcal{X})$, by Theorem 3.2 and independence of $\sigma(X_n)$ and $\mathcal{G}$,

$$\int f\, dP E h(X_n) = E f h(X_n) \to \int f \otimes h\, dP \otimes K \, .$$

In particular, $E h(X_n) \to \int h\, dP K$ and thus $E f h(X_n) \to \int f\, dP \int h\, dP K$. Corollary 3.3 yields $X_n \to P K$ $\mathcal{G}$-mixing.

(ii) $\Rightarrow$ (iii) is clear.

(iii) $\Rightarrow$ (ii). Assume $P^{X_n} \to \nu$ weakly. Then for $f \in \mathcal{L}^1(\mathcal{G}, P)$ and $h \in C_b(\mathcal{X})$,

$$E f h(X_n) = \int f\, dP E h(X_n) \to \int f\, dP \int h\, d\nu \, .$$

Corollary 3.3 yields $X_n \to \nu$ $\mathcal{G}$-mixing.

(ii) $\Rightarrow$ (i) is clear.

3.5. The implications (i) $\Rightarrow$ (ii) $\Rightarrow$ (iii) are obvious consequences of Corollary 3.3 and the Portmanteau theorem.

(iii) $\Rightarrow$ (i). Using the Portmanteau theorem again, we have

$$\lim_{n \to \infty} P(X_n \in B) = \nu(B)$$

and

$$\lim_{n \to \infty} P(\{X_n \in B\} \cap \{X_k \in B\}) = \nu(B) P(X_k \in B)$$

for every $k \in \mathbb{N}$ and $B \in \mathcal{B}(\mathcal{X})$ with $\nu(\partial B) = 0$. It remains to show that this implies

$$\lim_{n \to \infty} P(\{X_n \in B\} \cap F) = \nu(B) P(F)$$

for every $F \in \mathcal{F}$ and $B \in \mathcal{B}(\mathcal{X})$ with $\nu(\partial B) = 0$. The assertion (i) then follows from the Portmanteau theorem and Corollary 3.3.

In order to prove the above limiting relation, fix $B \in \mathcal{B}(\mathcal{X})$ with $\nu(\partial B) = 0$ and let $F_n := \{X_n \in B\}$. One checks that

$$\mathcal{L} := \left\{ f \in \mathcal{L}^2(P) : \lim_{n \to \infty} \int 1_{F_n} f \, dP = \nu(B) \int f \, dP \right\}$$

is a closed vector subspace of $\mathcal{L}^2(P)$ containing 1_Ω and 1_{F_k} for every $k \in \mathbb{N}$. Consequently, the closed linear span of $\{1_{F_k} : k \in \mathbb{N}\} \cup \{1_\Omega\}$ in $\mathcal{L}^2(P)$, denoted by $\mathcal{L}_1$, satisfies $\mathcal{L}_1 \subset \mathcal{L}$. Now let $F \in \mathcal{F}$ and let $1_F = f_1 + f_2$ with $f_1 \in \mathcal{L}_1$ and $f_2 = 1_F - f_1$ belonging to the orthogonal complement of $\mathcal{L}_1$ (f_1 is the P-almost surely unique best approximation to 1_F from $\mathcal{L}_1$). Then we obtain

$$P(F_n \cap F) = \int 1_{F_n} f_1 \, dP \to \nu(B) \int f_1 \, dP = \nu(B) P(F) .$$

(i) $\Rightarrow$ (iv) follows from Corollary 3.3.

(iv) $\Rightarrow$ (ii). In view of Proposition 3.4 (c) and $P(\nu \otimes \delta_{X_k}) = \nu \otimes P^{X_k}$, we have $X_n \to \nu \, \sigma(X_k)$-mixing for every $k \in \mathbb{N}$. The assertion now follows from Corollary 3.3.

3.6. Assume $X_n \to X$ in probability for some $(\mathcal{X}, \mathcal{B}(\mathcal{X}))$-valued random variable X. Then by Corollary 3.6, $X_n \to \delta_X$ stably. This implies $\nu = \delta_X$ almost surely, hence, ν is a Dirac measure.

3.7. (a) ([50], Lemma IX.6.5) Let $k \in \mathbb{N}$. There exists a compact set $A \subset \mathcal{X}$ such that $P(X \notin A) \le 1/k$. Here we use that $\mathcal{X}$ is polish. Choose $x_1, \ldots, x_p \in A$ such that $A \subset \bigcup_{i=1}^p \{d(\cdot, x_i) < 1/k\}$. ($d$ is a metric on $\mathcal{X}$ inducing the topology.) Since

$$\left\{ d(X_n, X) \ge \frac{3}{k} \right\} \cap \{X \in A\} \subset \bigcup_{i=1}^p \left\{ d(x_i, X) < \frac{1}{k}, d(x_i, X_n) > \frac{2}{k} \right\}$$

and setting $h_i(x) := k\left[(d(x_i, x) - 1/k)^+ \wedge 1\right]$ we obtain

$$P\left(d(X_n, X) \ge \frac{3}{k}\right) \le P(X \notin A) + P\left(d(X_n, X) \ge \frac{3}{k}, X \in A\right)$$

$$\le \frac{1}{k} + \sum_{i=1}^p E 1_{\{d(x_i, X) < 1/k\}} 1_{\{d(x_i, X_n) > 2/k\}}$$

$$\le \frac{1}{k} + \sum_{i=1}^p E 1_{\{d(x_i, X) < 1/k\}} h_i(X_n) .$$

Using $h_i \in C_b(\mathcal{X})$ and $\{d(x_i, X) < 1/k\} \in \mathcal{G}$, the last sum above converges to $\sum_{i=1}^{p} E1_{\{d(x_i,X)<1/k\}}h_i(X) = 0$, hence, $\limsup_{n\to\infty} P(d(X_n, X) \geq 3/k) \leq 1/k$. This yields $X_n \to X$ in probability as $n \to \infty$.

(b) By assumption and Theorem 3.2 we have $(X_n, X) \to \delta_X \otimes \delta_X$ $\mathcal{G}$-stably so that $(X_n, X) \xrightarrow{d} P(\delta_X \otimes \delta_X) = P^{(X,X)}$. Let d be a metric on $\mathcal{X}$ inducing the topology. Using the continuity of $d : \mathcal{X} \times \mathcal{X} \to \mathbb{R}_+$, this yields $Ed(X_n, X) \wedge 1 \to Ed(X, X) \wedge 1 = 0$.

3.8. For $\varepsilon > 0$, we have

$$
\begin{aligned}
E\left(d\left(X_{n,r}, Y_n\right) \wedge 1\right) &= \int_{\{d(X_{n,r},Y_n)\leq\varepsilon\}} d\left(X_{n,r}, Y_n\right) \wedge 1 \, dP \\
&\quad + \int_{\{d(X_{n,r},Y_n)>\varepsilon\}} d\left(X_{n,r}, Y_n\right) \wedge 1 \, dP \\
&\leq \varepsilon + P\left(d\left(X_{n,r}, Y_n\right) > \varepsilon\right)
\end{aligned}
$$

and for $\varepsilon \in (0, 1)$,

$$
P\left(d\left(X_{n,r}, Y_n\right) > \varepsilon\right) = P\left(d\left(X_{n,r}, Y_n\right) \wedge 1 > \varepsilon\right) \leq \varepsilon^{-1} E\left(d\left(X_{n,r}, Y_n\right) \wedge 1\right).
$$

This yields the assertion.

Based on this formulation of condition (iii) one can also prove Theorem 3.10 as follows.

For every bounded Lipschitz function $h : \mathcal{X} \to \mathbb{R}$ with Lipschitz constant $L \in \mathbb{R}_+$ and $F \in \mathcal{G}$ with $P(F) > 0$, we have

$$
\begin{aligned}
\left|\int h(Y_n) \, dP_F - \int h \, dP_F K\right| &\leq \int \left|h(Y_n) - h(X_{n,r})\right| dP_F \\
&\quad + \left|\int h(X_{n,r}) \, dP_F - \int h \, dP_F K_r\right| \\
&\quad + \left|\int h \, dP_F K_r - \int h \, dP_F K\right|
\end{aligned}
$$

and moreover,

$$
\begin{aligned}
\int \left|h(Y_n) - h(X_{n,r})\right| dP_F &\leq \int L d\left(X_{n,r}, Y_n\right) \wedge 2 \|h\|_{\sup} \, dP_F \\
&\leq \frac{L \vee 2 \|h\|_{\sup}}{P(F)} \int d\left(X_{n,r}, Y_n\right) \wedge 1 \, dP.
\end{aligned}
$$

We obtain

$$\lim_{n \to \infty} \left| \int h\,(Y_n)\, dP_F - \int h\, dP_F K \right| = 0$$

and hence, $P_F^{Y_n} \to P_F K$ weakly (cf. [26], Theorem 11.3.3).

3.9. (i) $\Rightarrow$ (iii) follows from Theorem 3.2 or Proposition 3.12.

(iii) $\Rightarrow$ (ii). Let $F \in \mathcal{E}$ with $P\,(F) > 0$ so that $F \in \sigma\,(X_k)$ for some $k \in \mathbb{N}$. Assume $(X_n, X_k) \xrightarrow{d} \mu$ as $n \to \infty$ for some $\mu \in \mathcal{M}^1\,(\mathcal{X} \times \mathcal{X})$. By Proposition 3.4 (a), (b), there exists a subsequence (X_m) of (X_n) and $K \in \mathcal{K}^1\,(\sigma\,(X_k)\,, P)$ such that $X_m \to K\ \sigma\,(X_k)$-stably. Theorem 3.2 yields $(X_m, X_k) \xrightarrow{d} P\,(K \otimes \delta_{X_k})$ as $m \to \infty$ so that $\mu = P\,(K \otimes \delta_{X_k})$. Now it follows from Proposition 3.4 (c) that $X_n \to K$ $\sigma\,(X_k)$-stably. Consequently, by Theorem 3.2, $P_F^{X_n} \to P_F K$ weakly.

(ii) $\Rightarrow$ (i). Assume $P_F^{X_n} \to \nu_F$ weakly for some $\nu_F \in \mathcal{M}^1\,(\mathcal{X})$ and every $F \in \mathcal{E} := \bigcup_{n=1}^{\infty} \sigma\,(X_n)$ with $P\,(F) > 0$. In view of Proposition 3.4 (a), there exists a subsequence (X_m) of (X_n) and $K \in \mathcal{K}^1$ such that $X_m \to K$ stably. This implies $P_F^{X_m} \to P_F K$ weakly for every $F \in \mathcal{F}$ with $P\,(F) > 0$ so that $\nu_F = P_F K$ for every $F \in \mathcal{E}$ with $P\,(F) > 0$. One checks that

$$\mathcal{L} := \left\{ f \in \mathcal{L}^2\,(P) : \lim_{n \to \infty} E f\,h\,(X_n) = \int f \otimes h\, dP \otimes K \text{ for every } h \in C_b\,(\mathcal{X}) \right\}$$

is a closed vector subspace of $\mathcal{L}^2\,(P)$ containing 1_F, $F \in \mathcal{E}$. Consequently, the closed linear span of $\{1_F : F \in \mathcal{E}\}$ in $\mathcal{L}^2\,(P)$, denoted by $\mathcal{L}_1$, satisfies $\mathcal{L}_1 \subset \mathcal{L}$. Now let $F \in \mathcal{F}$ with $P\,(F) > 0$ and let $1_F = f_1 + f_2$ with $f_1 \in \mathcal{L}_1$ and $f_2 = 1_F - f_1$ belonging to the orthogonal complement of $\mathcal{L}_1$. Since step functions are dense in $\mathcal{L}^p$-spaces, we have $\mathcal{L}^2\,(\sigma\,(X_k)\,, P) \subset \mathcal{L}_1 \subset \mathcal{L}$ for every $k \in \mathbb{N}$. Hence, for every $h \in C_b\,(\mathcal{X})$,

$$\int 1_F h\,(X_n)\, dP = \int f_1 h\,(X_n)\, dP \to \int f_1 \otimes h\, dP \otimes K\,.$$

The assertion follows from Proposition 3.12.

3.10. Let

$$X_t^n := \frac{1}{\sigma \sqrt{n}} \left(\sum_{j=1}^{[nt]} Z_j + (nt - [nt])\, Z_{[nt]+1} \right)\,, \quad t \geq 0\,.$$

By Example 3.14, $X^n \to \nu = P^W$ mixing in $C\,(\mathbb{R}_+)$, where $W = (W_t)_{t \geq 0}$ denotes a Brownian motion. Using the continuity of the restriction map $C\,(\mathbb{R}_+) \to C\,([0, 1])$, $x \mapsto x|\,[0, 1]$, we get

$$\left(X_t^n\right)_{t \in [0,1]} \to P^{(W_t)_{t \in [0,1]}} \quad \text{mixing in } C\,([0, 1])\,.$$

Define $g : C([0, 1]) \to \mathbb{R}$ by $g(x) := \max_{t \in [0,1]} x(t)$. Then g is continuous and hence by Theorem 3.7 (c),

$$\max_{t \in [0,1]} X_t^n \to P^{\max_{t \in [0,1]} W_t} \quad \text{mixing}.$$

Finally observe that $P^{\max_{t \in [0,1]} W_t} = \mu$ ([51], Proposition 13.13) and

$$\max_{t \in [0,1]} X_t^n = \frac{1}{\sigma \sqrt{n}} \max_{0 \le j \le n} \sum_{i=1}^{j} Z_j.$$

3.11. The "if" part. By Lemma A.5, we have $P^{(X,Y)} = P\left(P^{(X,Y)|\mathcal{G}}\right) = P\left(P^{X|\mathcal{G}} \otimes \delta_Y\right)$. Consequently, Proposition 3.4 (c) yields $X_n \to X$ $\mathcal{G}$-stably.

The "only if" part follows from Theorem 3.17.

3.12. An application of Theorem 3.17 with $\mathcal{E} = \bigcup_{k=1}^{\infty} \sigma(X_1, \ldots, X_k)$ shows that $X_n \to X$ $\mathcal{G}$-stably if and only if $X_n \to X$ $\sigma(X_1, \ldots, X_k)$-stably for every $k \in \mathbb{N}$. The assertion follows from Exercise 3.11.

4.1. Check the proof of Proposition 4.5.

4.2. Let d be a metric on $\mathcal{X}$ inducing the topology. For $\varepsilon > 0$ and $k \in \mathbb{N}$, we have

$$P\left(d\left(X_{\tau_n}, X\right) > \varepsilon\right) \le P\left(\tau_n < k\right) + P\left(d\left(X_{\tau_n}, X\right) > \varepsilon, \tau_n \ge k\right)$$

$$= P\left(\tau_n < k\right) + \sum_{j=k}^{\infty} P\left(d\left(X_j, X\right) > \varepsilon, \tau_n = j\right)$$

$$\le P\left(\tau_n < k\right) + \sum_{j=k}^{\infty} P\left(\sup_{m \ge k} d\left(X_m, X\right) > \varepsilon, \tau_n = j\right)$$

$$\le P\left(\tau_n < k\right) + P\left(\sup_{m \ge k} d\left(X_m, X\right) > \varepsilon\right).$$

Since $\lim_{n \to \infty} P\left(\tau_n < k\right) = 0$ and $\lim_{k \to \infty} P\left(\sup_{m \ge k} d\left(X_m, X\right) > \varepsilon\right) = 0$, we obtain $\lim_{n \to \infty} P\left(d\left(X_{\tau_n}, X\right) > \varepsilon\right) = 0$.

4.3. Recall that $P|\mathcal{H}$ is *purely atomic* for a sub-σ-field $\mathcal{H}$ of $\mathcal{F}$ if there exists a (possibly finite) sequence $(F_j)_{j \ge 1}$ of $P|\mathcal{H}$-atoms such that $P\left(\bigcup_{j \ge 1} F_j\right) = 1$, where $F \in \mathcal{H}$ is called a $P|\mathcal{H}$-*atom* if $P(F) > 0$ and every $H \in \mathcal{H}$ with $H \subset F$ satisfies $P(H) = 0$ or $P(H) = P(F)$. If $F, G \in \mathcal{H}$ are $P|\mathcal{H}$-atoms, then $P(F \cap G) = 0$ or $P(F \Delta G) = 0$.

The "only if" part. Assume $X_n \to K$ stably for some $K \in \mathcal{K}^1$. Assume that $\tau_n \to \infty$ in probability and $P|\mathcal{H}$ is purely atomic, where $\mathcal{H} := \sigma(\tau_n, n \geq 1)$. Let $(F_j)_{j \geq 1}$ be a sequence of $P|\mathcal{H}$-atoms satisfying $P\left(\bigcup_{j \geq 1} F_j\right) = 1$ and $P(F_j \cap F_k) = 0$ for $j \neq k$. Since $P(F_j) = \sum_{k=1}^{\infty} P(F_j \cap \{\tau_n = k\})$, there exist $k_{n,j} \in \mathbb{N}$ such that $P(F_j) = P(F_j \cap \{\tau_n = k_{n,j}\})$, that is, $F_j \subset \{\tau_n = k_{n,j}\}$ P-almost surely. Then $k_{n,j} \to \infty$ as $n \to \infty$. Consequently, for every $F \in \mathcal{F}$, $h \in C_b(\mathcal{X})$ and every j,

$$\int_F h(X_{\tau_n}) \, dP_{F_j} = \int_F h(X_{k_{n,j}}) \, dP_{F_j} \to \int 1_F \otimes h \, dP_{F_j} \otimes K \quad \text{as } n \to \infty$$

so that by Theorem 3.2, $X_{\tau_n} \to K$ stably under P_{F_j}. Using $P = \sum_{j \geq 1} P(F_j) P_{F_j}$, Proposition 3.11 yields $X_{\tau_n} \to K$ stably (under P) and hence, $X_{\tau_n} \xrightarrow{d} \nu := PK$.

The "if"-part follows as in Remark 4.7 (b) because $P|\mathcal{H}$ is purely atomic for every finite sub-σ-field $\mathcal{H} \subset \mathcal{F}$.

4.4. By the subsequent Corollary 5.9 (see also Example 4.3) we have $X^n \to P^W$ mixing in $C(\mathbb{R}_+)$. The assertion follows from Theorem 4.6 and Remark 4.7 (a).

4.5. The classical central limit theorem yields

$$P\left(\sum_{j=1}^{n} Z_j > 0\right) = (X_1^n > 0) \to 1 - \Phi(0) = \frac{1}{2}$$

and

$$P\left(\sum_{j=1}^{n} Z_j \leq 0\right) = (X_1^n \leq 0) \to \Phi(0) = \frac{1}{2},$$

where $X_1^n = \sigma^{-1} n^{-1/2} \sum_{j=1}^{n} Z_j$ and Φ denotes the distribution function of $N(0, 1)$, so that $\tau_n/n \xrightarrow{d} \frac{1}{2}(\delta_1 + \delta_2)$ (and $\frac{1}{2}(\delta_1 + \delta_2)(0, \infty) = 1$). On the other hand,

$$P(X_1^{\tau_n} \leq 0) = P\left(X_1^{2n} \leq 0, \sum_{j=1}^{n} Z_j \leq 0\right)$$

$$= P\left(\frac{1}{\sigma\sqrt{n}} \sum_{j=1}^{2n} Z_j \leq 0, \frac{1}{\sigma\sqrt{n}} \sum_{j=1}^{n} Z_j \leq 0\right)$$

$$= P^{\left(\sigma^{-1} n^{-1/2} \sum_{j=1}^{n} Z_j, \sigma^{-1} n^{-1/2} \sum_{j=n+1}^{2n} Z_j\right)} (C),$$

where $C = \{(x, y) \in \mathbb{R}^2 : x \leq 0, x + y \leq 0\}$, and hence, by the central limit theorem and Fubini's theorem, as $n \to \infty$

$$P\left(X_1^{T_n} \leq 0\right) = P^{\sigma^{-1}n^{-1/2}\sum_{j=1}^n Z_j} \otimes P^{\sigma^{-1}n^{-1/2}\sum_{j=1}^n Z_j}(C)$$

$$\rightarrow N(0,1) \otimes N(0,1)(C) = \int N(0,1)(C_y)\, dN(0,1)(y)$$

$$= \frac{1}{4} + \int_0^\infty \Phi(-y)\, dN(0,1)(y) = \frac{1}{4} + \int_{-\infty}^0 \Phi(y)\, dN(0,1)(y)$$

$$= \frac{3}{8} \neq \Phi(0),$$

where the last equation follows from integration by parts.

4.6. Let $w_0(\mathcal{X})$ denote the topology on $\mathcal{M}^1(\mathcal{X})$ generated by the maps $\nu \mapsto \int h_{B,n}\, d\nu, B \in \mathcal{B}, n \in \mathbb{N}$. In order to show that $w_0(\mathcal{X})$ coincides with the weak topology on $\mathcal{M}^1(\mathcal{X})$, we have to show that the map $\nu \mapsto \int h\, d\nu$ is $w_0(\mathcal{X})$-continuous for every $h \in C_b(\mathcal{X})$. Let $(\nu_\alpha)_\alpha$ be a net in $\mathcal{M}^1(\mathcal{X})$ and $\nu \in \mathcal{M}^1(\mathcal{X})$ such that $\nu_\alpha \rightarrow \nu$ with respect to $w_0(\mathcal{X})$. Let $O \subset \mathcal{X}$ be open. Choose $B_k \in \mathcal{B}$ such that $B_k \uparrow O$ as $k \rightarrow \infty$. Then $h_{B_k,m} \uparrow 1_{B_k}$ as $m \rightarrow \infty$ for every $k \in \mathbb{N}$. Therefore, using the monotone convergence theorem,

$$\liminf_\alpha \nu_\alpha(O) \geq \liminf_\alpha \int h_{B_k,m}\, d\nu_\alpha = \int h_{B_k,m}\, d\nu \quad \text{for every } k, m \in \mathbb{N},$$

$$\lim_{m \rightarrow \infty} \int h_{B_k,m}\, d\nu = \nu(B_k) \quad \text{and} \quad \lim_{k \rightarrow \infty} \nu(B_k) = \nu(O),$$

which yields $\liminf_\alpha \nu_\alpha(O) \geq \nu(O)$. Consequently, by the Portmanteau theorem, $\nu_\alpha \rightarrow \nu$ weakly, that is, $\int h\, d\nu_\alpha \rightarrow \int h\, d\nu$ for every $h \in C_b(\mathcal{X})$.

4.7. Since $\left(P^{X_n}\right)_{n \geq 1}$ is tight, $(X_n)_{n \geq 1}$ has a stably convergent subsequence by Proposition 3.4 (a). Let (X_k) be any subsequence of (X_n) with $X_k \rightarrow H$ stably for some $H \in \mathcal{K}^1$. By Theorem 4.9, there exists a subsequence (X_m) of (X_k) such that almost surely,

$$\frac{1}{r} \sum_{m=1}^r \delta_{X_m(\omega)} \rightarrow H(\omega, \cdot) \quad \text{weakly as } r \rightarrow \infty.$$

Hence, $H = K$ almost surely. Thus all subsequences of (X_n) which converge stably, converge stably to K. So the original sequence must converge stably to K.

5.1. Check the proof of Proposition 5.1.

5.2. Check the proof of Proposition 5.3.

5.3. Apply Proposition 5.3 with $f_n : \mathcal{Z}^n \rightarrow \mathbb{R}, f_n(z_1, \ldots, z_n) := \sum_{j=1}^n g(z_j)$.

6.1. For $n \in \mathbb{N}$ and $0 \leq k \leq k_n$, define $Y_{nk} := \sum_{j=1}^k X_{nj}$ with $Y_{n0} = 0$. Then $(Y_{nk})_{0 \leq k \leq k_n}$ is a nonnegative submartingale (with respect to the filtration

$(\mathcal{F}_{nk})_{0 \leq k \leq k_n}$ with compensator $A_{nk} = \sum_{j=1}^{k} E\left(X_{nj} | \mathcal{F}_{n,j-1}\right)$. For $\varepsilon, \delta > 0$ the Lenglart inequality of Theorem A.8 (a) yields

$$P\left(Y_{nk_n} \geq \varepsilon\right) \leq \frac{\delta}{\varepsilon} + P\left(A_{nk_n} > \delta\right).$$

Letting n tend to infinity and then letting δ tend to zero gives the assertion.

6.2. One checks that for every $n \in \mathbb{N}$ there exists an $r_n \in \mathbb{N}$, $r_n \geq n$ such that

$$\sum_{j > r_n} X_{nj} \to 0 \quad \text{in probability as } n \to \infty$$

and

$$\sum_{j=1}^{r_n} E\left(X_{nj}^2 | \mathcal{F}_{n,j-1}\right) \to \eta^2 \quad \text{in probability as } n \to \infty.$$

The σ-field $\mathcal{G}$ from Theorem 6.1 which takes the form $\mathcal{G} = \sigma\left(\bigcup_{n=1}^{\infty} \mathcal{G}_{n\infty}\right)$, where $\mathcal{G}_{n\infty} = \sigma\left(\bigcup_{j=0}^{\infty} \mathcal{G}_{nj}\right)$, coincides with the σ-field $\sigma\left(\bigcup_{n=1}^{\infty} \mathcal{G}_{nr_n}\right)$. Now apply Theorem 6.1 to the array $(X_{nk})_{1 \leq k \leq r_n, n \in \mathbb{N}}$ and $(\mathcal{F}_{nk})_{0 \leq k \leq r_n, n \in \mathbb{N}}$ and Theorem 3.7 (a) to get the assertion.

6.3. By Theorem 6.1, Remark 6.2 and Proposition 3.5 we have

$$\sum_{j=1}^{k_n} X_{nj} \to N\left(0, \eta^2\right) \quad \text{stably as } n \to \infty.$$

The assertion follows from Theorem 4.1.

6.4. Let $A_n = \bigcap_{k=1}^{n} \left\{X_{nk} = \frac{1}{n}\right\}$. Then $P\left(A_n\right) = 1 - \frac{1}{n}$ and

$$A_n \subset \left\{\max_{1 \leq k \leq n} |X_{nk}| = \frac{1}{n}\right\} \cap \left\{\sum_{k=1}^{n} X_{nk}^2 = \frac{1}{n}\right\} \cap \left\{\sum_{k=1}^{n} X_{nk} = 1\right\}$$

so that $\max_{1 \leq k \leq n} |X_{nk}| \to 0$ and $\sum_{k=1}^{n} X_{nk}^2 \to 0$ in probability (that is (R) with $\eta^2 = 0$), but $\sum_{k=1}^{n} X_{nk} \to 1$ in probability.

6.5. We have $E\left(X_n | \mathcal{F}_{n-1}\right) = 1_{\{X_0 \neq 0\}} E\left(Y_n | \mathcal{F}_{n-1}\right) = 0$ for $n \geq 1$ so that M is a martingale. Moreover,

$$\langle M \rangle_n = \sum_{j=1}^{n} E\left(X_j^2 | \mathcal{F}_{j-1}\right) = \sum_{j=1}^{n} \sigma^2 1_{\{X_0 \neq 0\}} = n\sigma^2 1_{\{X_0 \neq 0\}} = n\sigma^2 X_0^2$$

which implies $\langle M \rangle_n / n = \sigma^2 X_0^2$ and thus condition (N_{a_n}) is satisfied with $a_n = \sqrt{n}$. The conditional Lindeberg condition (CLB_{a_n}) is satisfied because

$$\frac{1}{n} \sum_{j=1}^{n} E\left(X_j^2 \mathbb{1}_{\{|X_j| \geq \varepsilon \sqrt{n}\}} | \mathcal{F}_{j-1} \right) = \frac{1}{n} \sum_{j=1}^{n} \mathbb{1}_{\{X_0 \neq 0\}} E Y_1^2 \mathbb{1}_{\{|Y_1| \geq \varepsilon \sqrt{n}\}}$$

$$= \mathbb{1}_{\{X_0 \neq 0\}} E Y_1^2 \mathbb{1}_{\{|Y_1| \geq \varepsilon \sqrt{n}\}} \to 0$$

on Ω as $n \to \infty$. Hence, Theorem 6.23 yields $M_n / \sqrt{n} \to N\left(0, \sigma^2 X_0^2\right)$ stably. In particular, $M_n / \sqrt{n} \xrightarrow{d} PN\left(0, \sigma^2 X_0^2\right) = P\left(X_0 = 0\right) \delta_0 + P\left(X_0 \neq 0\right) N\left(0, \sigma^2\right)$.

6.6. Let $Z_0 := 0$, $\mathcal{F}_n := \sigma\left(Z_0, Z_1, \ldots, Z_n\right)$, $\mathbb{F} = \left(\mathcal{F}_n\right)_{n \geq 0}$, $U_n := \sum_{j=1}^{n} Z_j / j$ with $U_0 = 0$ and $M_n := \sum_{j=1}^{n} U_{j-1} Z_j$ with $M_0 = M_1 = 0$. Then U and M are square integrable $\mathbb{F}$-martingales and U is $\mathcal{L}^2$-bounded because $E U_n^2 = \sum_{j=1}^{n} \sigma^2 / j^2 \leq \sum_{j=1}^{\infty} \sigma^2 / j^2 < \infty$. The martingale convergence theorem yields $U_n \to V = \sum_{j=1}^{\infty} Z_j / j$ almost surely. We have $\langle M \rangle_n = \sum_{j=1}^{n} E\left(U_{j-1}^2 Z_j^2 | \mathcal{F}_{j-1}\right) = \sigma^2 \sum_{j=1}^{n} U_{j-1}^2$ and hence, by the Toeplitz Lemma 6.28 (b), $\frac{1}{n} \langle M \rangle_n \to \sigma^2 V^2$ almost surely. Moreover, the conditional Lyapunov condition $(CLY_{a_n, p})$ with $a_n = \sqrt{n}$ and $p > 2$ is satisfied because

$$\frac{1}{n^{p/2}} \sum_{j=1}^{n} E\left(|U_{j-1} Z_j|^p | \mathcal{F}_{j-1}\right) = \frac{1}{n^{p/2}} \sum_{j=1}^{n} |U_{j-1}|^p E |Z_j|^p$$

$$= \frac{E |Z_1|^p}{n^{p/2-1}} \frac{1}{n} \sum_{j=1}^{n} |U_{j-1}|^p \to 0 \quad \text{a.s.}$$

using $\sum_{j=1}^{n} |U_{j-1}|^p / n \to |V|^p$ almost surely which follows again from the Toeplitz lemma. The assertion now follows from Theorem 6.23 and Remark 6.25.

6.7. For $n, k \in \mathbb{N}$, define $X_{nk} := a_n X_{n+k}$ and for $n \in \mathbb{N}$, $k \in \mathbb{N}_0$, $\mathcal{F}_{nk} := \mathcal{F}_{n+k}$. Then the nesting condition is obviously satisfied. Apply Exercise 6.2.

6.8. The stopping time τ_n is almost surely finite because $\langle M \rangle_\infty = \infty$ almost surely. Consider the arrays $X_{nk} = \Delta M_k / \sqrt{n}$, $k, n \in \mathbb{N}$, and $\mathcal{F}_{nk} = \mathcal{F}_k$, $k \in \mathbb{N}_0$, $n \in \mathbb{N}$. Then (X_{nk}) is a square integrable martingale difference array adapted to the nested array $(\mathcal{F}_{nk})$. We have for every $n \in \mathbb{N}$,

$$\sum_{k=1}^{\tau_n} E\left(X_{nk}^2 | \mathcal{F}_{n,k-1}\right) = \frac{1}{n} \sum_{k=1}^{\tau_n} E\left((\Delta M_k)^2 | \mathcal{F}_{k-1}\right) = \frac{1}{n} \langle M \rangle_{\tau_n}$$

and

$$1 \leq \frac{1}{n} \langle M \rangle_{\tau_n} \leq \frac{1}{n} \left(n + c^2 \right) = 1 + \frac{c^2}{n} .$$

Hence

$$\sum_{k=1}^{\tau_n} E \left(X_{nk}^2 | \mathcal{F}_{k-1} \right) \to 1 \quad \text{a.s. as } n \to \infty$$

and (CLB$_{\tau_n}$) is satisfied because $|X_{nk}| \leq c/\sqrt{n}$. Consequently, by Corollary 6.4,

$$\frac{1}{\sqrt{n}} M_{\tau_n} = \frac{M_0}{\sqrt{n}} + \sum_{k=1}^{\tau_n} X_{nk} \to N(0, 1) \quad \text{mixing as } n \to \infty .$$

6.9. Since $(Y_n - (r + s + mn) X_\infty) / \sqrt{n} = (r + s + mn) (X_n - X_\infty) / \sqrt{n}$ and $(r + s + mn) / \sqrt{n} \sim m\sqrt{n}$ as $n \to \infty$, the assertion follows from Example 6.30.

6.10. Let $\mathcal{G} := \sigma (X_n, n \geq 1)$ and $\mathcal{E} := \bigcup_{k=1}^\infty \sigma (X_1, \ldots, X_k)$. Then $\mathcal{E}$ is a field with $\sigma (\mathcal{E}) = \mathcal{G}$. If $h \in C_b (\mathcal{X})$ and $F \in \sigma (X_1, \ldots, X_k)$ for some $k \in \mathbb{N}$ with $P(F) > 0$, then for $n > k$

$$E 1_F h (X_n) = E E (1_F h (X_n) | X_1, \ldots, X_k, \mathcal{T}_X)$$
$$= E (1_F E (h (X_n) | X_1, \ldots, X_k, \mathcal{T}_X)) .$$

Since $\sigma (X_n)$ and $\sigma (X_1, \ldots, X_k)$ are conditionally independent given $\mathcal{T}_X$, we have

$$E (h (X_n) | X_1, \ldots, X_k, \mathcal{T}_X) = E (h (X_n) | \mathcal{T}_X) = E (h (X_1) \mathcal{T}_X)$$

([17], Theorem 7.3.1) so that $E 1_F h (X_n) \to E (1_F E (h (X_1) | \mathcal{T}_X))$. The assertion follows from Theorem 3.2 and Proposition 3.5.

8.1. The Cauchy-distribution $\mu = C (0, b)$ satisfies $\int \log^+ |x| \, d\mu (x) < \infty$. Moreover, since $\sum_{j=0}^\infty p^{-j/2} = \sqrt{p} / (\sqrt{p} - 1)$, the distribution of $\sum_{j=0}^\infty p^{-j/2} Z_j$ for an independent and identically distributed sequence $(Z_j)_{j \geq 0}$ of $C (0, b)$-distributed random variables is $C \left(0, b\sqrt{p} / (\sqrt{p} - 1) \right)$. Thus the assertion follows from Theorem 8.2.

9.1. We have $E \log^+ |Z_1| < \infty$ and $P^{Z_1} = C (0, b)$ is symmetric around zero. Since $\sum_{j=1}^\infty |\vartheta|^{-j} = 1/ (|\vartheta| - 1)$, the distribution of $(\vartheta^2 - 1)^{1/2} \sum_{j=1}^\infty \vartheta^{-j} Z_j$ for an independent and identically distributed sequence $(Z_j)_{j \geq 1}$ of $C (0, b)$-distributed random variables is $C \left(0, b (\vartheta^2 - 1)^{1/2} / (|\vartheta| - 1) \right)$. The assertion follows from Theorem 9.2. In the more general case $P^{Z_1} = S_\alpha (b)$, the distribution of $(\vartheta^2 - 1)^{1/2} \sum_{j=1}^\infty \vartheta^{-j} Z_j$ is $S_\alpha \left(b (\vartheta^2 - 1)^{\alpha/2} / (|\vartheta|^\alpha - 1) \right)$.

9.2. Since P^{Z_1} is symmetric around zero, the distribution P^U is also symmetric around zero. Hence, by Lemma A.5 (c), $P^{(U/Y)|\mathcal{F}_\infty} = P^{-(U/Y)|\mathcal{F}_\infty}$ so that Theorem 9.2 yields $\vartheta^n\left(\widehat{\vartheta}_n - \vartheta\right) \to \left(\vartheta^2 - 1\right) U/Y\ \mathcal{F}_\infty$-stably and $-\vartheta^n\left(\widehat{\vartheta}_n - \vartheta\right) \to \left(\vartheta^2 - 1\right) U/Y\ \mathcal{F}_\infty$-stably. This implies the assertion.

9.3. We have

$$\sum_{j=0}^n X_j = \sum_{j=0}^n \left(X_0 + \sum_{i=1}^j Z_i \right) = (n+1) X_0 + \sum_{j=1}^{n-1} \left(\sum_{i=1}^j Z_i \right) + \sum_{i=1}^n Z_i\,.$$

As in the proof of Theorem 9.3 one shows that

$$n^{-3/2} \sum_{j=1}^{n-1} \left(\sum_{i=1}^j Z_i \right) \to \sigma \int_0^1 W_t\, dt \quad \mathcal{F}_\infty\text{-mixing}\,.$$

Using $n^{-3/2}(n+1) X_0 \to 0$ almost surely, $n^{-3/2} \sum_{i=1}^n Z_i \to 0$ almost surely and Theorem 3.18 (a), the assertion follows.

9.4. From $Z_j = X_j - \vartheta X_{j-1}$ we obtain $Z_j^2 \le 2X_j^2 + 2\vartheta^2 X_{j-1}$ for every $j \in \mathbb{N}$ so that $\sum_{j=1}^n Z_j^2 \le 2\left(1+\vartheta^2\right) \sum_{j=0}^n X_j^2$ for every $n \in \mathbb{N}$ which, in view of $Z_1 \in \mathcal{L}^2$ with $E\left(Z_1^2\right) > 0$, implies $A_n = \sum_{j=1}^n X_{j-1}^2 \to \infty$ almost surely as $n \to \infty$ by Kolmogorov's strong law of large numbers. This yields $\tau_c < \infty$ almost surely for every $\gamma > 0$ and $c \in \mathbb{N}$.

For the proof of

$$\frac{X_n^2}{A_n} \to 0 \quad \text{a.s. as } n \to \infty$$

for all $|\vartheta| \le 1$ we set, for $n \ge 2$,

$$R_n := -\frac{2\vartheta B_n + X_0^2}{A_n} + 1 - \vartheta^2 = -\frac{2\vartheta \sum_{j=1}^n X_{j-1} Z_j + X_0^2}{\sum_{j=1}^n X_{j-1}^2} + 1 - \vartheta^2\,.$$

Moreover, we set, for $n \in \mathbb{N}$ and $\lambda > 0$

$$T_n := \frac{1}{\sqrt{n}} \max_{1 \le j \le n} |Z_j| \quad \text{and} \quad S_n(\lambda) := \frac{1}{\lambda n} \sum_{j=1}^n Z_j^2 - 1\,.$$

For every $\lambda \in \left(0, \sigma^2\right)$ and $m \in \mathbb{N}$ and all sufficiently large $n \in \mathbb{N}$ we will show that

$$\frac{X_n^2}{A_n} \le \frac{R_n}{S_n(\lambda)} + \frac{1}{m\left(1 - mT_n/\sqrt{\lambda}\right)^2}\,,$$

where $S_n(\lambda) > 0$ and $mT_n/\sqrt{\lambda} < 1$ is true almost surely for all sufficiently large n because $S_n(\lambda) \to \sigma^2/\lambda - 1 > 0$ and $T_n \to 0$ almost surely as $n \to \infty$ by Kolmogorov's strong law of large numbers. For the proof of this inequality, note that $X_j^2 = \vartheta^2 X_{j-1}^2 + 2\vartheta X_{j-1} Z_j + Z_j^2$ for all $j \in \mathbb{N}$ so that

$$X_n^2 + A_n - X_0^2 = \sum_{j=1}^{n} X_j^2 = \vartheta^2 A_n + 2\vartheta B_n + \sum_{j=1}^{n} Z_j^2$$

for all $n \in \mathbb{N}$, which, by rearranging terms, yields for $n \geq 2$

$$R_n A_n = \sum_{j=1}^{n} Z_j^2 - X_n^2 .$$

If $X_n^2 \leq \lambda n$, then

$$R_n A_n \geq \sum_{j=1}^{n} Z_j^2 - \lambda n = \lambda n S_n(\lambda) \geq X_n^2 S_n(\lambda)$$

so that

$$\frac{X_n^2}{A_n} \leq \frac{R_n}{S_n(\lambda)} .$$

Therefore, it remains to consider the case $X_n^2 \geq \lambda n$. For every $n \in \mathbb{N}$ we have, because $|\vartheta| \leq 1$,

$$|X_{n-1}| \geq |\vartheta| \, |X_{n-1}| = |X_n - Z_n| \geq |X_n| - |Z_n|$$

which gives, inductively, for all $1 \leq j \leq n$,

$$|X_{n-j}| \geq |X_n| - \sum_{k=0}^{j-1} |Z_{n-k}|$$

so that, for all $m, n \in \mathbb{N}$ with $n \geq m$

$$\min_{1 \leq j \leq m} |X_{n-j}| \geq |X_n| - \sum_{k=0}^{m-1} |Z_{n-k}| = |X_n| \left(1 - \frac{1}{|X_n|} \sum_{k=0}^{m-1} |Z_{n-k}| \right) .$$

Moreover, $X_n^2 \geq \lambda n$ implies

$$\frac{1}{|X_n|} \sum_{k=0}^{m-1} |Z_{n-k}| \leq \frac{m}{\sqrt{\lambda n}} \max_{0 \leq k \leq m-1} |Z_{n-k}| \leq \frac{m}{\sqrt{\lambda}} T_n$$

so that

$$1 - \frac{1}{|X_n|} \sum_{k=0}^{m-1} |Z_{n-k}| \geq 1 - \frac{m}{\sqrt{\lambda}} T_n > 0$$

for all sufficiently large n and therefore

$$\min_{1 \leq j \leq m} X_{n-j}^2 \geq X_n^2 \left(1 - \frac{m}{\sqrt{\lambda}} T_n\right)^2.$$

This implies

$$A_n \geq m \min_{1 \leq j \leq m} X_{j-1}^2 \geq m X_n^2 \left(1 - \frac{m}{\sqrt{\lambda}} T_n\right)^2$$

so that

$$\frac{X_n^2}{A_n} \leq \frac{1}{m \left(1 - m T_n / \sqrt{\lambda}\right)^2}.$$

This completes the proof of the inequality

$$\frac{X_n^2}{A_n} \leq \frac{R_n}{S_n(\lambda)} + \frac{1}{m \left(1 - m T_n / \sqrt{\lambda}\right)^2},$$

for all sufficiently large n.

Note that B is a square integrable martingale w.r.t. $\mathbb{F}$ and

$$\langle B \rangle_n = \sigma^2 A_n \to \infty \quad \text{a.s. as } n \to \infty.$$

Therefore, $R_n \to 1 - \vartheta^2$ almost surely as $n \to \infty$ by the strong law of large numbers A.9 for martingales. As noted above, $S_n(\lambda) \to \sigma^2/\lambda - 1$ and $T_n \to 0$ almost surely. Consequently, the right-hand side of the last inequality converges almost surely to $(1 - \vartheta^2) / (\sigma^2/\lambda - 1) + 1/m$, where $\lambda \in (0, \sigma^2)$ and $m \in \mathbb{N}$ are arbitrary. This implies

$$\frac{X_n^2}{A_n} \to 0 \quad \text{a.s. as } n \to \infty.$$

Let $\gamma > 0$ be fixed. Clearly, $\tau_c \to \infty$ almost surely as $c \to \infty$, and by definition of τ_c,

$$c\gamma \leq A_{\tau_c} = \sum_{j=1}^{\tau_c - 1} X_{j-1}^2 + X_{\tau_c - 1}^2 < c\gamma + X_{\tau_c - 1}^2$$

which in view of $X^2_{\tau_c-1}/A_{\tau_c} \to 0$ almost surely as $c \to \infty$ implies

$$\frac{1}{c}A_{\tau_c} \to \gamma \quad \text{a.s. as } c \to \infty .$$

In the next step we will show that

$$\frac{1}{\sqrt{c}}B_{\tau_c} \to \sigma\sqrt{\gamma}Z \quad \mathcal{F}_\infty\text{-mixing as } c \to \infty$$

by an application of Corollary 6.4. For all $c, j \in \mathbb{N}$ we set $X_{c,j} := X_{j-1}Z_j/\sqrt{c}$, and for all $c \in \mathbb{N}$ and $j \in \mathbb{N}_0$ we set $\mathcal{F}_{c,j} := \mathcal{F}_j$. Then $(X_{c,j})_{c,j\in\mathbb{N}}$ is a square integrable martingale difference array w.r.t. the nested array $(\mathcal{F}_{c,j})_{c\in\mathbb{N},j\in\mathbb{N}_0}$ of σ-fields, and for every $c \in \mathbb{N}$ the random variable τ_c is by construction an almost surely finite stopping time w.r.t. $(\mathcal{F}_{c,j})_{j\in\mathbb{N}_0}$. We have

$$\sum_{j=1}^{\tau_c} E\left(X^2_{c,j}|\mathcal{F}_{c,j-1}\right) = \frac{1}{c}\sum_{j=1}^{\tau_c} X^2_{j-1}E\left(Z^2_j|\mathcal{F}_{j-1}\right) = \frac{\sigma^2}{c}A_{\tau_c} \to \sigma^2\gamma$$

almost surely as $c \to \infty$ so that condition (N_{τ_n}) in Corollary 6.4 is satisfied with $\tau_n = \tau_c$ and the constant random variable $\eta^2 = \sigma^2\gamma$. To verify the Lindeberg condition (CLB_{τ_n}) with $\tau_n = \tau_c$ we write, for all $c \in \mathbb{N}$ and $\varepsilon, M > 0$,

$$\sum_{j=1}^{\tau_c} E\left(X^2_{c,j}1_{\{|X_{c,j}|\geq\varepsilon\}}|\mathcal{F}_{c,j-1}\right)$$

$$= \frac{1}{c}\sum_{j=1}^{\tau_c} X^2_{j-1}E\left(Z^2_j1_{\{|X_{j-1}Z_j|\geq\varepsilon\sqrt{c}\}\cap\{|Z_j|\leq M\}}|\mathcal{F}_{j-1}\right)$$

$$+ \frac{1}{c}\sum_{j=1}^{\tau_c} X^2_{j-1}E\left(Z^2_j1_{\{|X_{j-1}Z_j|\geq\varepsilon\sqrt{c}\}\cap\{|Z_j|>M\}}|\mathcal{F}_{j-1}\right)$$

$$\leq \frac{M^2}{c}\sum_{j=1}^{\tau_c} X^2_{j-1}1_{\{|X_{j-1}|\geq\varepsilon\sqrt{c}/M\}} + \left(\frac{1}{c}\sum_{j=1}^{\tau_c} X^2_{j-1}\right)E\left(Z^2_11_{\{|Z_1|>M\}}\right)$$

$$= I_c(M) + II_c(M) ,$$

say. To verify that $I_c(M)$ converges to zero in probability as $c \to \infty$ for every $M > 0$, we first show that

$$\frac{1}{c}\max_{1\leq j\leq\tau_c} X^2_{j-1} \to 0 \quad \text{a.s. as } c \to \infty .$$

For this, let $\delta > 0$. With probability one there exists an $n_\delta \in \mathbb{N}$ with $X_n^2/A_n \leq \delta$ for all $n \geq n_\delta$ and a $c_\delta \in \mathbb{N}$ with $\tau_c > n_\delta$ for all $c \geq c_\delta$. Therefore, with probability one for all $c \geq c_\delta$

$$\frac{1}{c} \max_{1 \leq j \leq \tau_c} X_{j-1}^2 \leq \frac{1}{c} \max_{1 \leq j \leq n_\delta} X_{j-1}^2 + \frac{1}{c} \max_{n_\delta < j \leq \tau_c} \frac{X_{j-1}^2}{A_{j-1}} A_{j-1} \leq \frac{1}{c} \max_{1 \leq j \leq n_\delta} X_{j-1}^2 + \frac{\delta}{c} A_{\tau_c}.$$

The first summand on the right-hand side of this inequality converges to zero almost surely as $c \to \infty$ and the second one to $\delta\gamma$ which, since $\delta > 0$ is arbitrary, concludes the proof. Now the inequality

$$P\left(\frac{M^2}{c} \sum_{j=1}^{\tau_c} X_{j-1}^2 1_{\{|X_{j-1}| \geq \varepsilon\sqrt{c}/M\}} \geq \delta \right) \leq P\left(\frac{1}{\sqrt{c}} \max_{1 \leq j \leq \tau_c} |X_{j-1}| \geq \varepsilon/M \right),$$

which holds for all $\delta > 0$, shows that $I_c(M) \to 0$ in probability as $c \to \infty$ for all $M > 0$. Clearly, $II_c(M) \to \gamma E\left(Z_1^2 1_{\{|Z_1| > M\}}\right)$ almost surely as $c \to \infty$ for every $M > 0$ where $E\left(Z_1^2 1_{\{|Z_1| > M\}}\right) \to 0$ as $M \to \infty$ because $Z_1 \in \mathcal{L}^2$. This completes the proof of

$$\sum_{j=1}^{\tau_c} E\left(X_{c,j}^2 1_{\{|X_{c,j}| \geq \varepsilon\}} | \mathcal{F}_{c,j-1} \right) \to 0 \quad \text{in probability as } c \to \infty$$

for every $\varepsilon > 0$. Now by Corollary 6.4

$$\frac{1}{\sqrt{c}} B_{\tau_c} = \sum_{j=1}^{\tau_c} X_{c,j} \to \sigma\sqrt{\gamma} N \quad \mathcal{F}_\infty\text{-mixing as } c \to \infty,$$

where $P^N = N(0,1)$ and N is independent of $\mathcal{F}_\infty$. For every $c \in \mathbb{N}$ we have

$$\widehat{\vartheta}_{\tau_c} - \vartheta = \frac{B_{\tau_c}}{A_{\tau_c}}$$

so that we have shown

$$\frac{1}{\sqrt{c}} \left(\sum_{j=1}^{\tau_c} X_{j-1}^2 \right) \left(\widehat{\vartheta}_{\tau_c} - \vartheta \right) = \frac{1}{\sqrt{c}} B_{\tau_c} \to \sigma\sqrt{\gamma} N$$

$\mathcal{F}_\infty$-mixing as $c \to \infty$, which in view of $A_{\tau_c}/c \to \gamma$ almost surely as $c \to \infty$ implies both

$$\left(\sum_{j=1}^{\tau_c} X_{j-1}^2\right)^{1/2} (\widehat{\vartheta}_{\tau_c} - \vartheta) \to \sigma N \quad \mathcal{F}_\infty\text{-mixing as } c \to \infty$$

and

$$c^{1/2} (\widehat{\vartheta}_{\tau_c} - \vartheta) \to \frac{\sigma}{\sqrt{\gamma}} N \quad \mathcal{F}_\infty\text{-mixing as } c \to \infty.$$

10.1. For every fixed $k \in \mathbb{N}_0$, set

$$V_{k,n} := \sum_{j=1}^{X_{n-1}} \left(1_{\{Y_{nj}=k\}} - p_k\right), \quad n \in \mathbb{N}.$$

Then $\left(V_{k,n}\right)_{n\in\mathbb{N}}$ is a martingale difference sequence w.r.t. $\mathbb{F} = (\mathcal{F}_n)_{n\geq0}$: Clearly, $V_{k,n}$ is $\mathcal{F}_n$-measurable for all $n \in \mathbb{N}$, and $\left|V_{k,n}\right| \leq X_{n-1}$ so that $V_{k,n} \in \mathcal{L}^1(P)$ and

$$E\left(V_{k,n}|\mathcal{F}_{n-1}\right) = \sum_{j=1}^{X_{n-1}} E\left(1_{\{Y_{nj}=k\}} - p_k|\mathcal{F}_{n-1}\right) = 0$$

because $E\left(1_{\{Y_{nj}=k\}}|\mathcal{F}_{n-1}\right) = p_k$ by independence of Y_{nj} and $\mathcal{F}_{n-1}$. Consequently,

$$M_n^{(k)} := \sum_{i=1}^{n} V_{k,i}, \quad n \in \mathbb{N},$$

defines an $\mathbb{F}$-martingale $M^{(k)} = \left(M_n^{(k)}\right)_{n\geq0}$ (with $M_0^{(k)} = 0$) for which

$$\widehat{p}_{k,n} - p_k = \frac{1}{Z_n} \sum_{i=1}^{n} \sum_{j=1}^{X_{i-1}} \left(1_{\{Y_{ij}=k\}} - p_k\right) = \frac{M_n^{(k)}}{Z_n}.$$

Since we also assume $Y_{11} \in \mathcal{L}^2$, the martingale $M^{(k)}$ is square integrable with quadratic characteristic

$$\left\langle M^{(k)} \right\rangle_n = \sum_{i=1}^{n} E\left(V_{k,i}^2|\mathcal{F}_{i-1}\right), \quad n \in \mathbb{N},$$

where

$$E\left(V_{k,i}^2 \middle| \mathcal{F}_{i-1}\right) = E\left(\left[\sum_{j=1}^{X_{i-1}} \left(1_{\{Y_{ij}=k\}} - p_k\right)\right]^2 \middle| \mathcal{F}_{i-1}\right)$$

$$= \sum_{j,m=1}^{X_{i-1}} E\left(\left(1_{\{Y_{ij}=k\}} - p_k\right)\left(1_{\{Y_{im}=k\}} - p_k\right) \middle| \mathcal{F}_{i-1}\right)$$

$$= \sum_{j=1}^{X_{i-1}} E\left(\left(1_{\{Y_{ij}=k\}} - p_k\right)^2\right) = p_k\left(1 - p_k\right) X_{i-1}$$

by independence of Y_{ij}, Y_{im} and $\mathcal{F}_{i-1}$ and independence of Y_{ij} and Y_{im} for $j \neq m$. Hence

$$\left\langle M^{(k)} \right\rangle_n = p_k\left(1 - p_k\right) Z_n .$$

If $p_k = 0$, then clearly $\widehat{p}_{k,n} = 0$ for all $n \in \mathbb{N}$, and both assertions are trivial. Therefore, assume $p_k > 0$ from now on. Then $p_k\left(1 - p_k\right) > 0$ and

$$\left\langle M^{(k)} \right\rangle_n \to \infty \quad \text{a.s. on } \left\{\lim_{n\to\infty} X_n = \infty\right\} = M_+ .$$

The strong law of large numbers for $\mathcal{L}^2$-martingales of Theorem A.9 implies

$$\frac{M_n^{(k)}}{\left\langle M^{(k)} \right\rangle_n} \to 0 \quad \text{a.s. as } n \to \infty \text{ on } M_+$$

which because

$$\widehat{p}_{k,n} - p_k = p_k\left(1 - p_k\right)\frac{M_n^{(k)}}{\left\langle M^{(k)} \right\rangle_n}$$

implies $\widehat{p}_{k,n} \to p_k$ almost surely as $n \to \infty$ on M_+.

To prove the stable limit theorem for $\widehat{p}_{k,n}$, we will apply Theorem 8.2 in combination with Corollary 8.5 and Remark 8.6 to $X = M^{(k)}$ and $A = \left\langle M^{(k)} \right\rangle$ with $G = \Omega$ and $a_n = \alpha^{n/2}$. According to Remark 8.6 we only have to verify conditions (i), (iii), and (iv) in Theorem 8.2.

As to condition (i), we have

$$\frac{\left\langle M^{(k)} \right\rangle_n}{\alpha^n} \to \frac{p_k\left(1 - p_k\right)}{\alpha - 1} M_\infty \quad \text{a.s. as } n \to \infty$$

and $P\left(p_k\left(1 - p_k\right) M_\infty / \left(\alpha - 1\right) > 0\right) = P\left(M_+\right) > 0$, so that the condition is satisfied with $\eta = \left(p_k\left(1 - p_k\right) M_\infty / \left(\alpha - 1\right)\right)^{1/2}$.

Clearly, for all $n, r \in \mathbb{N}$ with $n > r$,

$$\frac{\alpha^{n-r}}{\alpha^n} = \frac{1}{\alpha^r},$$

so that condition (iii) of Theorem 8.2 is satisfied with $p = \alpha \in (1, \infty)$.

It remains to prove condition (iv). For this, we set

$$W_{nj}^{(k)} := \frac{1_{\{Y_{nj}=k\}} - p_k}{(p_k (1 - p_k))^{1/2}}$$

and note that

$$\frac{\Delta M_n^{(k)}}{\langle M^{(k)} \rangle_n^{1/2}} = \frac{V_{k,n}}{(p_k (1 - p_k))^{1/2} Z_n^{1/2}} = \frac{1}{Z_n^{1/2}} \sum_{j=1}^{X_{n-1}} W_{nj}^{(k)}.$$

Let ϕ_k denote the characteristic function of the (normalized) random variable $W_{11}^{(k)}$. Then

$$E_P \left(\exp \left(it \frac{\Delta M_n^{(k)}}{\langle M^{(k)} \rangle_n^{1/2}} \right) \Big| \mathcal{F}_{n-1} \right) = E_P \left(\exp \left(it \frac{1}{Z_n^{1/2}} \sum_{j=1}^{X_{n-1}} W_{nj}^{(k)} \right) \Big| \mathcal{F}_{n-1} \right)$$

$$= \phi_k \left(\frac{t}{Z_n^{1/2}} \right)^{X_{n-1}}$$

because Z_n and X_{n-1} are measurable w.r.t. $\mathcal{F}_{n-1}$ and the random variables $W_{nj}^{(k)}$ are independent and identically distributed with characteristic function ϕ_k. The classical central limit theorem for sums of independent and identically distributed random variables yields

$$\phi_k \left(\frac{x}{\sqrt{n}} \right)^n \to \exp \left(-\frac{1}{2} x^2 \right) \quad \text{as } n \to \infty$$

uniformly in $x \in \mathbb{R}$ on compact intervals. Setting $x = t X_{n-1}^{1/2} / Z_n^{1/2}$ and $n = X_{n-1}$ we get

$$\phi_k \left(\frac{t}{Z_n^{1/2}} \right)^{X_{n-1}} = \phi_k \left(\frac{t X_{n-1}^{1/2}}{Z_n^{1/2}} \frac{1}{X_{n-1}^{1/2}} \right)^{X_{n-1}} \to \exp \left(-\frac{1}{2} t^2 \frac{\alpha - 1}{\alpha} \right)$$

almost surely on M_+ as $n \to \infty$ because

$$\frac{t X_{n-1}^{1/2}}{Z_n^{1/2}} \to t \left(\frac{\alpha - 1}{\alpha} \right)^{1/2}.$$

Consequently, condition (iv) of Theorem 8.2 is satisfied for $\mu = N(0, b)$ with $b = (\alpha - 1)/\alpha$. Now Corollary 8.5 implies

$$Z_n^{1/2} \frac{\widehat{p}_{k,n} - p_k}{(p_k(1 - p_k))^{1/2}} = \frac{M_n^{(k)}}{\langle M^{(k)} \rangle_n^{1/2}} \to N \quad \mathcal{F}_\infty\text{-mixing under } P_{M_+}$$

and

$$\frac{Z_n}{\alpha^{n/2}} \left(\widehat{p}_{k,n} - p_k\right) \to \left(\frac{p_k(1 - p_k)}{\alpha - 1}\right)^{1/2} M_\infty^{1/2} N \quad \mathcal{F}_\infty\text{-stably under } P_{M_+},$$

where N is independent of $\mathcal{F}_\infty$ and $P^N = N(0, 1)$, which because $Z_n/\alpha^n \to M_\infty/(\alpha - 1)$ almost surely is equivalent to

$$\frac{\alpha^{n/2}}{(\alpha - 1)^{1/2}} \left(\widehat{p}_{k,n} - p_k\right) \to (p_k(1 - p_k))^{1/2} M_\infty^{-1/2} N \quad \mathcal{F}_\infty\text{-stably under } P_{M_+}.$$

10.2. A little algebra gives, with $\widehat{p}_{n,k}$ denoting the estimator from Exercise 10.1,

$$\widehat{p}_{n,k} - \widehat{p}_{k,n-1} = \left(\frac{Z_{n-1}}{Z_n} - 1\right)\left(\widehat{p}_{k,n-1} - p_k\right) + \frac{X_{n-1}}{Z_n}\left(\widetilde{p}_{k,n} - p_k\right),$$

which by strong consistency of $\widehat{p}_{k,n}$ and

$$\frac{Z_{n-1}}{Z_n} \to \frac{1}{\alpha} \quad \text{a.s. as } n \to \infty \text{ on } M_+$$

as well as

$$\frac{X_{n-1}}{Z_n} \to \frac{\alpha - 1}{\alpha} \quad \text{a.s. as } n \to \infty \text{ on } M_+$$

yields $\widetilde{p}_{k,n} \to p_k$ almost surely as $n \to \infty$ on M_+.

Replacing the random variables $Y_{nj} - \alpha$ by $1_{\{Y_{nj}=k\}} - p_k$ in Theorem 10.1 we obtain

$$\frac{1}{\alpha^{(n-1)/2}} \sum_{j=1}^{X_{n-1}} \left(1_{\{Y_{nj}=k\}} - p_k\right) \to (p_k(1 - p_k))^{1/2} M_\infty^{1/2} N$$

$\mathcal{F}_\infty$-stably as $n \to \infty$, where $P^N = N(0, 1)$ and N is P-independent of $\mathcal{F}_\infty$. This gives

$$X_{n-1}^{1/2} \left(\widetilde{p}_{k,n} - p_k\right) \to (p_k(1 - p_k))^{1/2} N \quad \mathcal{F}_\infty\text{-mixing under } P_{M_+} \text{ as } n \to \infty.$$

Abbreviations of Formulas

© Springer International Publishing Switzerland 2015
E. Häusler and H. Luschgy, *Stable Convergence and Stable Limit Theorems*,
Probability Theory and Stochastic Modelling 74,
DOI 10.1007/978-3-319-18329-9

Notation Index

a.s.	Almost surely
$\mathcal{B}(\mathcal{X})$	Borel σ-field
$C(0, b)$	Cauchy distribution, 153
$C_b(\mathcal{X})$	Space of continuous, bounded functions on $\mathcal{X}$, 11
$C(\mathbb{R}_+)$	Space of continuous functions on $\mathbb{R}_+$
δ_x	Dirac-measure
δ_X	Dirac-kernel, 21
∂B	Topological boundary
ΔX_n	Increments, 146, 192
$\frac{dQ}{dP}$	P-density of Q
EX	Expectation
$E(X|\mathcal{G})$	Conditional expectation
$E(K|\mathcal{G})$	Conditional expectation, 15
$\mathbb{F} = (\mathcal{F}_n)_{n \in I}$	Filtration, 63, 192
$\mathcal{F}_\infty$	68, 108
$\mathcal{F}_\infty^n$	64
$f \otimes h$	Tensor product, 12
$\mathcal{I}_X$	Invariant σ-field, 111
$K_1 \otimes K_2$	189
$\mathcal{K}^1 = \mathcal{K}^1(\mathcal{F}) = \mathcal{K}^1(\mathcal{F}, \mathcal{X})$	Markov kernels, 12
$\mathcal{K}^1(\mathcal{G}) = \mathcal{K}^1(\mathcal{G}, \mathcal{X})$	$\mathcal{G}$-measurable Markov kernels, 13
$K^1(P), K^1(\mathcal{G}, P)$	P-equivalence classes of Markov kernels, 18
$\mathcal{L}^p(P)$	p-integrable functions
$\mathcal{L}^p(\mathcal{G}, P)$	$\mathcal{G}$-measurable, p-integrable functions
λ	Lebesgue measure
$\mathcal{M}^1(\mathcal{X})$	Probability measures on $\mathcal{B}(\mathcal{X})$, 11
$N(0, \sigma^2)$	Normal distributions
$\mathbb{N}, \mathbb{N}_0$	Natural numbers, $\mathbb{N} \cup \{0\}$
P^X	Distribution of X, image measure
$P^{X|\mathcal{G}}$	Conditional distribution, 21, 188
$P_F = \frac{P(\cdot \cap F)}{P(F)}$	13
$Q \otimes K$	12
QK	12

© Springer International Publishing Switzerland 2015
E. Häusler and H. Luschgy, *Stable Convergence and Stable Limit Theorems*,
Probability Theory and Stochastic Modelling 74,
DOI 10.1007/978-3-319-18329-9

References

1. Aït-Sahalia,Y., Jacod, J., *High-Frequency Financial Econometrics*. Princeton University Press, Princeton, 2014.
2. Aldous, D.J., Weak convergence of randomly indexed sequences of random variables. *Math. Proc. Cambridge Phil. Soc.* 83 (1978), 117–126.
3. Aldous, D.J., Exchangeability and related topics. *École d'Été de Probabilités de Saint-Flour XIII*, 1–198, Lecture Notes in Math. 1117, Springer, Berlin, 1985.
4. Aldous, D.J., Eagleson, G.K., On mixing and stability of limit theorems. *Ann. Probab.* 6 (1978), 325–331.
5. Anderson, T.W., On asymptotic distributions of estimates of parameters of stochastic difference equations. *Ann. Math. Statist.* 30 (1959), 676–687.
6. Arouna, B., Adaptive Monte Carlo method, a variance reduction technique. *Monte Carlo Methods and Appl.* 10 (2004), 1–24.
7. Balder, E.J., Lectures on Young measures. Cahiers du CEREMADE, 1995–17, Université Paris-Dauphine.
8. Billingsley, P., The Lindeberg-Lévy theorem for martingales. *Proc. Amer. Math. Soc. 12* (1961), 788–792.
9. Billingsley, P., *Convergence of Probability Measures*. Wiley, New York, 1968.
10. Bingham, N.H., Goldie, C.M., Teugels, J.L., *Regular Variation*. Cambridge University Press, Cambridge, 1987.
11. Blackwell, D., Freedman, D., The tail σ-field of a Markov chain and a theorem of Orey. *Ann. Math. Statist.* 35 (1964), 1291–1295.
12. Brown, B.M., Martingale central limit theorems. *Ann. Math. Statist.* 42 (1971), 59–66.
13. Castaing, C., Raynaud de Fitte, P., Valadier, M., *Young Measures on Topological Spaces. With Applications in Control Theory and Probability Theory.* Kluwer Academic Publishers, Dordrecht, 2004.
14. Chatterji, S.D., Les martingales et leurs application analytiques. In: Ecole d'Eté de Probabilités: Processus Stochastiques, 1971, 27–164 (J.L. Bretagnolle et al., eds.). Lecture Notes in Math. 307, Springer, Berlin, 1973.
15. Chatterji, S.D., A principle of subsequences in probability theory: the central limit theorem. *Advances in Math.* 13 (1974), 31–54.
16. Cheng, T.L., Chow, Y.S., On stable convergence in the central limit theorem. *Statist. Probab. Letters* 57 (2002), 307–313.
17. Chow, Y.S., Teicher, H., *Probability Theory*. Springer, New York, 1978. (Third edition 1997).

© Springer International Publishing Switzerland 2015

E. Häusler and H. Luschgy, *Stable Convergence and Stable Limit Theorems*,
Probability Theory and Stochastic Modelling 74,
DOI 10.1007/978-3-319-18329-9

18. Crauel, H., *Random Probability Measures on Polish Spaces*. Taylor & Francis, London, 2002.
19. Crimaldi, I., An almost sure conditional convergence result and an application to a generalized Pólya urn. *International Math. Forum* 4 (23) (2009), 1139–1156.
20. Crimaldi, I., Letta, G., Pratelli, L., A strong form of stable convergence. *Séminaire de Probabilités XL*, 203–225, Lecture Notes in Math 1899, Springer, Berlin, 2007.
21. Csörgő, M., Fischler, R., Some examples and results in the theory of mixing and random-sum central limit theorems. *Periodica Math. Hung.* 3 (1973), 41–57.
22. Dedecker, J., Merlevède, F., Necessary and sufficient conditions for the conditional central limit theorem. *Ann. Probab.* 30 (2002), 1044–1081.
23. Dellacherie, C., Meyer, P.A., *Probabilities and Potential B*. North Holland, Amsterdam, 1982.
24. Dickey, D.A., Fuller, W.A., Distribution of the estimators for autoregressive time series with a unit root. *J. Amer. Statist. Association* 74 (1979), 927–931.
25. Dion, J.-P., Estimation of the mean and the initial probabilities of a branching process. *J. Appl. Probab.* 11 (1974), 687–694.
26. Dudley, R.M., *Real Analysis and Probability*. Cambridge University Press, Cambridge, 2002.
27. Eagleson, G.K., On Gordin's central limit theorem for stationary processes. *J. Appl. Probab.* 12 (1975), 176–179.
28. Eagleson, G.K., Some simple conditions for limit theorems to be mixing. *Theory Probab. Appl.* 21 (1976), 637–643.
29. Farrell, R.H., Weak limits of sequences of Bayes procedures in estimation theory. *Proc. Fifth Berkeley Symp. Math. Statist. Probab.* 1, Univ. of California (1967), 83–111.
30. Fischler, R.M., Borel-Cantelli type theorems for mixing sets. *Acta Math. Acad. Sci. Hungar.* 18 (1967), 67–69.
31. Fischler, R.M., Stable sequences of random variables and the weak convergence of the associated empirical measures. *Sankhyā Ser. A* 33 (1971), 67–72.
32. Fischler, R.M., Convergence faible avec indices aléatoires. *Ann. Inst. Henri Poincaré* 12 (1976), 391–399.
33. Florescu, L.C., Godet-Thobie, C., *Young Measures and Compactness in Measure Spaces*. De Gruyter, Berlin, 2012.
34. Gaenssler, P., Haeusler, E., On martingale central limit theory. In: Dependence in Probability and Statistics. A Survey of Recent Results, 303–334 (E. Eberlein and M.S. Taqqu, eds.). Progress in Probability and Statistics 11, Birkhäuser, Basel, 1986.
35. Gänssler, P., Stute, W., *Wahrscheinlichkeitstheorie*. Springer, Berlin, 1977.
36. Gordin, M.I., The central limit theorem for stationary processes. (In Russian) *Dokl. Akad. Nauk. S.S.S.R.* 118 (1969), 739 – 741. English translation: *Soviet Math. Dokl.* 10 (1969), 1174–1176.
37. Gordin, M.I., A note on the martingale method of proving the central limit theorem for stationary sequences. *J. Math. Sciences* 133 (2006), 1277–1281.
38. Gordin, M., Peligrad, M., On the functional central limit theorem via martingale approximation. *Bernoulli* 17 (2011), 424–440.
39. Grübel, R., Kabluchko, Z., A functional central limit theorem for branching random walks, almost sure weak convergence, and an application to random trees. arXiv:1410.0469v1 [math PR], 2014.
40. Guttorp, P., *Statistical Inference for Branching Processes*. Wiley, New York, 1991.
41. Hall, P., Heyde, C.C., *Martingale Limit Theory and Its Application*. Academic Press, New York, 1980.
42. Harris, T.E., Branching processes. *Ann. Math. Statist.* 19 (1948), 474–494.
43. Heyde, C.C., Extension of a result of Seneta for the supercritical Galton-Watson process. *Ann. Math. Statist.* 41 (1970), 739–742.
44. Heyde, C.C., Remarks on efficiency in estimation for branching processes. *Biometrika* 62 (1975), 49–55.
45. Ibragimov, I.A., A central limit theorem for a class of dependent random variables. *Theory Probab. Appl.* 8 (1963), 83–89.

46. Jacod, J., On continuous conditional Gaussian martingales and stable convergence in law. *Séminaire de Probab. XXXI*, 232–246, Lecture Notes in Math. 1655, Springer, Berlin, 1997.

47. Jacod, J., On processes with conditional independent increments and stable convergence in law. *Séminaire de Probab. XXXVI*, 383–401, Lecture Notes in Math. 1801, Springer, Berlin, 2003.

48. Jacod, J., Memin, J., Sur un type de convergence intermédiaire entre la convergence en loi et la convergence en probabilité. *Séminaire de Probab. XV*, 529–546, Lecture Notes in Math. 850, Springer, Berlin, 1981. Corrections: *Séminaire Probab. XVII*, 509–511, Lecture Notes in Math. 986, Springer, Berlin, 1983.

49. Jacod, J., Protter, P., *Discretization of Processes*. Springer, Berlin, 2012.

50. Jacod, J., Shiryaev, A.N., *Limit Theorems for Stochastic Processes, Second Edition*. Springer, Berlin, 2003.

51. Kallenberg, O., *Foundations of Modern Probability, Second Edition*. Springer, New York, 2002.

52. Kallenberg, O., *Probabilistic Symmetries and Invariance Principles*. Springer, New York, 2005.

53. Klenke, A., *Probability Theory, Second Edition*. Springer, London, 2014.

54. Koul, H.L., Pflug, G.Ch., Weakly adaptive estimators in explosive autoregression. *Ann. Statist.* 18 (1990), 939–960.

55. Lai, T.L., Siegmund, D., Fixed accuracy estimation of an autoregressive parameter. *Ann. Statist.* 11 (1983), 478–485.

56. Leon, J.R., Ludena, C., Stable convergence of certain functionals of diffusions driven by fBm. *Stochastic Analysis and Applications* 22 (2004), 289–314.

57. Letta, G., Convergence stable et applications. *Atti Sem. Mat. Fis. Univ. Modena, Supplemento al Vol. XLVI* (1998), 191–211.

58. Letta, G., Pratelli, L., Convergence stable vers un noyau gaussien. *Rendiconti Academia Nazionale della Scienze detta dei XL, Memorie di Matematica e Applicazioni* 114 (1996), 205–213.

59. Lindberg, C., Rootzén, H., Error distributions for random grid approximations of multidimensional stochastic integrals. *Ann. Appl. Probab.* 23 (2013), 834–857.

60. Liptser, R.S., Shiryaev, A.N., *Theory of Martingales*. Kluwer Academic Publishers, Dordrecht, 1989.

61. Luschgy, H., Elimination of randomization and Hunt-Stein type theorems in invariant statistical decision problems. *Statistics* 18 (1987), 99–111.

62. Luschgy, H., Integral representation in the set of transition kernels. *Probab. Math. Statist.* 10 (1989), 75–92.

63. Luschgy, H., Asymptotic inference for semimartingale models with singular parameter points. *J. Statist. Plann. Inference* 39 (1994), 155–186.

64. Luschgy, H., *Martingale in diskreter Zeit*. Springer, Berlin, 2012.

65. Luschgy, H., Mussmann, D., A characterization of weakly dominated statistical experiments by compactness of the set of decision rules. *Sankhyā* 49 (1987), 388–394.

66. Merlevède, F., Peligrad, M., Utev, S., Recent advances in invariance principles for stationary sequences. *Probability Surveys* 3 (2006), 1–36.

67. Nagaev, A.V., On estimating the number of direct descendents of a particle in a branching process. *Theory Probab. Appl.* 12 (1967), 314–320.

68. Pagès, G., *Introduction to Numerical Probability for Finance*. Université Pierre et Marie Curie, 2015 (Forthcoming).

69. Parthasarathy, K.R., *Probability Measures on Metric Spaces*. Academic Press, New York, 1967.

70. Peccati, G., Taqqu, M.S., Stable convergence of L^2-generalized stochastic integrals and the principle of conditioning. *Electron. J. Probab.* 12 (2007), 447–480.

71. Peccati, G., Taqqu, M.S., Stable convergence of multiple Wiener-Ito integrals. *J. Theoret. Probab.* 21 (2008), 527–570.

72. Peligrad, M., Conditional central limit theorem via martingale approximation. arXiv:1101.0174v2[math.PR], 2011.

73. Phillips, P.C.B., Towards a unified asymptotic theory for autoregression. *Biometrika* 74 (1987), 535–547.

74. Pollard, D., *Convergence of Stochastic Processes,* Springer, New York, 1984.

75. Rényi, A., Contributions to the theory of independent random variables (In Russian, English Summary). *Acta Math. Acad. Sci. Hungar.* 1 (1950), 99–108.

76. Rényi, A., On mixing sequences of sets. *Acta Math. Acad. Sci. Hungar.* 9 (1958), 215–228.

77. Rényi, A., On stable sequences of events. *Sankhyā Ser. A* 25 (1963), 293–302.

78. Rényi, A., Révész, P., On mixing sequences of random variables. *Acta Math. Acad. Sci. Hungar.* 9 (1958), 389–393.

79. Rootzén, H., Some properties of convergence in distribution of sums and maxima of dependent random variables. *Z. Wahrscheinlichkeitstheorie verw. Gebiete* 29 (1974), 295–307.

80. Rootzén, H., Fluctuations of sequences which converge in distribution. *Ann. Probab.* 4 (1976), 456–463.

81. Rootzén, H., On the functional central limit theorem for martingales, II. *Z. Wahrscheinlichkeitstheorie verw. Gebiete* 51 (1980), 79–93.

82. Rootzén, H., Central limit theory for martingales via random change of time. In: Essays in honour of Carl Gustav Esséen, 154–190 (L. Holst and A. Gut, eds.). Uppsala University, 1983.

83. Sainte-Beuve, M.-F., On the extension of von Neumann-Aumann's theorem. *J. Funct. Anal.* 17 (1974), 112–129.

84. Salem, R., Zygmund, A., On lacunary trigonometric series. *Proc. Nat. Acad. Sci. U.S.A.* 33 (1947), 333–338.

85. Salem, R., Zygmund, A., On lacunary trigonometric series II. *Proc. Nat. Acad. Sci. U.S.A.* 34 (1948), 54–62.

86. Schaefer, H.H., *Topological Vector Spaces.* Springer, New York, 1971.

87. Schäl, M., On dynamic programming: Compactness in the space of policies. *Stoch. Processes and Their Applications* 3 (1975), 345–364.

88. Scott, D.J., A central limit theorem for martingales and an application to branching processes. *Stoch. Proc. Appl.* 6 (1978), 241–252.

89. Shiryaev, A.N., Spokoiny, V.G., *Statistical Experiments and Decisions.* World Scientific, Singapore, 2000.

90. Smith, J.C., On the asymptotic distribution of the sums of Rademacher functions. *Bull. Amer. Math. Soc.* 51 (1945), 941–944.

91. Stroock, D.W., *An Introduction to the Theory of Large Deviations.* Springer, New York, 1984.

92. Sucheston, L., On mixing and the zero-one law. *J. Math. Analysis and Applications* 6 (1963), 447–456.

93. Takahashi, S., On the asymptotic distribution of the sum of independent random variables. *Proc. Japan Acad.* 27 (1951), 393–400.

94. Takahashi, S., On the central limit theorem. *Tohoku Math. J.* 3 (1951), 316–321.

95. Touati, A., Two theorems on convergence in distribution for stochastic integrals and statistical applications. *Theory Probab. Appl.* 38 (1993), 95–117.

96. Wellner, J.A., A martingale inequality for the empirical process. *Ann. Probab.* 5 (1977), 303–308.

97. White, J.S., The limiting distribution of the serial correlation coefficient in the explosive case. *Ann. Math. Statist.* 29 (1958), 1188–1197.

98. White, J.S., The limiting distribution of the serial correlation coefficient in the explosive case II. *Ann. Math. Statist.* 30 (1959), 831–834.

99. van Zanten, H., A multivariate central limit theorem for continuous local martingales. *Statistics & Probability Letters* 50 (2000), 229–235.

100. Zweimüller, R., Mixing limit theorems for ergodic transformations. *J. Theoret. Probab.* 20 (2007), 1059–1071.

Index

A
Adapted, 192
Approximation, 29, 35
Atom, 203
Autoregressive process of order one
    critical, 168
    ergodic, 160
    explosive, 156, 163

B
Borel-Cantelli feature, 43
Brown's inequality, 194

C
Chow's SLLN, 193
Classical
    Lindeberg condition, 85
    Lyapunov condition, 86
    stable CLT, 30
    stable functional CLT, 31, 143
    stable functional random-sum CLT, 48
Compensator, 192
Conditional
    distribution, 21, 188
    expectation, 15
    Gaussian increments, 153
    Lindeberg condition, 68, 70, 123, 139
    Lyapunov condition of order p, 85, 110
    probability measure, 13
Convergence
    in distribution, 1, 2
    in probability, 26
    mixing, 22, 33

    stable, 21, 22, 33
    weak, 11, 13
Convergence determining, 50

D
δ-method, 51
Differences, 192
Dirac-kernel, 21
Discrete rule of de l'Hospital, 113
Disintegration, 190
Distribution, 188

E
Empirical measure theorem, 49
Ergodic process, 57
Estimator
    adaptive, 113
    conditional least squares, 177
    conditional moment, 175
    least squares, 159
    moment, 174
Exchangeable process, 112, 143

F
Filtration, 192

G
Galton-Watson branching process, 173
    supercritical, 174
Gauss-kernel, 22

© Springer International Publishing Switzerland 2015
E. Häusler and H. Luschgy, *Stable Convergence and Stable Limit Theorems*,
Probability Theory and Stochastic Modelling 74,
DOI 10.1007/978-3-319-18329-9

Printed in the United States
By Bookmasters